T0206932

Lecture Notes in Computer Science 12276

More information about this series at http://www.springer.com/series/7409

Josep Domingo-Ferrer ·
Krishnamurty Muralidhar (Eds.)

Privacy in Statistical Databases

UNESCO Chair in Data Privacy, International Conference, PSD 2020
Tarragona, Spain, September 23–25, 2020
Proceedings

Editors
Josep Domingo-Ferrer
Rovira i Virgili University
Tarragona, Catalonia, Spain

Krishnamurty Muralidhar
University of Oklahoma
Norman, OK, USA

ISSN 0302-9743 ISSN 1611-3349 (electronic)
Lecture Notes in Computer Science
ISBN 978-3-030-57520-5 ISBN 978-3-030-57521-2 (eBook)
https://doi.org/10.1007/978-3-030-57521-2

LNCS Sublibrary: SL3 – Information Systems and Applications, incl. Internet/Web, and HCI

This Springer imprint is published by the registered company Springer Nature Switzerland AG
The registered company address is: Gewerbestrasse 11, 6330 Cham, Switzerland

Preface

Privacy in statistical databases is a discipline whose purpose is to provide solutions to the tension between the social, political, economic, and corporate demands of accurate information, and the legal and ethical obligation to protect the privacy of the various parties involved. In particular, the need to enforce the EU General Data Protection Regulation (GDPR) in our world of big data has made this tension all the more pressing. Stakeholders include the subjects, sometimes a.k.a. respondents (the individuals and enterprises to which the data refer), the data controllers (those organizations collecting, curating, and to some extent sharing or releasing the data) and the users (the ones querying the database or the search engine, who would like their queries to stay confidential). Beyond law and ethics, there are also practical reasons for data controllers to invest in subject privacy: if individual subjects feel their privacy is guaranteed, they are likely to provide more accurate responses. Data controller privacy is primarily motivated by practical considerations: if an enterprise collects data at its own expense and responsibility, it may wish to minimize leakage of those data to other enterprises (even to those with whom joint data exploitation is planned). Finally, user privacy results in increased user satisfaction, even if it may curtail the ability of the data controller to profile users.

There are at least two traditions in statistical database privacy, both of which started in the 1970s: the first one stems from official statistics, where the discipline is also known as statistical disclosure control (SDC) or statistical disclosure limitation (SDL), and the second one originates from computer science and database technology. In official statistics, the basic concern is subject privacy. In computer science, the initial motivation was also subject privacy but, from 2000 onwards, growing attention has been devoted to controller privacy (privacy-preserving data mining) and user privacy (private information retrieval). In the last few years, the interest and the achievements of computer scientists in the topic have substantially increased, as reflected in the contents of this volume. At the same time, the generalization of big data is challenging privacy technologies in many ways: this volume also contains recent research aimed at tackling some of these challenges.

Privacy in Statistical Databases 2020 (PSD 2020) was held in Tarragona, Catalonia, Spain, under the sponsorship of the UNESCO Chair in Data Privacy, which has provided a stable umbrella for the PSD biennial conference series since 2008. Previous PSD conferences were held in various locations around the Mediterranean, and had their proceedings published by Springer in the LNCS series: PSD 2018, Valencia, LNCS 11126; PSD 2016, Dubrovnik, LNCS 9867; PSD 2014, Eivissa, LNCS 8744; PSD 2012, Palermo, LNCS 7556; PSD 2010, Corfu, LNCS 6344; PSD 2008, Istanbul, LNCS 5262; PSD 2006 (the final conference of the Eurostat-funded CENEX-SDC project), Rome, LNCS 4302; and PSD 2004 (the final conference of the European FP5 CASC project) Barcelona, LNCS 3050. The nine PSD conferences held so far are a follow-up of a series of high-quality technical conferences on SDC which started

22 years ago with Statistical Data Protection-SDP 1998, held in Lisbon in 1998 with proceedings published by OPOCE, and continued with the AMRADS project SDC workshop, held in Luxemburg in 2001 with proceedings published by Springer in LNCS 2316.

The PSD 2020 Program Committee accepted for publication in this volume 25 papers out of 49 submissions. Furthermore, 10 of the above submissions were reviewed for short oral presentation at the conference. Papers came from 14 different countries and 4 different continents. Each submitted paper received at least two reviews. The revised versions of the 25 accepted papers in this volume are a fine blend of contributions from official statistics and computer science. Covered topics include privacy models, microdata protection, protection of statistical tables, protection of interactive and mobility databases, record linkage and alternative methods, synthetic data, data quality, and case studies.

We are indebted to many people. Firstly, to the Organization Committee for making the conference possible, and especially to Jesús Manjón, who helped prepare these proceedings. In evaluating the papers we were assisted by the Program Committee and by Weiyi Xia, Zhiyu Wan, Chao Yan, and Jeremy Seeman as external reviewers. We also wish to thank all the authors of submitted papers and we apologize for possible omissions.

July 2020

Josep Domingo-Ferrer
Krishnamurty Muralidhar

Organization

Program Committee

Jane Bambauer	University of Arizona, USA
Bettina Berendt	Technical University of Berlin, Germany
Elisa Bertino	CERIAS, Purdue University, USA
Aleksandra Bujnowska	EUROSTAT, EU
Jordi Castro	Polytechnical University of Catalonia, Spain
Anne-Sophie Charest	Université Laval, Canada
Chris Clifton	Purdue University, USA
Graham Cormode	University of Warwick, USA
Josep Domingo-Ferrer	Universitat Rovira i Virgili, Catalonia, Spain
Jörg Drechsler	IAB, Germany
Khaled El Emam	University of Ottawa, Canada
Mark Elliot	The University of Manchester, UK
Sébastien Gambs	Université du Québec à Montréal, Canada
Sarah Giessing	Destatis, Germany
Sara Hajian	Nets Group, Denmark
Hiroaki Kikuchi	Meiji University, Japan
Bradley Malin	Vanderbilt University, USA
Laura McKenna	Census Bureau, USA
Anna Monreale	Università di Pisa, Italy
Krishnamurty Muralidhar	University of Oklahoma, USA
Anna Oganyan	National Center for Health Statistics, USA
David Rebollo-Monedero	Universitat Rovira i Virgili, Catalonia, Spain
Jerome Reiter	Duke University, USA
Yosef Rinott	Hebrew University, Israel
Steven Ruggles	University of Minnesota, USA
Nicolas Ruiz	OECD, Universitat Rovira i Virgili, Catalonia, Spain
Pierangela Samarati	Università di Milano, Italy
David Sánchez	Universitat Rovira i Virgili, Catalonia, Spain
Eric Schulte-Nordholt	Statistics Netherlands, The Netherlands
Natalie Shlomo	The University of Manchester, UK
Aleksandra Slavković	Penn State University, USA
Jordi Soria-Comas	Catalan Data Protection Authority, Catalonia, Spain
Tamir Tassa	The Open University, Israel
Vicenç Torra	Umeå University, Sweden
Lars Vilhuber	Cornell University, USA
Peter-Paul de Wolf	Statistics Netherlands, The Netherlands

Program Chair

Josep Domingo-Ferrer	UNESCO Chair in Data Privacy, Universitat Rovira i Virgili, Catalonia, Spain

General Chair

Krishnamurty Muralidhar	University of Oklahoma, USA

Organization Committee

Joaquín García-Alfaro	Télécom SudParis, France
Jesús Manjón	Universitat Rovira i Virgili, Catalonia, Spain
Romina Russo	Universitat Rovira i Virgili, Catalonia, Spain

Contents

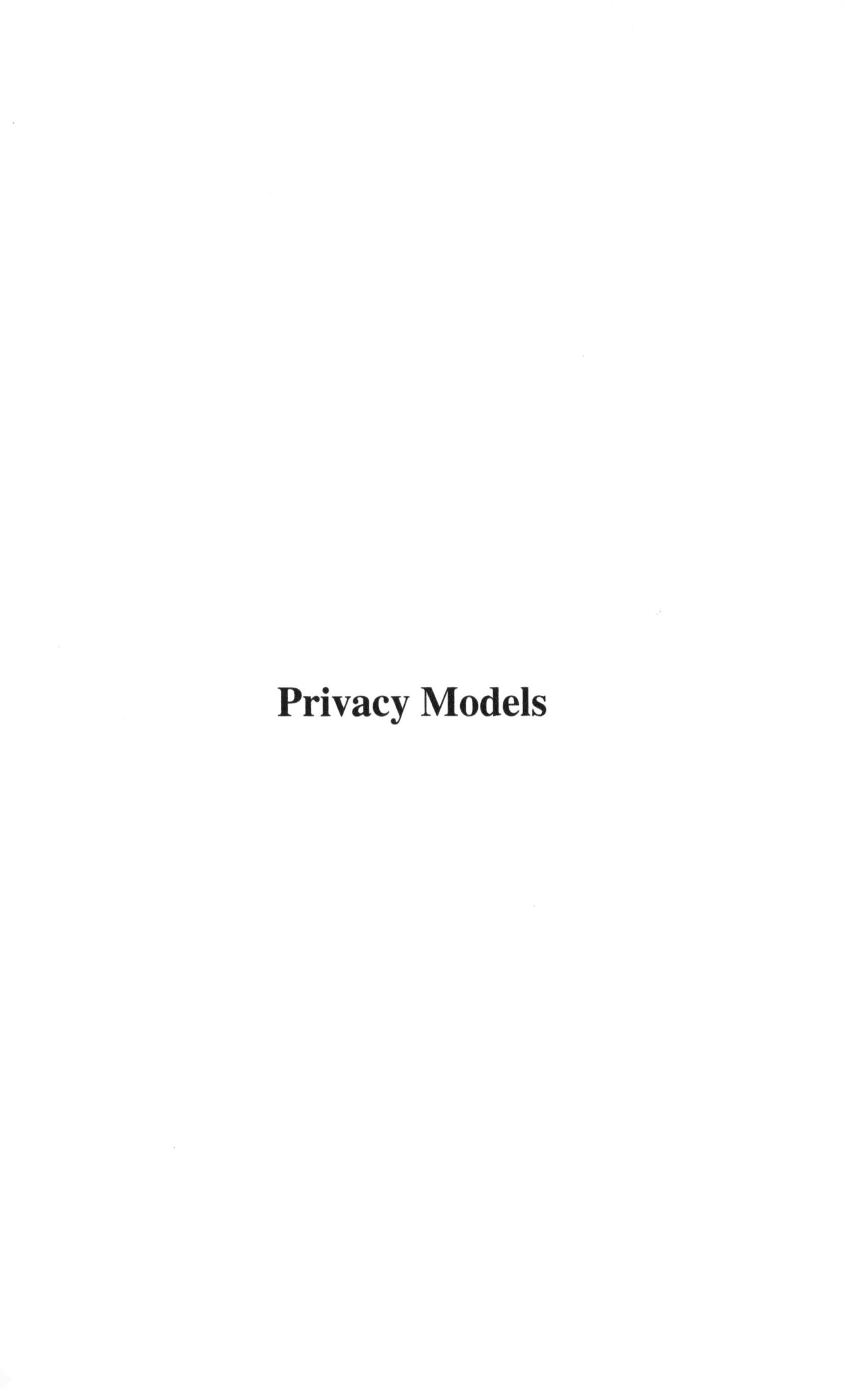

Privacy Models

$P_{\alpha,\beta}$-Privacy: A Composable Formulation of Privacy Guarantees for Data Publishing Based on Permutation

Nicolas Ruiz(✉)

Departament d'Enginyeria Informàtica i Matemàtiques, Universitat Rovira i Virgili, Av. Països Catalans 26, 43007 Tarragona, Catalonia, Spain
nicolas.ruiz@urv.cat

Abstract. Methods for Privacy-Preserving Data Publishing (PPDP) have been recently shown to be equivalent to essentially performing some permutations of the original data. This insight, called the permutation paradigm, establishes a common ground upon which any method can be evaluated *ex-post*, but can also be viewed as a general *ex-ante* method in itself, where data are anonymized with the injection of suitable permutation matrices. It remains to develop around this paradigm a formal privacy model based on permutation. Such model should be sufficiently intuitive to allow non-experts to understand what it really entails for privacy to permute, in the same way that the privacy principles lying behind k-anonymity and differential privacy can be grasp by most. Moreover, similarly to differential privacy this model should ideally exhibit simple composition properties, which are highly handy in practice. Based on these requirements, this paper proposes a new privacy model for PPDP called $P_{\alpha,\beta}$-privacy. Using for benchmark a one-time pad, an absolutely secure encryption method, this model conveys a reasonably intuitive meaning of the privacy guarantees brought by permutation, can be used *ex-ante* or *ex-post*, and exhibits simple composition properties. We illustrate the application of this new model using an empirical example.

Keywords: Privacy-Preserving Data Publishing · Permutation paradigm · Privacy model · One-time pad

1 Introduction

Statistical disclosure control (SDC), that is anonymization techniques for Privacy-Preserving data publishing (PPDP), has a rich history in providing the analytical apparatus through which the privacy/information trade-off can be assessed and implemented [1]. The overall goal of SDC is to provide to data releasers tools for modifying the original data set in some way that reduces disclosure risk while altering as little as possible the information that it contains. Overall, SDC techniques can be classified into two main approaches:

- *Privacy-first*: the method is applied with the primary goal of complying with some privacy models, judged as acceptable and under which data exchange can take place.

J. Domingo-Ferrer and K. Muralidhar (Eds.): PSD 2020, LNCS 12276, pp. 3–20, 2020.
https://doi.org/10.1007/978-3-030-57521-2_1

- *Utility-first*: the method is applied with the primary goal of complying with some pre-requisites on the level of information, judged as valuable enough to make data exchange worthwhile.

Over the years, SDC has burgeoned in many directions. Such diversity is undoubtedly useful but has one major drawback: a lack of a comprehensive view across the relative performances of the techniques available in different context. A step toward the resolution of this limitation has been recently proposed [2, 3], by establishing that any anonymization method can be viewed as functionally equivalent to a permutation of the original data, plus eventually a small noise addition. This insight, called the permutation paradigm, unambiguously establishes a common ground upon which any anonymization method can be used. This paradigm was originally considered by its authors as a way to evaluate any method applied to any data, but can in fact also be viewed as a general anonymization method where the original data are anonymized with the injection of permutation matrices [4].

While appealing, it remains however to develop around those insights a formal privacy model. Clearly, differentially private or k-anonymous published data sets can be seen through the lens of permutation [5]. Thus, it is always possible to assess *a posteriori* which parametrizations of permutations are compliant with k-anonymity or differential privacy. Nevertheless, given the fact that the very nature of SDC is permutation, it seems natural enough to require a privacy model based on permutation directly, not least for the practical purpose of being able to define privacy guarantees *a priori*, build the according permutation matrices, and anonymize the original data with them.

At a general level, a privacy model is a condition, based on one or a set of parameters, that guarantees an upper-bound on the disclosure risk in the case of an attack from an intruder [6], and which must be linked to a large extent to a set of tools able to enforce this model. Moreover, to be widely used it should comply with a non-formal but nonetheless essential requirement, in the sense that it must be easy to understand by non-experts. The two current mainstream privacy models, k-anonymity and ε-differential privacy, fulfil these two conditions [1]. In particular, they provide intuitive guarantees that can be readily understood by a large audience: in k-anonymity, the intuition is that an individual is hidden in a crowd of k identical individuals; in ε-differential privacy, the intuition is, at least for small ε, that the presence or absence of one individual cannot be noticed from the anonymized data.

An additional and desirable requirement for a privacy model relates to its composability. If several blocks of an anonymized data set are proven to fulfill the privacy model selected, each for some parameters' levels, then it seems natural, and very convenient in practice, to require that the privacy guarantees fulfil by the overall anonymized data set can be assessed from the guarantees of the blocks, and that the parameters' levels of the overall data set be expressed as a function of the levels of each block. Composability could appear stringent at first glance, but it is a condition also satisfied by differential privacy [7], which allows notably splitting computational workloads into independent chunks that can be independently anonymized.

Over the years, several privacy models have been proposed, all different along the type and strength of the privacy guarantee that they propose [1]. To the best of the author's knowledge, only one privacy model based on permutation is available in the literature, which in fact has been proposed in the same paper establishing the permutation paradigm [3]. This model, called *(d, v, f)-permuted privacy*, relies on the identification of constraints on absolute permutation distances plus a diversity criterion. Being essentially an *ex-post* privacy model, (d, v, f)-permuted privacy offers the possibility of verifiability by each subject, which is an appealing property, and obviously also by the data protector. However, its elaborate definition clearly lacks of intuitiveness and is also not composable. Moreover, it does not lean against some intuitive benchmark to assess the privacy guarantees.

Based on this state of affairs, the purpose of this paper is to establish a new privacy model based on permutation called $P_{\alpha,\beta}$-privacy, that we believe can be easily understood by the layman, that is composable, and can be used both *ex-ante* (for the calibration of permutation matrices) or *ex-post* (for assessing the privacy guarantees of any SDC method on any data set from the permutation perspective). Moreover, this model leans back against the equivalent in anonymization of a *one-time pad*, a procedure which offers absolute privacy guarantees. The rest of the paper proceeds as follow. Section 2 gives some background elements on the permutation paradigm and its consequences, needed later on. Section 3 introduces $P_{\alpha,\beta}$-privacy and identifies its key functioning and properties. Section 4 presents some empirical results based on this new model. Conclusions and future research directions are gathered in Sect. 5.

2 Background Elements

2.1 The Permutation-Based Paradigm in Anonymization

The permutation paradigm in data anonymization starts from the observation that any anonymized data set can be viewed as a permutation of the original data plus a non-rank perturbative noise addition. Thus, it establishes that all masking methods can be thought of in terms of a single ingredient, i.e. permutation. This result clearly has far-reaching conceptual and practical consequences, in the sense that it provides a single and easily understandable reading key, independent of the model parameters, the risk measures or the specific characteristics of the data, to interpret the utility/protection outcome of an anonymization procedure.

To illustrate this equivalence, we introduce a toy example which consists (without loss of generality) of five records and three attributes $X = (X_1, X_2, X_3)$ generated by sampling $N(10, 10^2)$, $N(100, 40^2)$ and $N(1000, 2000^2)$ distributions, respectively. Noise is then added to obtain $Y = (Y_1, Y_2, Y_3)$, the three anonymized version of the attributes, from $N(0, 5^2)$, $N(0, 20^2)$ and $N(0, 1000^2)$ distributions, respectively. One can see that the masking procedure generates a permutation of the records of the original data (Table 1).

Now, as the attributes' values of a data set can always be ranked (see below), it is always possible to derive a data set Z that contains the attributes X_1, X_2 and X_3, but ordered according to the ranks of Y_1, Y_2 and Y_3, respectively, i.e. in Table 1 re-ordering (X_1, X_2, X_3) according to (Y_{1R}, Y_{2R}, Y_{3R}). This can be done following the post-masking reverse procedure outlined in [2]. Finally, the masked data Y can be fully reconstituted by adding small noises to each observation in each attribute. By construction, Z has the same marginal distributions as X, which is an appealing property. Moreover, under a

Table 1. An illustration of the permutation paradigm

Original dataset X			Masked dataset Y		
X_1	X_2	X_3	Y_1	Y_2	Y_3
13	135	3707	8	160	3248
20	52	826	20	57	822
2	123	-1317	-1	122	248
15	165	2419	18	135	597
29	160	-1008	29	164	-1927
Rank of the original attribute			**Rank of the masked attribute**		
X_{1R}	X_{2R}	X_{3R}	Y_{1R}	Y_{2R}	Y_{3R}
4	3	1	4	2	1
2	5	3	2	5	2
5	4	5	5	4	4
3	1	2	3	3	3
1	2	4	1	1	5

maximum-knowledge intruder model of disclosure risk evaluation, the small noise addition turns out to be irrelevant [8]: re-identification via record linkage can only come from permutation, as by construction noise addition cannot alter ranks. Reverse mapping thus establishes permutation as the overarching principle of data anonymization, allowing the functioning of any method to be viewed as the outcome of a permutation of the original data, independently of how the method operates. This functional equivalence leads to the following result (see [9] for the original and full proposal):

Result 1: *For a dataset X with n records and p attributes (X_1, ..., X_p), its anonymized version Y can always be written, regardless of the anonymization methods used, as:*

$$\mathrm{Y} = \left(P_1X_1, \ldots, P_pX_p\right) + \mathrm{E} \tag{1}$$

where $P_1 = A_1^T D_1 A_1, \ldots, P_p = A_p^T D_p A_p$ is a set of p permutation matrices and E is a matrix of small noises. $A_1, \ldots, A_P$ is a set of p permutation matrices that sort the attributes in increasing order, $A_1^T, \ldots A_p^T$ a set of p transposed permutation matrices that put back the attribute in the original order, and $D_1, \ldots D_P$ is a set of permutation matrices for anonymizing the data.

This proposition characterizes permutation matrices as an encompassing tool for data anonymization. Proceeding attribute by attribute, each is first permuted to appear in increasing order, then the key is injected, and finally it is re-ordered back to its original form by applying the inverse of the first step (which in the case of a permutation matrix is simply its transpose). This formalizes the common basis of comparison for different mechanisms that the permutation paradigm originally proposed, but is also an anonymization method in itself. Whatever the differences in the natures of the methods to be compared and the distributional features of the original data, the methods can fundamentally always be viewed as the application of different permutation matrices to the original data.

2.2 Ranks as the Most Basic Level of Information for Anonymization

An immediate consequence of the permutation paradigm is that ranks constitute the most basic level of information, necessary and sufficient, to conduct anonymization. Data sets come in a variety of structures and shapes. They can be made of only numerical or categorical attributes, or more often as a mix of numerical, categorical and nominal attributes, to which different anonymization methods can be applied [10]. However, whatever the types of attributes it is always possible to rank individuals. This is obvious in the cases of numerical or categorical attributes. In the specific case of nominal attributes this may require some involved methods to come up with a ranking, as for example when an attribute lists some medical terms in plain text, but it is still always possible [11]. Moreover, the permutation-based paradigm made it clear that any anonymization method, whatever its operating principle or the data upon which it is applied, can be thought of in terms of permutation. As we saw above, the small noises are necessary to recreate strictly the anonymized version of the numerical attributes for some methods, but in terms of privacy and information properties they are not required. As intuitively any information loss should come hand in hand with a security improvement, and as small noises do not qualify as such, at a fundamental level they are cosmetic and can be discarded. In particular, the term E in *Eq. (1)* can be ignored. Moreover, by construction these noises do not exist in the case of ordinal or nominal attributes.

Rank permutation has thus been shown to be the essential principle lying behind anonymization. Then, knowing the ranks of each individual is the only information that is truly necessary. As far as performing anonymization is concerned, one can put the underlying metrics of each attribute aside, stripping down the data set to its most basic level, and works with the ranks to anonymize.

To formalize this and introduce some elements that will be used in the rest of this paper, let $N = \{1, \ldots n\}$ denotes a finite set of individuals with $n \geq 2$, let $P = \{1, \ldots, \mathrm{p}\}$ denotes a finite set of attributes with $p \geq 2$. $\forall j \in P$, let V^j be a non-empty set of strictly positive integers with $\#V^j \geq 2$. Moreover, let $R = \left(r_{ij}\right)_{(n,p)}$ be a matrix of size $n \times p$ with $r_{ij} \in V^j\, \forall i \in N, \forall j \in P$. The matrix R is the most basic structure for a data set and is strictly sufficient to conduct anonymization. Its building blocks are made of rank strings, i.e. the succession of ranks with which each individual is contributing to the data. One way to see this is:

$$R = \begin{pmatrix} r_{11} & r_{12} & r_{13} & \cdots & r_{1p} \\ r_{21} & r_{22} & r_{23} & \cdots & r_{2p} \\ r_{31} & r_{32} & r_{33} & \cdots & r_{3p} \\ \vdots & \vdots & \vdots & \vdots & \vdots \\ r_{n1} & r_{n2} & r_{n3} & \cdots & r_{np} \end{pmatrix} = \begin{pmatrix} r_{1.} \\ r_{2.} \\ r_{3.} \\ \vdots \\ r_{n.} \end{pmatrix}$$
$$= \left(r_{.1}\; r_{..2}\; r_{.3} \ldots r_{.p}\right) \tag{2}$$

The matrix R can be represented either as a collection of rank strings at the individual level, i.e. each individual i has a rank in each attributes j $\forall j \in P$, or a collection of rank strings at the attributes level, i.e. each attribute j is ranking the individual i $\forall i \in N$. Thus, a data set in its most basic form can be represented by either a collection of stacked individual rank strings $r_{i.}$, or a collection of collated attribute rank strings $r_{.j}$.

An individual rank string $r_{i.}$ is the contribution of individual i in term of information to the data set. From this perspective, which starts by considering first the individual, it is clear that anonymization must alter individual rank strings. If this is not the case, then the individual is left untouched and is not protected. In what follows, we use the comparison of individual rank strings before and after anonymization as the starting point to build a new privacy model based on permutation.

3 $P_{\alpha,\beta}$-Privacy

The fact that anonymization appears to rely on the single principle of permutation can be phrased under the following casual statement: *"to be protected, become someone else"*. Indeed, for an individual to be protected she must inherit the attributes' values of other individuals in the original data set, for some attributes if not all. This is obviously what permutation is about.

Note that a trivial consequence is that one cannot be anonymized alone. For anonymization to take place, other individuals are necessary. For instance, if some noise is added to some of the attributes' values of one individual taken in isolation, but that the noise magnitude is not enough to switch ranks with other individuals, then anonymization did not really take place and in *Eq. (1)* $D_1, \ldots, D_p$ will turn to be the identity matrices. In that case, no permutation happened. Thus, to be anonymized you need to consider at the same time other individuals and exchange values with them.

Because anonymization means for an individual to inherit the values of others, this raises two natural questions: *"are these changes common in what defines me?"* and *"if some changes happened, are they important? What are their depths?"*. We will call these two casual questions the "frequency" and the "deepness" questions, respectively. The "frequency" question may be of lesser relevancy for data sets with a small number of attributes. However, when this number is very large all attributes may not be altered by anonymization and this question may turn relevant.

Naturally, an individual will rightly feel more protected than otherwise the more attributes have been changed and if her new values for the modified attributes come from individuals that are far from her in the original data set. Thus, in principle protection happens, and can be gauged, according to the frequency of the transformation of an individual along her attributes, as well as the depth of these changes.

Along these two intuitive considerations we will consider a third, may be slightly less intuitive but we believe nonetheless relevant, one: *"do my modified characteristics receive the same treatment? Among my values that have been changed, has everything been changed the same way?"*. We will call it the "balance in treatment" question. Indeed, among the attributes that have been changed, some may have been changed more deeply than others may. Thus, an individual may be interested in knowing if all her modified attributes received the same treatment or not. For example, if a large number of attributes are deeply changed the same way, then the individual can reasonably consider herself more protected in comparison to a situation where some attributes have been changed deeply and other not so much.

At this stage, we are fully aware that the reader may be surprised by such casual formulations and considerations in a supposedly scientific paper. However, we believe that they are key. As we outlined above, the success of mainstream privacy models such as ε-differential privacy and k-anonymity relies on the fact that a solid analytical framework can be expressed as casual principles understandable by most. And insofar as anonymization is a field practiced by experts who deal with questions that non-experts are very interested in, any privacy model should have an easy interpretation. As a result, our objective is to build a privacy model to reply to the casual questions asked above. Thus, this paper is about a formal treatment of some casual questions, an approach that is generally better than the opposite.

3.1 The P_α Aggregator

Consider the following quantity:

$$z_{ij} = \frac{abs\left(r_{ij} - r'_{ij}\right)}{\max\left(r_{1j}, \ldots, r_{nj}\right) - 1} \tag{3}$$

where r_{ij} is the rank of individual i in attribute j in the *original* data set, as defined in *Eq. (2)*, and r'_{ij} is the rank of individual i in attribute j in the *anonymized* data set. Thus, z_{ij} is simply a measure of absolute rank differences between the original and the anonymized data set for the individual i in attribute j, *normalized*, without loss of generality, by the maximum number of ranks that an individual can be switched in the attribute. For a numerical attribute collected over n individuals, generally $\max\left(r_{1j}, \ldots, r_{nj}\right) = n$, but for an ordinal attribute it can be much smaller, and the normalization is a way to align the possibility of rank changes on a comparable scale, here expressed in percentages.

Next, we introduce the following aggregator:

$$\mathrm{P}_{\alpha,i}\left(z_{i1}, \ldots, z_{ip}\right) = \frac{1}{p}\sum_{j=1}^{q}\left(z_{ij}\right)^{\alpha} \forall\alpha \geq 0 \tag{4}$$

Remark that $\mathrm{P}_{\alpha,i}(.)$ is a norm on a Fréchet space [12]. It aggregates the normalized absolute rank distances of the q attributes that have been modified by the anonymization process, thus $q \leq p$. By construction, $\mathrm{P}_{\alpha,i}\left(z_{i1}, \ldots, z_{ip}\right)$ is bounded between zero and one.

Let $C_i = q/p$ be the share of attributes that have been modified for individual i, and $D_i = \sum_{j=1}^{q} z_{ij}/q$ the average of the absolute normalized rank distances taken over the modified attributes. Clearly, we have:

$$\mathrm{P}_{0,i}\left(z_{i1}, \ldots, z_{ip}\right) = \frac{q}{p} = C_i \tag{5}$$

and

$$\mathrm{P}_{1,i}\left(z_{i1}, \ldots, z_{ip}\right) = C_i \times D_i \tag{6}$$

For $\alpha = 0$, $\mathrm{P}_{\alpha,i}(.)$ brings a reply to the "frequency" question about how many attributes have been changed, as a percentage of all the attributes in the data set. For $\alpha = 1$, it brings

a reply to the "deepness" question, by measuring the average depth of the changes taken over all the attributes, which can be decomposed further as the product of the share of the changes multiplied by the depth of the changes *only* for the attributes that have been modified. The case $\alpha = 2$ squares the normalized absolute rank differences and thus weighs the rank differences by the rank differences. It can be straightforwardly demonstrated that for $\alpha = 2$, the following result holds:
Result 2: *For* $\alpha = 2$, $\mathrm{P}_{\alpha,i}(z_{i1}, \ldots, z_{im})$ *satisfies the following decomposition:*

$$\mathrm{P}_{2,i}\left(z_{i1}, \ldots, z_{ip}\right) = C_i\left[D_i^2 + (1 - D_i)^2 V_i^2\right] \tag{7}$$

where V_i^2 *is the squared coefficient of variation of the absolute normalized rank distances among the attributes that have been modified.*

Suppose that all modified attributes have the same absolute normalized rank distances. Then $\mathrm{P}_{2,i}\left(z_{i1}, \ldots, z_{ip}\right) = C_i D_i^2$ as $V_i^2 = 0$. However, if the modified attributes did not receive the same treatment, then $\mathrm{P}_{2,i}\left(z_{i1}, \ldots, z_{ip}\right)$ is larger by $C_i(1 - D_i)^2 V_i^2$. When C_i and D_i are held constant, then $\mathrm{P}_{2,i}\left(z_{i1}, \ldots, z_{ip}\right)$ varies with V_i^2, the difference in treatment on the changed attributes. Thus, $\mathrm{P}_{2,i}\left(z_{i1}, \ldots, z_{ip}\right)$ and its decomposition brings a reply to the "balance in treatment", allowing to evaluate if each attributes have been treated the same way or not.

Finally, assume that the p attributes are divided into L collections of attributes $l = 1, \ldots, L$ of size $p^{(l)}$ and denotes by $z^{(l)}$ the vector of normalized absolute rank differences for the lth collection. Then we have the following result:
Result 3: *For an individual i and any set of attributes broken down into subgroups with associated normalized absolute rank differences vectors* $z^{(1)}, \ldots, z^{(L)}$:

$$\mathrm{P}_{\alpha,i}\left(z_{i1}, \ldots, z_{ip}\right) = \sum_{l=1}^{L} \frac{p^{(l)}}{p} \mathrm{P}_{\alpha,i}\left(z^{(l)}\right) \forall \alpha \geq 0 \tag{8}$$

Following *Eq. (8)*, $\mathrm{P}_{\alpha,i}\left(z_{i1}, \ldots, z_{ip}\right)$ is additively decomposable with attribute share weights for all positive α.

To summarize, the additively decomposable aggregator $\mathrm{P}_{\alpha}(.)$ allows evaluating different quantities to reply to the casual questions asked above, by simply making vary one parameter. Note that the cases $\alpha \geq 2$ can also be considered. In fact, as α tends to infinity, the value of the largest normalized absolute rank difference is all that matters. However, the larger the α, the smaller $\mathrm{P}_{\alpha,i}\left(z_{i1}, \ldots, z_{ip}\right)$ for a given distribution $\left(z_{i1}, \ldots, z_{ip}\right)$. As a result, and taking into account some tentative concerns about the "psychology of numbers", in the rest of this paper we will focus mainly on $\alpha = 0, 1, 2$. As we saw, those cases display some interesting decomposition properties, and will yield some "healthier-looking" values with magnitude easy to grasp and relatively intuitive.

3.2 $P_{\alpha,\beta}$-Privacy at the Individual Level

To bring meaningful replies to the casual questions asked above, one needs a proper benchmark. For instance, one can reply to the "deepness" question by computing $\mathrm{P}_{1,i}\left(z_{i1}, \ldots, z_{ip}\right)$ and the higher the value, the better the protection. However, up to which level can it be really qualified as better? Of course, if $P_{1,i}(z_{i1}, \ldots, z_{ip})$ is close to 100% undoubtedly strong anonymization has been applied and one can reasonably think that in that case the task of an intruder for re-identifying individual i will be arduous. However, this cannot be the case for all individuals in the data set, as not all can be permuted maximally. Moreover, we believe as desirable to reason with a benchmark that brings some sense and intuition to the whole exercise. In this paper, we propose to reply to the following question: *"how do my anonymized values differ compare to absolutely secure modifications?"*.

Coming back to *Eq. (1)*, one can observe that, theoretically speaking, and if the permutation matrices are truly selected randomly, it is a *one-time pad* [13]. Indeed, considering the attributes as the plaintext, the anonymized attributes as the ciphertext and the permutation matrices as encryption keys, which are of the *same size* than the plaintext, then for a *truly random* generation of keys *Eq. (1)* describes a one-time pad [14]. Of course, while in cryptography a truly random generation of keys is acceptable, in data anonymization it is not. In addition to providing some privacy guarantees to individuals, anonymized data should also meet data users' needs by offering some information. As a result, some structures and constraints must be applied to the permutation keys for the released data to be meaningful. The fact that in data anonymization the keys selection must be guided with both protection and information in mind precludes randomly generating them.

Still, while not acceptable in data anonymization it provides an interesting benchmark. Being information-theoretically secure, a one-time pad is impossible to decrypt as the ciphertext provides absolutely no information to an intruder about the plaintext (except the length of the message) and a ciphertext can be translated into any plaintext of the same length, with all being equally likely. In data anonymization, this means that if $D_1, \ldots, D_p$ are truly random, then an intruder can never plausibly claim to have re-identified any individual or learn some of her attributes' values. It must be noted that, without being named, one-time pads have been used in one occurrence in the SDC literature to conduct attacks by a maximum-knowledge intruder [8].

The simple idea of $P_{\alpha,\beta}$-privacy is thus to compare P_α's values with the ones achieved by a one-time pad. Stated otherwise, $P_{\alpha,\beta}$-privacy aims at delivering the following statement: *"you have been changed at P_α%, and this change is at most β percentage points different compared to the case of an absolutely secure change"*. We believe such statement vehicles an intuitive understanding for most people. To formalize it, we introduce the following quantity:

$$z_{ij}^S = \frac{abs\left(r_{ij} - r_{ij}^S\right)}{\max\left(r_{1j}, \ldots, r_{nj}\right) - 1} \tag{9}$$

where r_{ij}^S is the rank of individual i in attribute j in the one-time pad anonymized data set, i.e. where all attributes have been permuted randomly. The definition of $P_{\alpha,\beta}$-privacy follows:

Definition 1: *An individual i is $P_{\alpha,\beta}$-private if:*

$$\left|P_{\alpha,i}\left(z_{i1}, \ldots, z_{ip}\right) - P_{\alpha,i}\left(z_{i1}^S, \ldots, z_{ip}^S\right)\right| \leq \beta \tag{10}$$

$P_{\alpha,\beta}$-privacy requires that the difference between $\mathrm{P}_{\alpha,i}(z_{i1},\ldots,z_{ip})$ and $\mathrm{P}_{\alpha,i}\left(z_{i1}^{S},\ldots,z_{ip}^{S}\right)$ must be contained in an interval centered around zero, and for which the narrowness is driven by β. Stated otherwise, it requires a degree of proximity between the changes incurred, in terms of permutation, by an anonymization method and a one-time pad. The proximity is controlled by the parameter β, with by construction $0 \leq \beta \leq 1$, and which can be interpreted as a *maximum tolerance to departure from absolute security*: the smaller (resp. larger) is β, the greater (resp. lesser) is the required proximity. For instance, if $\mathrm{P}_{0,i}(z_{i1},\ldots,z_{ip}) = 0.4$ and $\mathrm{P}_{0,i}\left(z_{i1}^{S},\ldots,z_{ip}^{S}\right) = 0.5$, then the individual i is $P_{0,0.1}$-private. In plain words, the individual can make the following claim: *"40% of my attributes have been changed, which is at most 10% points different from the situation offering me absolute security"*. If $\mathrm{P}_{1,i}(z_{i1},\ldots,z_{ip}) = 0.25$ and $\mathrm{P}_{1,i}\left(z_{i1}^{S},\ldots,z_{ip}^{S}\right) = 0.40$, then the individual is $P_{1,0.15}$-private and the claim becomes: *"I have been transformed into other individuals that were originally and on average 25% far from me; this transformation departs by at most 15 percentage points from a transformation offering me absolute security"*.

β also offers a direct link to the amount of information provided by the anonymized data set. Clearly, the narrower it is, the likelier the permutation matrices are to be generated randomly. For an anonymized data set to be useful and worthwhile disseminating, it must preserve information to a certain extent. Stated otherwise, the permutations applied to the original data must be constrained, e.g. it is forbidden to permute some attributes more than a certain distance, which means that one must depart from randomness. Thus, the narrower is β, the lesser information is preserved. An additional consequence is that $P_{\alpha,\beta}$*-privacy* can be used *ex-ante* to guide the calibration of permutation matrices.

Remark that we consider that $\mathrm{P}_{\alpha,i}(z_{i1},\ldots,z_{ip})$ can depart from $\mathrm{P}_{\alpha,i}\left(z_{i1}^{S},\ldots,z_{ip}^{S}\right)$ by above or below. Generally, applying random permutations at the data set level will tend to break the rank-dependencies among attributes. However, in practice all configurations can occur (even having as random permutations the identity matrices, leaving the data untouched [8]). This is even truer at the individual level, where the values of $\mathrm{P}_{\alpha,i}\left(z_{i1}^{S},\ldots,z_{ip}^{S}\right)$ will depend on the original rank strings and the redundancy in each rank. In the absence of clear directions for the changes for each individual after having been randomly permuted, we assess the possible changes from above or below. In addition, it is advisable in practice to generate several random permutations and take the average of $\mathrm{P}_{\alpha,i}\left(z_{i1}^{S},\ldots,z_{ip}^{S}\right)$ across the permutations to avoid the privacy assessment being driven by a peculiar random draw in the permutation matrices.

Now, from the additive decomposability of $\mathrm{P}_{\alpha,i}(z_{i1},\ldots,z_{ip})$ it can be demonstrated, after some manipulations, that:

Result 4: *For any set of p attributes broken down into subgroups with associated normalized absolute rank differences vectors $z^{(1)},\ldots,z^{(L)}$ and with each subgroups l being $P_{\alpha,\beta^{(l)}}$-private, i.e. $\left|\mathrm{P}_{\alpha,i}(z^{(l)}) - \mathrm{P}_{\alpha,i}(z^{S,(l)})\right| \leq \beta^{(l)}$, it holds that:*

$$\left|\mathrm{P}_{\alpha,i}\left(z^{(1)},\ldots,z^{(L)}\right) - \mathrm{P}_{\alpha,i}\left(z^{S,(1)},\ldots,z^{S,(L)}\right)\right| \leq \sum_{l=1}^{L}\frac{p^{(l)}}{p}\beta^{(l)} \tag{11}$$

$P_{\alpha,\beta}$-privacy is *horizontally composable.* If an individual i is anonymized separately over some different sets of attributes (each with different $P_{\alpha,\beta}$-privacy guarantees) and then the sets are merged, the overall set is also $P_{\alpha,\beta}$-private, with as guarantee the attribute shares weighted average of each sub-set's guarantee.

3.3 $P_{\alpha,\beta}$-Privacy at the Data Set Level

So far, we focused on the individuals. Indeed, the very nature of anonymization is about individuals and the protection of their information, so it seems only natural that privacy at the data set level must be designed from basic building blocks which are the assessment at the individual level. In fact, it can be proved that any coherent approach for evaluating some anonymization outcomes can only start from an assessment at the individual level [15].

To evaluate the privacy guarantees at the data set level, one possible path is to slot the $P_\alpha(.)$ aggregators at the individual level within a P_α aggregator that will deliver an evaluation at the data set level. That is, we propose the following, where D denotes the data set:

$$\begin{aligned}&\mathrm{P}_{\alpha,D}\big(\mathrm{P}_{\alpha,1}(z_{11},\ldots,z_{1p}),\ldots,\mathrm{P}_{\alpha,n}(z_{n1},\ldots,z_{np})\big)\\&\quad=\frac{1}{n}\sum_{i=1}^{s}\big(\mathrm{P}_{\alpha,i}(z_{i1},\ldots,z_{ip})\big)^{\alpha}\forall\alpha\geq 0\end{aligned}\tag{12}$$

$\mathrm{P}_{\alpha,D}(.)$ aggregates the individual $\mathrm{P}_{\alpha,i}(z_{i1},\ldots,z_{ip})$ of the s individuals that have been modified by the anonymization process, thus $s\leq n$. Its interpretation is the same than for the aggregator at the individual level, also inheriting *de facto* its properties outlined above. By construction, $\mathrm{P}_{\alpha,D}(.)$ is also bounded between zero and one.

For $\alpha=0$, $\mathrm{P}_{0,D}(.)$ measures the share of individuals that have been altered by the anonymization process. For $\alpha=1$, it measures the depth of the changes averaged over all individuals, which can be decomposed further as the product of the share of the modified individuals multiplied by the average depth of the changes *only* on the individuals that have been modified. The case $\alpha=2$ squares the $\mathrm{P}_{\alpha,i}(z_{i1},\ldots,z_{ip})$'s and thus weighs the $\mathrm{P}_{\alpha,i}(z_{i1},\ldots,z_{ip})$ by the $\mathrm{P}_{\alpha,i}(z_{i1},\ldots,z_{ip})$. Using the decomposition in *Eq. (7)*, one can assess if all individuals have been treated equally or not during the anonymization process.

We can now define $P_{\alpha,\beta}$-privacy at the data set level:

Definition 2: *A data set D is $P_{\alpha,\beta}$-private if*

$$\begin{aligned}&\big|\mathrm{P}_{\alpha,D}\big(\mathrm{P}_{\alpha,1}(z_{11},\ldots,z_{1p}),\ldots,\mathrm{P}_{\alpha,n}(z_{n1},\ldots,z_{np})\big)\\&\quad-\mathrm{P}_{\alpha,D}\Big(\mathrm{P}_{\alpha,1}\big(z_{11}^{S},\ldots,z_{1p}^{S}\big),\ldots,\mathrm{P}_{\alpha,n}\big(z_{n1}^{S},\ldots,z_{np}^{S}\big)\Big)\Big|\\&\quad\leq\beta\end{aligned}\tag{13}$$

Using the decomposability property of $P_\alpha(.)$ and assuming that we consider individuals equally for anonymization, i.e. they have the same weight and none deserves more care than another *ex-ante* anonymization, we can express the $P_{\alpha,\beta}$-privacy guarantee of a data set as a function of the $P_{\alpha,\beta}$-privacy guarantees at the individual level:

Result 4: *For a data set D made of n individuals and p attributes, if* $\left|\mathrm{P}_{\alpha,i}(z_{i1},\ldots,z_{ip})-\mathrm{P}_{\alpha,i}\left(z_{i1}^{S},\ldots,z_{ip}^{S}\right)\right|\leq\beta_i\ \forall i=1,\ldots,n$, *then it holds that:*

$$\begin{aligned}&\left|\mathrm{P}_{\alpha,D}(\mathrm{P}_{\alpha,1}(z_{11},\ldots,z_{1p}),\ldots,\mathrm{P}_{\alpha,n}(z_{n1},\ldots,z_{np}))-\right.\\&\left.\mathrm{P}_{\alpha,D}\left(\mathrm{P}_{\alpha,1}\left(z_{11}^{S},\ldots,z_{1p}^{S}\right),\ldots,\mathrm{P}_{\alpha,n}\left(z_{n1}^{S},\ldots,z_{np}^{S}\right)\right)\right|\leq\frac{1}{n}\sum\nolimits_{i=1}^{n}\beta_i\end{aligned}\tag{14}$$

The $P_{\alpha,\beta}$-privacy guarantee of a data set is the average of the $P_{\alpha,\beta}$-privacy guarantees at the individual level. A direct corollary of this result is that $P_{\alpha,\beta}$-privacy is *vertically* composable. For a given set of attributes, if one anonymizes separately different groups of individuals, whatever the method used in each, then the privacy guarantee of the stacked data set will simply be the group shares weighted average of each sub-data set's guarantee:

Result 5: *For a set of n individuals broken down into subgroups* $n^{(1)},\ldots,n^{(L)}$ *and with each observed over the same p attributes, if each sub-data sets l is* $P_{\alpha,\beta^{(l)}}$*-private, i.e.*

$$\begin{aligned}&\left|\mathrm{P}_{\alpha,D^{(l)}}\left(\mathrm{P}_{\alpha,1}(z_{11},\ldots,z_{1p}),\ldots,\mathrm{P}_{\alpha,n^{(l)}}\left(z_{n^{(l)}1},\ldots,z_{n^{(l)}p}\right)\right)-\right.\\&\left.\mathrm{P}_{\alpha,D^{(l)}}\left(\mathrm{P}_{\alpha,1}\left(z_{11}^{s},\ldots,z_{1p}^{s}\right),\ldots,\mathrm{P}_{\alpha,n^{(l)}}\left(z_{n^{(l)}1}^{s},\ldots,z_{n^{(l)}p}^{s}\right)\right)\right|\leq\beta^{(l)}\end{aligned}$$

, then it holds that:

$$\begin{aligned}&\left|\mathrm{P}_{\alpha,D}\left(\begin{array}{c}\mathrm{P}_{\alpha,D^{(1)}}\left(\mathrm{P}_{\alpha,1}(z_{11},\ldots,z_{1p}),\ldots,\mathrm{P}_{\alpha,n^{(1)}}\left(z_{n^{(1)}1},\ldots,z_{n^{(1)}p}\right)\right),\\\ldots,\mathrm{P}_{\alpha,D^{(L)}}\left(\mathrm{P}_{\alpha,1}(z_{11},\ldots,z_{1p}),\ldots,\mathrm{P}_{\alpha,n^{(L)}}\left(z_{n^{(L)}1},\ldots,z_{n^{(L)}p}\right)\right)\end{array}\right)\right.\\&\left.-\begin{array}{c}\mathrm{P}_{\alpha,D^{(1)}}\left(\mathrm{P}_{\alpha,1}\left(z_{11}^{s},\ldots,z_{1p}^{s}\right),\ldots,\mathrm{P}_{\alpha,n^{(1)}}\left(z_{n^{(1)}1}^{s},\ldots,z_{n^{(1)}p}^{s}\right)\right),\\\ldots,\mathrm{P}_{\alpha,D^{(L)}}\left(\mathrm{P}_{\alpha,1}\left(z_{11}^{s},\ldots,z_{1p}^{s}\right),\ldots,\mathrm{P}_{\alpha,n^{(L)}}\left(z_{n^{(L)}1}^{s},\ldots,z_{n^{(L)}p}^{s}\right)\right)\end{array}\right|\\&\leq\sum_{l=1}^{L}\frac{n^{(l)}}{n}\beta^{(l)}\end{aligned}\tag{15}$$

4 Empirical Illustrations

The objective of this section is to illustrate how $P_{\alpha,\beta}$-privacy operates. The experiment is based, without loss of generality, on a small data set of 20 observations and three attributes. In particular, this simple example will allow to display in a tractable way $P_{\alpha,\beta}$-privacy at the individual level.

We generate the original data set by sampling N(50, 10^2), N(500, 50^2) and N(2500, 250^2) distributions, respectively. All attributes are then anonymized with additive noise under three scenario with some standard deviations equal to 10%, 50% and 100% of the standard errors of the original values, respectively. Regarding the generation of the one-time pad, to avoid the problems mentioned above we generate five randomly permuted data sets and take the average of $\mathrm{P}_{\alpha,i}\left(z_{i1}^{S},\ldots,z_{ip}^{S}\right)$ over those sets. Tables 2 and 3 show the values obtained under the three scenario plus the one-time pads averaged values, Table 4 shows $P_{0,\beta}$-privacy guarantees and Table 5 $P_{1,\beta}$-privacy guarantees.

Table 2. P_α aggregators: scenario 1 & 2

ID	Scenario 1				Scenario 2			
	P0	P1	P2 of which V		P0	P1	P2 of which V	
1	33%	2%	0%	141%	100%	11%	1%	41%
2	67%	5%	0%	82%	33%	2%	0%	141%
3	67%	7%	1%	71%	67%	9%	1%	75%
4	33%	2%	0%	141%	33%	2%	0%	141%
5	67%	4%	0%	71%	100%	19%	5%	46%
6	67%	4%	0%	71%	100%	9%	1%	28%
7	33%	2%	0%	141%	33%	7%	1%	141%
8	0%	0%	0%	0%	67%	5%	0%	82%
9	0%	0%	0%	0%	67%	23%	13%	125%
10	33%	2%	0%	141%	100%	14%	2%	18%
11	0%	0%	0%	0%	67%	9%	2%	102%
12	33%	2%	0%	141%	100%	9%	1%	28%
13	67%	5%	0%	82%	67%	5%	0%	82%
14	67%	4%	0%	71%	67%	14%	3%	77%
15	0%	0%	0%	0%	33%	2%	0%	141%
16	33%	5%	1%	141%	67%	19%	6%	78%
17	33%	4%	0%	141%	0%	0%	0%	0%
18	67%	9%	2%	102%	67%	11%	2%	108%
19	67%	4%	0%	71%	67%	11%	2%	71%
20	33%	2%	0%	141%	67%	11%	2%	71%
Data set	**40%**	**3%**	**0%**	**88%**	**65%**	**9%**	**2%**	**80%**

As expected, the higher the noise injected, the higher P_α for all α, which is rather intuitive. Consider for example the first individual. Under scenario 1, she can claims that one-third of her attributes have been changed, while she became someone else by 2%, i.e. on average across her attributes she inherited values from other individuals that were originally 2% far from her. Under scenario 2 she can claim higher values; for instance, she inherited values from other individuals that were originally 11% far from her. Under scenario 3, the magnitudes are clearly more pronounced: all her attributes have been altered and she became someone else at 42%. Moreover, according to the coefficient of variation V, which following *Eq. (7)* is one of the components of P_2, the second and third scenario applies similar balanced treatment to each attribute.

Overall, the *average* individual in the data set can claim that 40%, 65% and 87% of her attributes have been changed and that with anonymization she inherited new values

Table 3. P_α aggregators: scenario 3 & one-time pads averaged values

ID	Scenario 3				One-time pads average	
	P0	P1	P2 of which V		P0	P1
1	100%	42%	18%	18%	100%	38%
2	100%	40%	22%	61%	87%	45%
3	100%	19%	5%	56%	93%	24%
4	100%	18%	3%	14%	93%	31%
5	100%	47%	34%	73%	93%	36%
6	0%	0%	0%	0%	100%	41%
7	100%	19%	4%	34%	93%	30%
8	67%	11%	2%	108%	93%	36%
9	100%	21%	8%	89%	100%	36%
10	100%	30%	11%	44%	93%	25%
11	100%	21%	6%	54%	93%	29%
12	100%	18%	3%	28%	100%	39%
13	100%	12%	2%	53%	93%	38%
14	67%	12%	2%	73%	100%	32%
15	67%	7%	1%	94%	100%	35%
16	100%	33%	12%	32%	100%	44%
17	67%	19%	6%	72%	100%	33%
18	67%	16%	4%	82%	93%	36%
19	100%	32%	15%	68%	93%	46%
20	100%	25%	9%	71%	87%	24%
Data set	**87%**	**22%**	**8%**	**56%**	**95%**	**35%**

coming from individuals that were originally at 3%, 9% and 22% far from her, according to scenario 1, 2 and 3, respectively.

By comparing these results to the outcomes obtained from a one-time pad, we can measure if they are indeed important in magnitude, and conclude on the $P_{\alpha,\beta}$-privacy guarantees of each scenario. For $P_{1,\beta}$-privacy, the modifications on the first individual are 38% points different from a situation of absolute security in scenario 1, 28% in scenario 2 and 4% in scenario 3 (Table 5). That is, in scenario 3 her inherited values depart only by 4% points from a situation where she would have been absolutely secure. Overall, with the third scenario she can make the following claim: *"all my attributes have been changed and I metamorphosed on average into individuals that were originally 42%*

Table 4. $P_{0,\beta}$-privacy

ID	Scenario 1	Scenario 2	Scenario 3
1	67%	0%	0%
2	20%	53%	13%
3	27%	27%	7%
4	60%	60%	7%
5	27%	7%	7%
6	33%	0%	100%
7	60%	60%	7%
8	93%	27%	27%
9	100%	33%	0%
10	60%	7%	7%
11	93%	27%	7%
12	67%	0%	0%
13	27%	27%	7%
14	33%	33%	33%
15	100%	67%	33%
16	67%	33%	0%
17	67%	100%	33%
18	27%	27%	27%
19	27%	27%	7%
20	53%	20%	13%
Data set	**55%**	**32%**	**17%**

far from me; such transformation departs only by 4% points from an ideal situation of absolute security". Regarding the privacy guarantees of the whole anonymized data set, here again they produce an ordering conform to intuition (Table 5): scenario 1 delivers $P_{1,0.33}$-privacy guarantees, scenario 2 $P_{1,0.26}$-privacy guarantees and scenario 3 $P_{1,0.15}$-privacy guarantees; for instance, in scenario 3 the anonymized data set is only 15% points different from a one-time pad.

Table 5. $P_{1,\beta}$-privacy

ID	Scenario 1	Scenario 2	Scenario 3
1	38%	28%	4%
2	43%	43%	5%
3	19%	15%	5%
4	27%	29%	13%
5	36%	17%	11%
6	36%	33%	41%
7	27%	23%	11%
8	27%	31%	26%
9	33%	13%	15%
10	23%	11%	5%
11	26%	21%	8%
12	39%	31%	22%
13	38%	33%	26%
14	32%	18%	20%
15	35%	34%	28%
16	44%	24%	10%
17	33%	33%	14%
18	36%	26%	20%
19	46%	35%	14%
20	24%	14%	0%
Data set	**33%**	**26%**	**15%**

5 Conclusions and Future Work

This paper introduced a new privacy model for data publishing based on permutation, $P_{\alpha,\beta}$-privacy. Following the insight proposed by the permutation paradigm, which showed that any SDC techniques can be thought of in terms of permutations, $P_{\alpha,\beta}$-privacy aims at providing a formal treatment of what it really means for privacy to be permuted and leans against an absolutely private benchmark. This treatment is also designed to be translatable into casual statements understandable by most people, so that a wide audience can grasp what $P_{\alpha,\beta}$-privacy means.

Moreover, and because it is based on permutations, it can be applied *ex-post* to assess the privacy guarantees of any data set anonymized through any methods, but can also be used *ex-ante* to tune permutation matrices. $P_{\alpha,\beta}$-privacy is also composable, both horizontally and vertically. This means that if two $P_{\alpha,\beta}$-private data sets comprising the same individuals (resp. attributes), and anonymized through different techniques, are merged (resp. stacked), the resulting overall data set will also be $P_{\alpha,\beta}$-private, with as parameters a function of the parameters of the two first data sets.

In $P_{\alpha,\beta}$-privacy, α stands for the different types of transformation that the information provided by an individual contributing to a data set undergoes. For $\alpha = 0$, one is monitoring the frequencies of the modified attributes. For $\alpha = 1$, one is measuring how an individual has been converted, as a result of permutations, into someone else and how far this someone else was in the original data. Finally, for $\alpha = 2$ more weight is given to the largest transformations, which allows to assess if the attributes' values of an individual received in fact the same transformation. As for β, it measures how an individual is far from absolute secure changes in her characteristics, as conveyed by a one-time pad, a well-known cryptographic protocol that provides absolute secrecy and that this paper brought to use in the context of anonymization.

We leave as future works the deeper exploration of $P_{\alpha,\beta}$-privacy. First of all, the generation of permutation matrices through the *ex-ante* setting of $P_{\alpha,\beta}$-privacy conditions should be explored. Second, establishing, through the lens of $P_{\alpha,\beta}$-privacy guarantees, an inventory of popular SDC methods under different parametrizations and data contexts, is warranted. Finally, the links between $P_{\alpha,\beta}$-privacy and k-anonymity and ε-differential privacy should be investigated, both theoretically and empirically.

References

1. Hundepool, A., et al.: Statistical Disclosure Control. Wiley, Hoboken (2012)
2. Muralidhar, K., Sarathy, R., Domingo-Ferrer, J.: Reverse mapping to preserve the marginal distributions of attributes in masked microdata. In: Domingo-Ferrer, J. (ed.) PSD 2014. LNCS, vol. 8744, pp. 105–116. Springer, Cham (2014). https://doi.org/10.1007/978-3-319-11257-2_9
3. Domingo-Ferrer, J., Muralidhar, K.: New directions in anonymization: permutation paradigm, verifiability by subjects and intruders, transparency to users. Inf. Sci. **337**, 11–24 (2016)
4. Ruiz, N.: On some consequences of the permutation paradigm for data anonymization: centrality of permutation matrices, universal measures of disclosure risk and information loss, evaluation by dominance. Inf. Sci. **430–431**, 620–633 (2018)
5. Domingo-Ferrer, J., Soria-Comas, J.: Connecting randomized response, post-randomization, differential privacy and t-closeness via deniability and permutation, CoRR abs/1803.02139 (2018)
6. Elliot, M., Domingo-Ferrer, J.: The Future of Statistical Disclosure Control, The National Statistician's Quality Review, December 2018
7. Dwork, C., Roth, A.: The algorithmic foundations of differential privacy. Theor. Comput. Sci. **9**(3–4), 211–407 (2014)
8. Domingo-Ferrer, J., Ricci, S., Soria-Comas, J.: Disclosure risk assessment via record linkage by a maximum-knowledge attacker. In: 13th Annual International Conference on Privacy, Security and Trust-PST 2015, Izmir, Turkey, September 2015
9. Ruiz, N.: A General cipher for individual data anonymization. Int. J. Fuzziness Uncertainty Knowl.- Syst. (2020). Accepted article in press 2020
10. Domingo-Ferrer, J., Sánchez, D., Rufian-Torrell, G.: Anonymization of nominal data based on semantic marginality. Inf. Sci. **242**, 35–48 (2013)
11. Rodriguez-Garcia, M., Batet, M., Sánchez, D.: Utility-preserving privacy protection of nominal data sets via semantic rank swapping. Inf. Fusion **45**, 282–295 (2019)

12. Dunford, N., Schwartz, J.T.: Linear Operators. Part I: General Theory. Interscience publishers Inc., New York (1958)
13. Shannon, C.E.: A mathematical theory of communication. Bell Syst. Tech. J. **27**, 379–423 (1948)
14. Stinson, D.R.: Cryptography: Theory and Practice, 3rd edn. Chapman and Hall/CRC, Boca Raton (2005)
15. Ruiz, N.: Anonymization as communication over a noisy channel and the trade-off between privacy and information (2020). Under review

ϵ-Differential Privacy for Microdata Releases Does Not Guarantee Confidentiality (Let Alone Utility)

Krishnamurty Muralidhar[1], Josep Domingo-Ferrer[2](✉), and Sergio Martínez[2]

[1] Department of Marketing and Supply Chain Management, University of Oklahoma, 307 West Brooks, Adams Hall Room 10, Norman, OK 73019, USA
krishm@ou.edu

[2] Department of Computer Engineering and Mathematics, Universitat Rovira i Virgili, CYBERCAT-Center for Cybersecurity Research of Catalonia UNESCO Chair in Data Privacy, Av. Països Catalans 26, 43007 Tarragona, Catalonia
{josep.domingo,sergio.martinezl}@urv.cat

Abstract. Differential privacy (DP) is a privacy model that was designed for interactive queries to databases. Its use has then been extended to other data release formats, including microdata. In this paper we show that setting a certain ϵ in DP does not determine the confidentiality offered by DP microdata, let alone their utility. Confidentiality refers to the difficulty of correctly matching original and anonymized data, and utility refers to anonymized data preserving the correlation structure of original data. Specifically, we present two methods for generating ϵ-differentially private microdata. One of them creates DP synthetic microdata from noise-added covariances. The other relies on adding noise to the cumulative distribution function. We present empirical work that compares the two new methods with DP microdata generation via prior microaggregation. The comparison is in terms of several confidentiality and utility metrics. Our experimental results indicate that different methods to enforce ϵ-DP lead to very different utility and confidentiality levels. Both confidentiality and utility seem rather dependent on the amount of permutation performed by the particular SDC method used to enforce DP. Thus suggests that DP is not a good privacy model for microdata releases.

Keywords: Anonymized microdata · Differential privacy · Synthetic data · Confidentiality · Analytical utility

1 Introduction

Traditional anonymization by national statistical institutes consists of applying a statistical disclosure control (SDC) method with a heuristic choice of parameters

J. Domingo-Ferrer and K. Muralidhar (Eds.): PSD 2020, LNCS 12276, pp. 21–31, 2020.
https://doi.org/10.1007/978-3-030-57521-2_2

and then assessing the disclosure risk and the analytical utility of the anonymized data. If the risk is deemed too high, the SDC method is run again with more stringent parameters, which is likely to reduce the risk and the utility as well.

Privacy models, originated in the computer science community, take a different view of anonymization. A privacy model is an *ex ante* parameterized condition that is meant to guarantee a pre-specified level of disclosure protection—that is, confidentiality—regardless of the impact on utility. If the utility loss is deemed too high, then the privacy model parameter must be made less strict. Privacy models are enforced using SDC methods whose parameters depend on the privacy model parameter. The earliest privacy model instance was k-anonymity [10], and the most talked about privacy model these days is differential privacy (DP, [4]). Privacy models are usually enforced by using one or more SDC methods: in the case of k-anonymity, one uses generalization, local suppression or microaggregation. In the case of DP, there are several options, the most usual being Laplace noise addition.

The initial formulation of DP was for the interactive setting. A randomized query function κ (that returns the query answer plus some noise) satisfies ϵ-DP if for all data sets D_1 and D_2 that differ in one record and all $S \subset Range(\kappa)$, it holds that $\Pr(\kappa(D_1) \in S) \leq \exp(\epsilon) \times \Pr(\kappa(D_2) \in S)$. In plain English, the presence or absence of any single record must not be noticeable from the query answers, up to an exponential factor ϵ (called the privacy budget). The smaller ϵ, the higher the protection. The most usual SDC method employed to enforce differential privacy is Laplace noise addition. The amount of noise depends on ϵ (the smaller ϵ, the more noise is needed) and, for fixed ϵ, it increases with the global sensitivity of the query (defined as the maximum variation of the query output when one record in the data set is changed, added or suppressed).

Differential privacy offers a neat privacy guarantee for interactive queries, at least for small values of ϵ. Unlike k-anonymity, its privacy guarantee does not require any assumptions on the intruder's background knowledge. Very soon, researchers proposed extensions of DP for the non-interactive setting, that is, to produce DP microdata sets that could be used for any analysis, rather than for a specific query. Based on that, Google, Apple and Facebook are currently using DP to anonymize microdata collection from their users, although in most cases with ϵ values much larger than 1 [6] (which is against the recommendations of [5]).

Unfortunately, as noted in [9], generating DP microdata is a very challenging task. A DP microdata set can be viewed as a collection of answers to identity queries, where an identity query is about the content of a specific record (*e.g.* tell me the content of the i-th record in the data set). Obviously, the sensitivity of an identity query is very high: if one record is changed the value of each attribute in the record can vary over the entire attribute domain. This means that a lot of noise is likely to be needed to produce DP microdata, which will result in poor utility. This should not be surprising, because by design, DP attempts to make the presence or absence of any single original record undetectable in the DP output, in this case, the DP microdata set.

The usual approach to obtain DP microdata is based on histogram queries [12,13]. In [11], a method to generate DP microdata that uses a prior microaggregation step was proposed. Microaggregation replaces groups of similar records by their average record. Since the average record is less sensitive than individual original records, if one takes the microaggregation output as the input to DP, the amount of Laplace noise required to enforce a certain ϵ is smaller than if taking the original data set as input. This yields DP microdata with higher utility than competing approaches.

1.1 Contribution and Plan of This Paper

Our aim in this paper is to demonstrate that a certain privacy budget ϵ can result in very different levels of confidentiality and utility. In fact, we show that the achieved confidentiality and utility depend on the particular SDC methods used to enforce DP.

Specifically, we present two new methods for generating DP microdata. One of them creates DP synthetic microdata from noise-added covariances. The other relies on adding noise to the cumulative distribution function (CDF). We present empirical work that compares the two new methods with the microaggregation-based method [11]. The comparison is in terms of several confidentiality and utility metrics. It becomes apparent that different methods to enforce ϵ-DP lead to very different utility and confidentiality levels.

Section 2 describes the synthetic data method. Section 3 describes the CDF-based method. Empirical work comparing the two new methods among them and with the microaggregation-based method is reported in Sect. 4. Conclusions and future research issues are gathered in Sect. 5.

2 A Method for Generating Synthetic DP Microdata

In this method, a DP synthetic microdata set is generated based on the original data. The approach is to add Laplace noise to: i) the sum of each attribute; ii) the sum of squared values of each attribute; and iii) the sum of the product of each pair of attributes. This allows obtaining DP versions of the attribute means and covariances. Finally, the synthetic microdata are obtained by sampling a multivariate normal distribute with parameters the DP mean vector and the DP covariance matrix. Therefore, *the synthetic data thus obtained are DP by construction.*

If the original data set has m attributes, there are m sums of attribute values, m sums of squared attribute values and $m(m-1)/2$ sums of products of pairs of attributes. Hence, the privacy budget ϵ must be divided among the total $2m + m(m-1)/2$ sums. Let $\epsilon^* = \epsilon/(2m + m(m-1)/2)$.

Let x_{ij}, for $i = 1, \ldots, n$ and $j = 1, \ldots, m$, represent the value of the j-th attribute in the i-th record of the original data set. Let μ_j and σ_{jj} denote, respectively, the mean and the variance of the j-th attribute, and let σ_{jk}, for $j \neq k$, represent the covariance between the j-th and the k-th attributes. On the other hand, let Δf_j represent the global sensitivity of the j-th attribute. Then Algorithm 1 formalizes the above-sketched method to generate DP synthetic microdata.

Algorithm 1. METHOD 1: DP SYNTHETIC MICRODATA GENERATION

Input: Original data set $\{x_{ij} : i = 1, \ldots, n; j = 1, \ldots, m\}$
Output: DP data set $\{x^*_{ij} : i = 1, \ldots, n; j = 1, \ldots, m\}$

```
1  for j = 1 to m do                          /* DP-perturb means and variances */
2      R_j = Σ_{i=1}^n x_ij + Laplace(0, Δf_j/ε*);
3      S_j = Σ_{i=1}^n x_ij^2 + Laplace(0, Δf_j^2/ε*);
4      μ*_j = R_j/n;
5      σ*_jj = (S_j − R_j^2/n)/(n−1);
6  for j = 1 to m do                          /* DP-perturb covariances */
7      for k = j + 1 to m do
8          T_jk = Σ_{i=1}^n (x_ij x_ik) + Laplace(0, Δf_j × Δf_k/ε*);
9          σ*_jk = (T_jk − R_j R_k/n)/(n−1);
10         σ*_kj = σ*_jk;
11 for i = 1 to n do                 /* Draw DP synthetic data from DP-normal */
12     (x*_i1, ..., x*_im) = Sample(N(μ*, σ*)), where μ* = (μ*_1, ..., μ*_m) and
       σ* = [σ*_ij]_{i,j=1,...m}.
```

Note 1. This method is problematic unless the number of attributes is really small. Indeed, for fixed ϵ the Laplace perturbation added to the variances and covariances quadratically grows with m, because this perturbation has privacy budget $\epsilon/(2m + m(m-1)/2)$. Thus, m does not need to be very large to risk getting a perturbed covariance matrix $\boldsymbol{\sigma}^*$ that is no longer positive definite, and hence not valid as a covariance matrix.

3 A CDF-Based Method to Obtain DP Microdata

This method follows the inspiration of [7] in that it anonymizes by sampling a distribution adjusted to the original data. Yet, unlike [7], we DP-perturb the distribution. In this way, rather than drawing from a multivariate normal distribution with DP-perturbed parameters as in Algorithm 1, we obtain DP microdata by: i) for each attribute, obtaining DP attribute values by sampling from a *univariate* normal distribution with DP-perturbed mean and variance; ii) replacing each original attribute value with a DP-attribute value whose rank is a DP-perturbed version of the rank of the original attribute value. The DP-perturbation of attribute values ensures that the original attribute values are unnoticeable in the DP microdata, whereas the DP-perturbation of ranks ensures that the rank correlation of attribute values within each record is altered enough for the resulting multivariate data set to be DP.

If we normalize ranks by dividing them by n, adding or suppressing one record will at most change the CDF (normalized rank) of any other record by $1/n$, and hence the global sensitivity of the CDF is $1/n$. Since records are assumed independent of each other, there is no sequential composition among records in the DP sense, and therefore the privacy budget ϵ does not need to be divided among the number of records. If there are m attributes, ϵ must just be divided among the m sums of attribute values, the m sums of squared attribute values and the m empirical CDFs of the attributes. This yields a budget $\epsilon^* = \epsilon/3m$ for the required Laplace perturbations.

Algorithm 2 formalizes the CDF-based method to generate DP microdata. Note that this approach is not synthetic, because each record in the DP data set results from a specific record in the original data set. For each attribute j, each original attribute value x_{ij} is replaced by a DP attribute value whose rank is DP-perturbed version of the rank of x_{ij}.

Algorithm 2. METHOD 2: CDF-BASED DP MICRODATA GENERATION

Input: Original data set $\{x_{ij} : i = 1, \ldots, n; j = 1, \ldots, m\}$
Output: DP data set $\{x^*_{ij} : i = 1, \ldots, n; j = 1, \ldots, m\}$

1 **for** $j = 1$ **to** m **do** `/* DP-perturb means and variances */`
2 $\quad R_j = \sum_{i=1}^{n} x_{ij} + \text{Laplace}\left(0, \frac{\Delta f_j}{\epsilon^*}\right)$;
3 $\quad S_j = \sum_{i=1}^{n} x_{ij}^2 + \text{Laplace}\left(0, \frac{\Delta f_j^2}{\epsilon^*}\right)$;
4 $\quad \mu^*_j = R_j/n$;
5 $\quad \sigma^*_{jj} = \frac{S_j - \frac{R_j^2}{n}}{n-1}$;
6 **for** $j = 1$ **to** m **do**
7 $\quad$ **for** $i = 1$ **to** n **do**
8 $\quad\quad y_i = \text{Sample}(N(\mu^*_j, \sigma^*_{jj}))$; `/* Generate DP attribute values */`
$\quad\quad c_i = \left(\frac{\text{Rank}(x_{ij})}{n}\right) + \text{Laplace}\left(0, \frac{1}{n\epsilon^*}\right)$; `/* Convert the rank of the original attribute value to real and DP-perturb it */`
9 $\quad$ **for** $i = 1$ **to** n **do** `/* Replace original attribute values by DP attribute values with DP-perturbed ranks */`
10 $\quad\quad x^*_{ij} = y_{[\text{Rank}(c_i)]}$, where $y_{[k]}$ stands for the value in $\{y_1, \ldots, y_n\}$ with rank k.

The following holds.

Proposition 1. *If the number of attributes is m is more than 3, for a given privacy budget ϵ the method of Algorithm 2 perturbs means and variances less than the method of Algorithm 1.*

Proof. In both algorithms, the perturbations of means and variances are directly proportional to the perturbations used to obtain R_j and S_j. On the other hand, the latter perturbations are inversely proportional to ϵ^*. In Algorithm 1 we have $\epsilon^* = \epsilon/(2m + m(m-1)/2)$, whereas in Algorithm 2 we have $\epsilon^* = \epsilon/3m$. Now $(2m + m(m-1)/2) > 3m$ if and only if $m > 3$.

Note that Proposition 1 does not necessarily imply that for $m > 3$ the utility of the output DP microdata is better in Algorithm 2 than in Algorithm 1, because the ways in which perturbed means and variances are used in both algorithms differ.

4 Empirical Work

We implemented the two proposed methods and we measured the analytical utility and the confidentiality they provide. Note that, although DP is a privacy model specifying an *ex ante* privacy condition with the ϵ budget, absolute unnoticeability of any particular record only holds when $\epsilon = 0$. For any other value of ϵ it makes sense to measure how protected against disclosure are the data, that is, what is the confidentiality level being achieved.

Further, to compare the two proposed methods against the state of the art, we included in the comparison the microaggregation-based DP microdata generation method [11].

4.1 Utility Metrics

We considered two metrics for generic analytical utility, which do not require assumptions on specific data uses.

The first one is the sum of squared errors SSE, defined as the sum of squares of attribute distances between records in the original data set and their versions in the DP data set. That is,

$$SSE = \sum_{i=1}^{n} \sum_{j=1}^{m} (x_{ij} - x^*_{ij})^2. \quad (1)$$

We took the squared Euclidean distance between x_{ij} and x^*_{ij} because our in experiments all attributes were numerical. For a version of SSE that works also with categorical data, see [11]. On the other hand, SSE needs to know which DP attribute value x^*_{ij} corresponds to each original attribute value x_{ij}. For that reason, SSE cannot be used to measure the utility of Method 1, because in that method the DP data are synthetic, which means that no correspondence can be established between original attribute values and DP attribute values.

The second utility metric is the one proposed in [2]:

$$UM(\mathbf{X}, \mathbf{Y}) = \begin{cases} 1 & \text{if } \hat{\lambda}_j^X = \hat{\lambda}_j^{Y|X} = 1/m \text{ for } j = 1, \ldots, m; \\ 1 - \min\left(1, \frac{\sum_{j=1}^{m} (\hat{\lambda}_j^X - \hat{\lambda}_j^{Y|X})^2}{\sum_{j=1}^{m} (\hat{\lambda}_j^X - 1/m)^2}\right) & \text{otherwise.} \end{cases} \quad (2)$$

In Expression (2), $\mathbf{X}$ is the original microdata set, $\mathbf{Y}$ is the DP microdata set, $\hat{\lambda}_j^X$ are the eigenvalues of the covariance matrix $\mathbf{C}_{XX}$ of $\mathbf{X}$ scaled so that they add to 1, and $\hat{\lambda}_j^{Y|X}$ are scaled versions of

$$\lambda_j^{Y|X} = (\mathbf{v}_j^X)^T \mathbf{C}_{YY} \mathbf{v}_j^X, \quad j = 1, \ldots, m,$$

where $\mathbf{C}_{YY}$ is the covariance matrix of $\mathbf{Y}$ and $\mathbf{v}_j^X$ is the j-th eigenvector of $\mathbf{C}_{XX}$.

The rationale of UM is as follows. Each eigenvalue $\hat{\lambda}_j^X$ represents the proportion of the variance of the attributes in $\mathbf{X}$ explained by the corresponding eigenvector $\mathbf{v}_j^X$. On the other hand each $\lambda_j^{Y|X}$ represents the proportion of the variance of the attributes in $\mathbf{Y}$ explained by $\mathbf{v}_j^X$. Then we have:

- The highest level of utility ($UM(\mathbf{X}, \mathbf{Y}) = 1$) occurs when $\hat{\lambda}_j^X = \hat{\lambda}_j^{Y|X}$ for $j = 1, \ldots, m$, which occurs when $\mathbf{C}_{XX} = \mathbf{C}_{YY}$.
- The lowest level of utility ($UM(\mathbf{X}, \mathbf{Y}) = 0$) occurs if $\hat{\lambda}_j^X$ and $\hat{\lambda}_j^{Y|X}$ differ at least as much as $\hat{\lambda}_j^X$ and the eigenvalues of an uncorrelated data set (which are $1/m$).

Note that an advantage of UM over SSE is that the former also applies for synthetic data, and hence for Method 1. However, both metrics view utility as the preservation of variability, more precisely as the preservation of the correlation structure of the original data set.

4.2 Confidentiality Metrics

We used four confidentiality metrics. First, the share of records in the original data set that can be correctly matched from the DP data set, that is, the proportion of correct record linkages

$$RL = \frac{\sum_{\mathbf{x}_i \in \mathbf{X}} \Pr(\mathbf{x}_i^*)}{n}, \tag{3}$$

where $\Pr(\mathbf{x}_i^*)$ is the correct record linkage probability for the i-th DP record $\mathbf{x}_i^*$. If the original record $\mathbf{x}_i$ from which $\mathbf{x}_i^*$ originates is not at minimum distance from $\mathbf{x}_i^*$, then $\Pr(\mathbf{x}_i^*) = 0$; if $\mathbf{x}_i$ is at minimum distance, then $\Pr(\mathbf{x}_i^*) = 1/M_i$, where M_i is the number of original records at minimum distance from $\mathbf{x}_i^*$.

The three other confidentiality metrics $CM1$, $CM2$ and $CM3$ are those proposed in [2], based on canonical correlations. We have:

$$CM1(\mathbf{X}, \mathbf{Y}) = 1 - \rho_1^2, \tag{4}$$

where ρ_1^2 is the largest canonical correlation between the *ranks* of attributes in $\mathbf{X}$ and $\mathbf{Y}$. The rationale is that:

- Top confidentiality ($CM1(\mathbf{X}, \mathbf{Y}) = 1$) is reached when the ranks of the attributes in $\mathbf{X}$ are independent of the ranks of attributes in $\mathbf{Y}$, in which case anonymization can be viewed as a random permutation.

- Zero confidentiality ($CM1(\mathbf{X}, \mathbf{Y}) = 0$) is achieved then the ranks are the same for *at least* one original attribute X^j and one DP attribute X^j. Note that this notion of confidentiality is quite strict: leaving a single attribute unprotected brings the confidentiality metric down to zero.

The next confidentiality metric is similar $CM1$ but it considers all canonical correlations:

$$CM2(\mathbf{X}, \mathbf{Y}) = \prod_{i=1}^{m}(1 - \rho_i^2) \left[= e^{-I(\mathbf{X};\mathbf{Y})}\right]. \tag{5}$$

The second equality between brackets in Expression (5) can only be guaranteed if the collated data sets $(\mathbf{X}, \mathbf{Y})$ follow an elliptically symmetrical distribution (a generalization of the multivariate Gaussian), in which case Expression (5) can be rewritten in terms of the mutual information $I(\mathbf{X}; \mathbf{Y})$ between the original and the DP data sets.

Regardless of the distributional assumptions, $CM2(\mathbf{X}, \mathbf{Y})$ can be computed from the canonical correlations and the following holds:

- Top confidentiality $CM2(\mathbf{X}, \mathbf{Y}) = 1$ is reached when the anonymized data set and the original data sets tell nothing about each other, which is the same as saying that mutual information between them is $I(\mathbf{X}; \mathbf{Y}) = 0$.
- Zero confidentiality $CM2(\mathbf{X}, \mathbf{Y}) = 0$ occurs if at least one of the canonical correlations is 1. This occurs if at least one original attribute is disclosed when releasing $\mathbf{Y}$. Since ρ_1 is the largest correlation, this means that we have $CM2(\mathbf{X}, \mathbf{Y}) = 0$ if and only if $\rho_1 = 1$, in which case we also have that the metric of Expression (4) is $CM1(\mathbf{X}, \mathbf{Y}) = 0$.

Note that RL, $CM1$ and $CM2$ cannot be applied to the DP synthetic data produced by Method 1, because the three metrics need to know the mapping between original and DP records. The last metric $CM3$ that we use is mapping-free and is intended for synthetic data (yet it should not be used when the mapping between original and DP records is known). Specifically, $CM3$ is derived from $CM2$ as follows:

$$CM3(\mathbf{X}, \mathbf{Y}) = \min_{1 \leq j \leq m} CM2(\mathbf{X}^{-j}, \mathbf{Y}^{-j}). \tag{6}$$

where $\mathbf{X}^{-j}$, resp. $\mathbf{Y}^{-j}$, is obtained from $\mathbf{X}$, resp. $\mathbf{Y}$, by sorting $\mathbf{X}$, resp. $\mathbf{Y}$, by its j-th attribute and suppressing the values of this attribute in the sorted data set.

The common principle of RL, $CM1$, $CM2$ and $CM3$ is to view confidentiality as permutation. The farther the anonymized values from the original values in value (for RL) or in rank (for the other metrics), the higher the confidentiality.

4.3 Results for Data Sets with Two Attributes

We considered five data sets with two numerical attributes X_1, X_2 and 10,000 records. In each data set, X_1 was drawn from a $N(50, 10)$ distribution and X_2

was also drawn from a $N(50, 10)$ distribution but in such a way that the expected correlation between X_1 and X_2 was 0.5.

For each data set, we ran the microaggregation-based DP microdata generation method of [11] with $\epsilon = 1$ (higher values are not recommended in [5]) and microaggregation group sizes $k = 250$, 500, 1000, 2000 and 3000. Since this method is not synthetic, for each resulting DP microdata set we computed utility metrics SSE and UM, and confidentiality metrics RL, $CM1$ and $CM2$. In Table 1 we report the values of those metrics for each value k averaged over the five data sets.

We then ran Method 1 for each data set with $\epsilon = 1$. Since it is a synthetic method, we computed utility metric UM and confidentiality metric $CM3$. In Table 1 we display those metrics averaged over the five data sets.

Finally, we ran Method 2 for each data set with $\epsilon = 1$. Since this method is not synthetic, we computed the same metrics as for the microaggregation-based method. Table 1 reports the averages for the five data sets.

Table 1. Empirical comparison of microaggregation-based DP generation, Method 1 and Method 2. In all cases $\epsilon = 1$ and all results are averages over five original data sets with the same distribution. "Micro*" denotes microaggregation-based DP microdata generation with $k = *$.

	SSE	UM	RL	CM1	CM2	CM3
Micro250	26833.22	0.647230517	0.00028	0.639798854	0.57703286	N/A
Micro500	3446.60	0.972413957	0.0008	0.226067842	0.112192366	N/A
Micro1000	2160.91	0.984149182	0.00096	0.164139854	0.057491616	N/A
Micro2000	2855.62	0.959011191	0.0005	0.214820038	0.10679245	N/A
Micro3000	3980.19	0.502650927	0.0003	0.197760886	0.197698071	N/A
Method1	N/A	0.992537652	N/A	N/A	N/A	0,935883394
Method2	49.74	0.981339047	0.1212	0.000492087	7.43603E-07	N/A

One thing that stands out in Table 1 is that utility metrics SSE and UM are consistent with each other. Higher values of SSE translate into lower values of UM, meaning less utility. Also, lower values of SSE result in higher values for the UM, meaning more utility. Thus, they capture the same utility notion and it is enough for us to consider one utility metric in what follows. We choose UM because it can be computed both for non-synthetic and synthetic data.

In terms of UM, we see in Table 1 that Method 1 achieves the highest utility, while offering a confidentiality metric $CM3$ that is also high, being close to 1. Thus, Method 1 seems the best performer.

Method 2 also offers high utility UM, but extremely low confidentiality in terms of $CM1$ and $CM2$. The DP data it produces turn out to be very similar to the original data.

The microaggregation-based DP microdata generation method can be seen to offer intermediate performance regarding the trade-off between utility and confidentiality. Whatever the choice of k, it achieves better confidentiality metrics $CM1$ and $CM2$ than Method 2, but its utility UM only beats Method 2 for $k = 1000$. Thus, microaggregation-based DP generation for $k = 1000$ is the second best performer.

The microaggregation-based DP method offers poorer utility for extreme values of k. The explanation is that for smaller k (250, 500) the prior microaggregation step does not reduce the sensitivity of the data as much as $k = 1000$ and hence still needs sustantial Laplace noise to attain DP with $\epsilon = 1$. On the other hand, for large $k = 2000, 3000$, averaging over such large groups causes a lot of information loss.

On the other hand, we can see that setting the same $\epsilon = 1$ for all methods can lead to very different confidentiality and utility levels.

4.4 Results for Data Sets with 10 Attributes

To check what is said in Note 1 and Proposition 1, we also tested Methods 1 and 2 for data sets with $m = 10$ attributes. We generated five data sets with normally distributed attributes and we took $\epsilon = 1$ as above. We kept running Method 1 for the five data sets until we got positive definite DP covariance matrices. The results were:

- As expected, the average utility achieved by Method 1 was extremely low, namely $UM = 0.00733781$. In contrast, the average confidentiality was high, $CM3 = 0.99752121$, even higher than for the two-attribute data sets.
- Method 2 yielded an extremely high average utility $UM = 0.99999618$. In contrast, confidentiality was as small as in the two-attribute case, with $CM1 = 0.00076754$ and $CM2$ =2.3331E-13.

5 Conclusions

We have compared three methods for generating DP microdata, two of them new. The three of them leverage different principles to generate DP microdata with $\epsilon = 1$. However, the confidentiality and utility levels they achieve for that value of ϵ are extremely different. Hence, setting a certain value of ϵ does *not* guarantee a certain level of confidentiality, let alone utility. The actual confidentiality and utility offered depend on the specific method used to enforce ϵ-DP. Our results complement those obtained in [8] for ϵ-DP synthetic data. In that paper, DP-synthetic data were generated with a single method but using several values of ϵ; it turned out that ϵ determined neither the protection against disclosure nor the utility of the synthetic data.

In our experiments, the methods that result in higher confidentiality seem to be those that operate a stronger permutation in terms of the permutation model of SDC [1]. Specifically Method 1, being synthetic, can be viewed as a random permutation of ranks, whereas Method 2 yields ranks for masked data that are very close to the ones of original data; this would explain the high confidentiality offered by Method 1 and the low confidentiality of Method 2.

In conclusion, the fact that parameter ϵ does not give any specific confidentiality guarantee for microdata releases suggests that DP should not be used to anonymize microdata. This adds to the arguments given in [3] in that sense.

Acknowledgments. Partial support to this work has been received from the European Commission (project H2020-871042 "SoBigData++"), the Government of Catalonia (ICREA Acadèmia Prize to J. Domingo-Ferrer and grant 2017 SGR 705), and from the Spanish Government (project RTI2018-095094-B-C21 "Consent"). The second and third authors are with the UNESCO Chair in Data Privacy, but the views in this paper are their own and are not necessarily shared by UNESCO.

References

1. Domingo-Ferrer, J., Muralidhar, K.: New directions in anonymization: permutation paradigm, verifiability by subjects and intruders, transparency to users. Inf. Sci. **337–338**, 11–24 (2016)
2. Domingo-Ferrer, J., Muralidhar, K., Bras-Amorós, M.: General confidentiality and utility metrics for privacy-preserving data publishing based on the permutation model. IEEE Trans. Dependable Secure Comput. (2020). To appear
3. Domingo-Ferrer, J., Sánchez, D., Blanco-Justicia, A.: The limits of differential privacy (and its misuse in data release and machine learning). Commun. ACM. To appear
4. Dwork, C.: Differential privacy. In: Bugliesi, M., Preneel, B., Sassone, V., Wegener, I. (eds.) ICALP 2006. LNCS, vol. 4052, pp. 1–12. Springer, Heidelberg (2006). https://doi.org/10.1007/11787006_1
5. Dwork, C.: A firm foundation for private data analysis. Commun. ACM **54**(1), 86–95 (2011)
6. Greenberg, A.: How one of Apple's key privacy safeguards falls short. Wired, 15 September 2017. https://www.wired.com/story/apple-differential-privacy-shortcomings/
7. Liew, C.K., Choi, U.J., Liew, C.J.: A data distortion by probability distribution. ACM Trans. Database Syst. **10**(3), 395–411 (1985)
8. McClure, D., Reiter, J.P.: Differential privacy and statistical disclosure risk measures: an investigation with binary synthetic data. Trans. Data Privacy **5**(3), 535–552 (2012)
9. Ruggles, S., Fitch, C., Magnuson, D., Schroeder, J.: Differential privacy and census data: implications for social and economic research. AEA Papers Proc. **109**, 403–408 (2019)
10. Samarati, P., Sweeney, L.: Protecting Privacy When Disclosing Information: k-Anonymity and Its Enforcement Through Generalization and Suppression. Technical report, SRI International (1998)
11. Soria-Comas, J., Domingo-Ferrer, J., Sánchez, D., Martínez, S.: Enhancing data utility in differential privacy via microaggregation-based k-anonymity. VLDB J. **23**(5), 771–794 (2014). https://doi.org/10.1007/s00778-014-0351-4
12. Xiao, Y., Xiong, L., Yuan, C.: Differentially private data release through multidimensional partitioning. In: Proceedings of the 7th VLDB Conference on Secure Data Management-SDM 2010, pp. 150–168 (2010)
13. Xu, J., Zhang, Z., Xiao, X., Yang, Y., Yu, G.: Differentially private histogram publication. In: IEEE International Conference on Data Engineering-ICDE 2012, pp. 32–43 (2012)

A Bayesian Nonparametric Approach to Differentially Private Data

Fadhel Ayed[1], Marco Battiston[2](✉), and Giuseppe Di Benedetto[1]

[1] Oxford University, Oxford OX1 4BH, UK
fadhel.ayed@gmail.com, benedett@stats.ox.ac.uk
[2] Lancaster University, Lancaster LA1 4YW, UK
m.battiston@lancaster.ac.uk

Abstract. The protection of private and sensitive data is an important problem of increasing interest due to the vast amount of personal data collected. *Differential Privacy* is arguably the most dominant approach to address privacy protection, and is currently implemented in both industry and government. In a decentralized paradigm, the sensitive information belonging to each individual will be locally transformed by a known privacy-maintaining mechanism Q. The objective of differential privacy is to allow an analyst to recover the distribution of the raw data, or some functionals of it, while only having access to the transformed data. In this work, we propose a Bayesian nonparametric methodology to perform inference on the distribution of the sensitive data, reformulating the differentially private estimation problem as a latent variable Dirichlet Process mixture model. This methodology has the advantage that it can be applied to any mechanism Q and works as a "black box" procedure, being able to estimate the distribution and functionals thereof using the same MCMC draws and with very little tuning. Also, being a fully non-parametric procedure, it requires very little assumptions on the distribution of the raw data. For the most popular mechanisms Q, like Laplace and Gaussian, we describe efficient specialized MCMC algorithms and provide theoretical guarantees. Experiments on both synthetic and real dataset show a good performance of the proposed method.

Keywords: Differential Privacy · Bayesian Nonparametrics · Exponential mechanism · Laplace noise · Dirichlet Process mixture model · Latent variables

1 Introduction

Recent years have been characterized by a remarkable rise in the size and quality of data collected online. As the volume of personal data held by organizations increases, the problem of preserving the privacy of individuals present in the data set becomes more and more important. In the computer science literature, *Differential Privacy* is currently the most dominant approach to privacy protection and has already been implemented in their protocols by many IT companies [7,8,17].

J. Domingo-Ferrer and K. Muralidhar (Eds.): PSD 2020, LNCS 12276, pp. 32–48, 2020.
https://doi.org/10.1007/978-3-030-57521-2_3

In the *non-interactive* model, the data curator holds some sensitive data $X_{1:n} = \{X_1 \dots, X_n\}$ about n individuals, which is of interest to end users in order to conduct statistical analysis with it and infer either its distribution P or some functionals of it, e.g. the population mean or median. Under the framework of differential privacy, the data curator applies a mechanism Q to the sensitive data set $X_{1:n}$ that returns a sanitized data set $Z_{1:n} = \{Z_1, \dots, Z_n\}$ to be released to the public. The end users have access only to $Z_{1:n}$, while the raw data $X_{1:n}$ is kept secret by the curator. Differential Privacy is a property of the mechanism Q that guarantees that the presence or absence of a specific individual in the raw data does not remarkably affect the output database $Z_{1:n}$. As a consequence, a malicious intruder who has access only to the transformed data is unlikely to guess correctly whether a specific individual is in the initial data set or not, and therefore her privacy will be preserved. Different mechanisms have been proposed in the literature for many statistical problems, e.g. in hypothesis testing [10,11,23], frequency tables [2,6,9,21] and network modeling [3,4,14,15]. The most applied mechanism Q is arguably addition of noise to the raw data, with Laplace or Gaussian noise being the most popular choices.

Inherent in any choice of the mechanism Q, there is a trade-off between privacy guarantees and statistical utility of the released data set. Specifically, as more noise is added to $X_{1:n}$, the achieved level of privacy will be higher, but so will the deterioration of the statistical utility of the released data set. In an interesting recent paper [5], the authors discuss minimax optimality within a framework of differential privacy. For many statistical tasks, optimal choices of the mechanism Q and estimators $T(Z_{1:n})$ that attain optimal minimax convergence rates under privacy constraints are presented. However, the choice of the optimal pair $(Q, T(Z_{1:n}))$ is strongly dependent on the particular task of interest, i.e. the particular functional $\theta(P)$ that the end user is interested in estimating. For two different tasks $\theta_1(P)$ and $\theta_2(P)$, e.g. mean and median estimation of P, the two optimal pairs can be completely different. Moreover, the choice of Q and T occurs separately, with Q being chosen by the curator, while T by the end users. In the non-interactive setting, where the sanitized data will be used by multiple end users for different tasks, these facts lead to important practical challenges: 1) the curator cannot choose a Q that is optimal for all possible tasks of interests; 2) the end user cannot directly apply the prescribed optimal T, unless the curator applied the mechanism Q targeted for her specific task. A further problem is that if the end user tries to apply some naive natural estimator T, without properly accounting for the transformation Q, her results could be very misleading.

In this work, we solve this practical problem by proposing a "black box" methodology to make inference on P or any functional $\theta(P)$ of the distribution of the raw data $X_{1:n}$, when we have access only to the sanitized data $Z_{1:n}$. In order to do so, we apply tools from Bayesian Nonparametrics and reformulate the problem as a latent variable mixture model, where the observed component is $Z_{1:n}$, the latent one is $X_{1:n}$ and we are interested in estimating the mixing measure P, i.e. the distribution of $X_{1:n}$. We adapt known Markov Chain Monte

Carlo (MCMC) algorithms for the Dirichlet Process Mixture (DPM) Model to our setting of differentially private data and show, both theoretically and empirically, how the proposed strategy is able to recover P. The proposed methodology has the following advantages: 1) it is fully nonparametric and it does not require any assumption on the distribution P of the raw data $X_{1:n}$; 2) it provides estimates of P, functionals of it and also credible intervals or variance estimates, all at once, using the same MCMC draws; 3) it is a black box procedure and does not require a careful analytic study of the problem and can be applied for every Q; 4) for the general case of Q in the exponential mechanism, it is shown that the proposed methodology satisfies theoretical guarantees for a large class of functions, including the important cases of Laplace and Gaussian noise addition.

This paper is organized as follows. Section 2 briefly reviews the literature on Differential Privacy and Bayesian Nonparametrics and discusses related work. Section 3 introduces the proposed approach based on Differential Privacy and Dirichlet Process Mixtures. In Sect. 3.1, we discuss the MCMC algorithm, while in Sect. 3.2 present the theoretical results for the proposed methodology. Finally, in Sect. 4 we present the empirical performances of the algorithm on one synthetic data set and two real data examples. The Appendix contains proofs and a specialized algorithm for Laplace noise.

2 Literature Review and Related Work

2.1 Differential Privacy

The notion of Differential Privacy was introduced in the computer science literature in [6] as a mathematical formalization of the idea that the presence or absence of an individual in the raw data should have a limited impact on the transformed data, in order for the latter to be considered privatized. Specifically, let $X_{1:n} = \{X_1 \ldots, X_n\}$ be a sample of observations, taking values in a state space $\mathcal{X}^n \subseteq \mathbb{R}^n$, that is transformed into a sanitized data set $Z_{1:n} = \{Z_1, \ldots, Z_n\}$, with $\mathcal{Z}^n \subseteq \mathbb{R}^n$ endowed with a sigma algebra $\sigma(\mathcal{Z}^n)$, through a *mechanism* Q, i.e. a conditional distribution of $Z_{1:n}$ given $X_{1:n}$. Then Differential Privacy is a property of Q that can be formalized as follows:

Definition 1 ([6]). *The mechanism Q satisfies α-*Differential Privacy *if*

$$\sup_{S \in \sigma(\mathcal{Z}^n)} \frac{Q(Z_{1:n} \in S | X_{1:n})}{Q(Z_{1:n} \in S | X'_{1:n})} \leq \exp(\alpha) \tag{1}$$

for all $X_{1:n}, X'_{1:n} \in \mathcal{X}^n$ s.t. $H(X_{1:n}, X'_{1:n}) = 1$, where H denotes the Hamming distance, $H(X_{1:n}, X'_{1:n}) = \sum_{i=1}^{n} \mathbb{I}(X_i \neq X'_i)$ and $\mathbb{I}$ is the indicator function of the event inside brackets.

For small values of α the right hand side of (1) is approximately 1. Therefore, if Q satisfies Differential Privacy, (1) guarantees that the output database $Z_{1:n}$

has basically the same probability of having been generated from either one of two *neighboring databases* $X_{1:n}$, $X'_{1:n}$, i.e. databases differing in only one entry. The two most common choices for the mechanism Q are addition of either Laplace or Gaussian noise. These two mechanisms are special cases of the general Exponential Mechanism, which will be reviewed in Subsect. 3.2, where theoretical guarantees for the proposed methodology will be provided for a large class of mechanisms, including both Laplace and Gaussian noise.

Differential Privacy has been studied in a wide range of problems, differing among them in the way data is collected and/or released to the end user. The two most important classifications are between Global vs Local privacy, and Interactive vs Non-Interactive models. In the *Global (or Centralized) model* of privacy, each individual sends her data to the data curator who privatizes the entire data set centrally. Alternatively, in the *Local (or Decentralized) model*, each user privatizes her own data before sending it to the data curator. In this latter model, data also remains secret to the possibly untrusted curator. In the *Non-Interactive (or Off-line) model*, the transformed data set $Z_{1:n}$ is released in one spot and each end user has access to it to perform her statistical analysis. In the *Interactive (or On-line) model* however, no data set is directly released to the public, but each end user can ask queries f about $X_{1:n}$ to the data holder who will reply with a noisy version of the true answer $f(X_{1:n})$. Even though all combinations of these classifications are of interest and have been studied in the literature, in our work we will focus on the Local Non-Interactive case, which is used in a large variety of applications.

2.2 Dirichlet Process Mixture Models

In Bayesian Nonparametrics, the most common approach to model observations from a continuous density is by convolving a Dirichlet Process with a known kernel K, e.g. a Normal kernel with unknown mean and variance. The resulting model is called *Dirichlet Process Mixture (DPM) Model* [16] (see also, [13], Chap. 5), and is formulated as follows

$$\begin{aligned} X_i|P &\sim \int K(X_i,\theta)P(d\theta) \qquad i=1,\dots,n \\ P &\sim DP(\epsilon, P_0). \end{aligned}$$

There is also an equivalent representation in terms of latent variables mixture models

$$\begin{aligned} X_i|\theta_i &\sim K(X_i,\theta_i) \qquad & i=1,\dots,n \\ \theta_i|P &\overset{iid}{\sim} P \qquad & i=1,\dots,n \\ P &\sim DP(\epsilon, P_0). \end{aligned}$$

The DPM model has been used in a variety of applications in statistics and machine learning both for density estimation and clustering. In density estimation, we are interested in estimating the density of observations through a mixture model, while in clustering we are interested in grouping observations into groups having similar distributions. For the latter tasks, the discreteness property of the Dirichlet Process turns out to be very convenient, since, being P almost surely discrete, we are going to observe ties in the latent variables θ_i. Two observations X_i and X_j having the same value of the latent variables θ_i and θ_j will be assigned to the same cluster and have the same distribution. There are many computational algorithms to perform inference in DPM models, including MCMC, variational and Newton recursion algorithms, see Chapter 5 of [13].

3 Methodology

The main idea of this work is to reformulate the statistical problem of estimating the distribution P of the raw data $X_{1:n}$, having access only to the sanitized data $Z_{1:n}$ obtained through a Differential Private mechanism Q, as a latent variables problem. The observed component is $Z_{1:n}$, the latent one $X_{1:n}$, Q is the likelihood of Z_i given X_i and the interest is in estimating the mixing measure P, which is endowed with a Dirichlet Process prior. Therefore, we are considering the following DPM model,

$$\begin{aligned} Z_i|X_i &\overset{ind}{\sim} Q(Z_i|X_i) \qquad & i=1,\ldots,n \\ X_i|P &\overset{iid}{\sim} P & i=1,\ldots,n \\ P &\sim DP(\epsilon, P_0) & \end{aligned} \tag{2}$$

The model can be easily extended to the Global Privacy Model by requiring a joint mechanism $Q(Z_{1:n}|X_{1:n})$, but, for easiness of presentation, in this work we will focus on the Local one, in which Q factorizes into the product of the conditionals of Z_i given X_i. In Subsect. 3.1, we describe computational algorithm to make inference on P, while in Subsect. 3.2, we provide theoretical guarantees on these estimates when Q belongs to the Exponential Mechanism for a large class of loss functions.

3.1 Computational Algorithms

The posterior distribution of DPM is a Mixture of Dirichlet Processes [1]. Closed form formulas for posterior quantities of interest can be derived analytically ([13], Proposition 5.2), but they involve summation over a large combinatorial number of terms, which becomes computationally prohibitive for any reasonable sample size. However, many computational algorithms have been proposed to solve this problem, see [13], Chapter 5. An estimate of the posterior mean of P,

$$\mathbb{E}(P|Z_{1:n}) = \int_{\mathcal{X}^n} \mathbb{E}(P|X_{1:n})\mathbb{P}(X_{1:n}|Z_{1:n})dX_{1:n}.$$

can be obtained following this two step procedure:

1. Sample M vectors $X_{1:n}^{(1)}, ..., X_{1:n}^{(M)}$ from $\mathbb{P}(X_{1:n}|Z_{1:n})$ through Algorithm 2 below;
2. Compute the empirical mean approximation

$$\int_{\mathcal{X}^n} \mathbb{E}(P|X_{1:n})\mathbb{P}(X_{1:n}|Z_{1:n})dX_{1:n} \approx \frac{1}{M}\sum_{m=1}^{M} \mathbb{E}(P|X_{1:n}^{(m)}) = \widehat{P}(Z_{1:n})$$

Using the posterior properties of the Dirichlet Process for $\mathbb{E}(P|X_{1:n}^{(m)})$, [13] Chapter 4, the estimator in 2. Can be more conveniently expressed as

$$\widehat{P}(Z_{1:n}) = \frac{\epsilon}{n+\epsilon}P_0 + \frac{n}{M(n+\epsilon)}\sum_{m=1}^{M}\sum_{i=1}^{n}\delta_{X_i^{(m)}}.$$

Step 1 can be carried over by approximating $\mathbb{P}(X_{1:n}|Z_{1:n})$ through a general MCMC algorithm for DPM models, ([19] Algorithm 8), detailed in Algorithm 2. Because $X_{1:n}$ are sampled from the discrete distribution P, they will display ties with positive probability. The algorithm resamples the distinct values $(X_1^*, \ldots, X_{K_n}^*)$ in $X_{1:n}$ and the classification variables $(c_1, \ldots, c_n)$ assigning each data point to one of these K_n distinct values separately. The parameter m is a free parameter that in our experiments has been set equal to 10, following [19]. If the hyperparameters of the models are endowed with hyperpriors, updates of ϵ and the parameters of P_0 are included in Algorithm 2 at the end of each iteration according to standard updating mechanisms, see [19] Section 7 or [13] Chapter 5.

In our experiments, we have observed that a *good initialization* of the chain can improve considerably the mixing of the Markov chain. As initialization of the algorithm, we suggest to use a generalization of k-means with a predetermined number of clusters S. Specifically, let S be a given number of clusters, let $X_{1:S}^*$ be the cluster centroids and $c_{1:n}$ the cluster allocations. Initialize Algorithm 2 using the following Algorithm 1.

Algorithm 1: Initialization of Algorithm 2

```
1 for j in 1:iterations do
2     for i in 1:n do
3         Update c_i = argmax_{1≤k≤S} Q(Z_i|X_k^*)
4     for k in 1:S do
5         Denote C_k = {Z_i | c_i = k}
6         Update X_k^* = E_{P_0}(X|C_k).
```

Algorithm 2 can be applied to any mechanism Q. When we are updating $X_{1:K_n}^*$, any update that leaves the distribution $\mathbb{P}(X_k^*|C_k)$ invariant can be used.

In general, this can be achieved by introducing a Metropolis-Hastings step. However, when possible, it is strongly recommended to update $X^*_{1:K_n}$ using directly a sample from $\mathbb{P}(X^*_k|C_k)$. This will improve the mixing of the Markov chain and produce a more efficient MCMC algorithm. In Appendix A, for the special case of *Laplace noise addition*, we derive the exact posterior distribution $\mathbb{P}(X^*_k|C_k)$ that can be expressed in terms of the base measure P_0 and can be simulated from in $O(n_k)$ operations, with n_k being the size of cluster k. This posterior distribution can be used in combination with Algorithm 2 to obtain more efficient a MCMC scheme for Laplace noise. Since the number of clusters of the DP scales as $\log n$, each iteration of the proposed algorithm scales in $O(n \log n)$ for the Laplace and Gaussian noises for common priors P_0.

Algorithm 2: General MCMC scheme [19]

1 **for** *t in 1:number of iterations* **do**
2 **for** *i in 1:n* **do**
3 Let K^- be the number of distinct c_j for $j \neq i$, labeled $\{1, \ldots, K^-\}$
4 Let $h = K^- + m$.
5 **if** $\exists j \neq i$ *such that* $c_i = c_j$ **then**
6 Draw $X^*_{K^-+1}, \ldots, X^*_h$ indep from P_0
7 **else**
8 Set $c_i = K^- + 1$
9 Draw $X^*_{K^-+2}, \ldots, X^*_h$ indep from P_0
10 Draw new value c_i from

$$\mathbb{P}(c_i = c \,|\, c_{-i}, Z_{1:n}, X^*_{1:K_n}) \propto \begin{cases} n_{-i,c} Q(Z_i, X^*_c) & \text{for } 1 \leq c \leq K^- \\ \frac{\epsilon}{m} Q(Z_i, X^*_c) & \text{for } K^- < c \leq h \end{cases}$$

where $n_{-i,c}$ is the number of $c_j = c$ for $j \neq i$
11 **for** *k in 1:K_n* **do**
12 Denote $C_k = \{Z_i \mid c_i = k\}$
13 Perform an update of $X^*_k|C_k$ that leaves the distribution $\mathbb{P}(X^*_k|C_k)$ invariant

3.2 Theoretical Guarantees

In this section, we will show that the proposed methodology is able to recover the true distribution for a large class of mechanisms. Specifically, we will consider the Exponential Mechanism proposed by [18] (see also [24] for a theoretical study of the Exponential Mechanism).

Definition 2 ([18]). *Let $L : \mathcal{X} \times \mathcal{Z} \to \mathbb{R}_+$ be a loss function. The* Local Exponential Mechanism *Q is defined as follows,*

$$Q(Z_{1:n}|X_{1:n}) \propto \exp\left(-\alpha \sum_{i=1}^{n} \frac{L(X_i, Z_i)}{2\Delta}\right) \tag{3}$$

where $\Delta = \sup_{X \neq X' \in \mathcal{X}} \sup_{Z \in \mathcal{Z}} |L(X, Z) - L(X', Z)|$.

The most popular examples of Exponential Mechanisms are the following:

1. Laplace Mechanism: $L(X, Z) = |X - Z|$;
2. Gaussian Noise: $L(X, Z) = (X - Z)^2$.

In Proposition 1, we will consider a larger class of loss functions, including both Gaussian and Laplace noise, having the following form, for some function ρ

$$L(X, Z) = \rho(X - Z),$$

and, within this class, we prove that the proposed methodology is able to recover the true distribution P_* that generated the data $X_{1:n}$, under the following mild assumptions:

1. The modulus of the Fourier transform of $\exp(-\alpha\rho/\Delta)$ is strictly positive;
2. $\mathcal{X} \subset \text{supp}(P_0) \subset [a, b]$, where $\text{supp}(P_0)$ is the support of the base measure P_0.

Remark 1. The first assumption ensures that the resulting privacy mechanism preserves enough information to be able to recover the true distribution (see proof of Proposition 1 in Appendix B). In order to have $\Delta < +\infty$, which is necessary for defining the Local Exponential Mechanism, it is common to suppose that $\mathcal{X} \subset [a, b]$ (see [5]). For example this assumption is necessary for the Laplace and Gaussian noises. Assumption 2 only ensures that the true distribution is in the Kullback-Leibler support of the Dirichlet Process Prior. This class corresponds to Exponential mechanisms with additive noise. A level of privacy α is then achieved with

$$Q(Z_{1:n}|X_{1:n}) \propto \exp\left(-\alpha \sum_{i=1}^{n} \frac{\rho(X_i - Z_i)}{\Delta}\right)$$

Proposition 1. *Let P_* the unknown true distribution of the unobserved sensitive data $X_{1:n}$. Then under the previous assumptions, $\mathbb{E}(P|Z_{1:n})$ will almost surely converge to P_* in Wasserstein distance, implying in particular convergence of all moments.*

In the case of Gaussian and Laplace kernels, results on rates of convergence of the proposed estimator can be derived from results of [20] and [12]. More specifically, logarithmic and polynomial rates can be derived for the Gaussian and Laplace mechanisms respectively. These rates are however not sharp, and obtaining better rates is still an open problem.

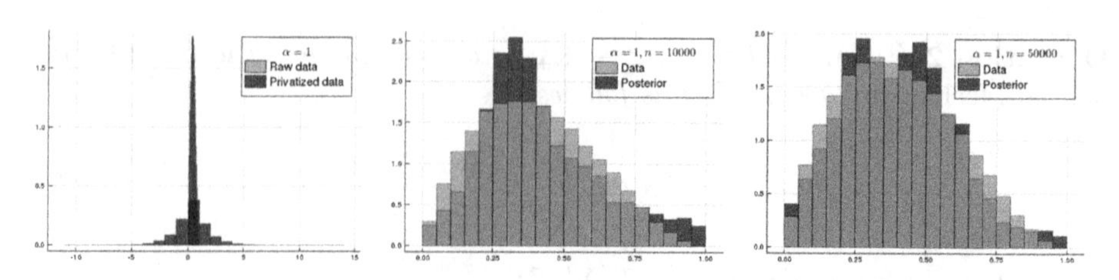

Fig. 1. $X_i \sim \text{Beta}(2, 3)$ and Z_i Laplace noise $\alpha = 1$. From left to right: a) Raw vs Privatized data; b) Estimated and True density of P_*, $n = 10.000$; c) Estimated and True density of P_*, $n = 50.000$.

Table 1. $X_i \sim \text{Beta}(2, 3)$ and Z_i Laplace noise $\alpha = 1$. Posterior estimates of $med_{P_*}(X)$, $\mathbb{E}(P_*)$, $\text{Var}(P_*)$ for sample sizes $n = 10.000, 25.000, 50.000$.

Statistic	$n = 10.000$	$n = 25.000$	$n = 50.000$	Raw data
Median	0.383 ± 0.011	0.382 ± 0.034	0.385 ± 0.016	0.386
Mean	0.402 ± 0.011	0.401 ± 0.007	0.399 ± 0.004	0.400
Variance	0.047 ± 0.009	0.043 ± 0.006	0.040 ± 0.003	0.040

4 Experiments

In this section, we will present experiments using one synthetic data set and two real data sets of Body Mass Index and California Salaries. In the experiments, we use a truncated Normal distribution as base measure P_0.

4.1 Beta Distribution

In this set of simulations, we consider $X_i \overset{iid}{\sim} \text{Beta}(2, 3)$ and $Z_i|X_i$ generated using the Laplace mechanism with $\alpha = 1$. Figure 1a) displays both raw and noisy data, from which it is clear how the noisy data is much over-dispersed compared to the original one. We are interested in recovering the distribution of $X_{1:n}$, its population mean, median and variance using the noisy data $Z_{1:n}$ and the proposed methodology. From Fig. 1b-c), we can see that the posterior approximations obtained through MCMC resemble quite accurately the true distribution of the data. Also from Table 1, the estimated posterior median, mean and variance follow very closely their corresponding true values, having also very narrow and accurate credible intervals.

4.2 Body Mass Index Data-Set

The first data set analyzed collects the Body Mass Index of a random sample of the approximately 6 million patient records from Medical Quality Improvement Consortium (MQIC) database. These records have been stripped of any personal identification and are freely available at https://www.visualizing.org/mqic-patient-data-100k-sample/. It is important in social and health studies to estimate the proportions p_{ob} and p_{uw} of people who suffer of either obesity (BMI $>$ 45) or underweight (BMI $<$ 16) conditions respectively. Statistical interest therefore lies in estimating accurately the probability mass of the tails of the distribution of the Body Mass Index. In this experiment, the privatized data is obtained through a Gaussian mechanism with $\alpha = 0.5$. This level of $\alpha = 0.5$ guarantees a very strong level of privacy. This fact is evident in Fig. 2a), in which the privatized data looks very over dispersed compared to the raw one. However, from Fig. 2b-c), we can see how the posterior approximation of the true distribution, obtained using the proposed Bayesian Nonparametric approach applied to the noisy data, seems very accurate. Table 2 displays the estimated and true values of mean, variance, median, p_{ob} and p_{uw}, including also the corresponding credible intervals, obtained using different samples of the noisy data $p = n/N$. The true values always fall inside the credible intervals, also when using only a small fraction of the noisy data.

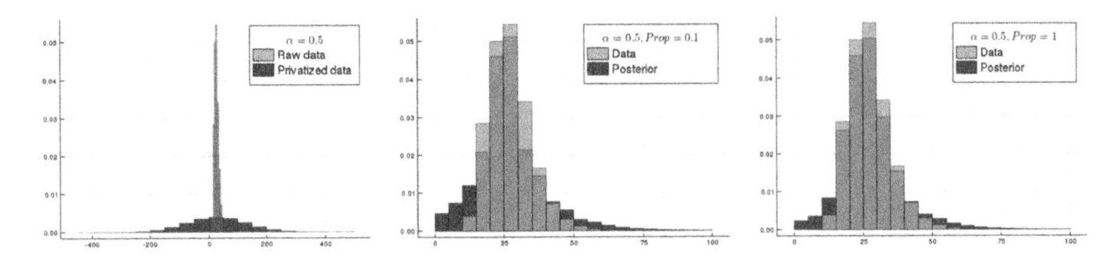

Fig. 2. Body Mass Index data-set, size N, Gaussian noise $\alpha = 0.5$. From left to right: a) Raw vs Privatized data; b) Estimated and True density of P_*, sample size $n = 0.1 * N$; c) Estimated and True density of P_*, $n = N$.

Table 2. Body Mass Index data-set. Guassian Mechanism, $\alpha = 0.5$. Posterior estimates and credible intervals $P_*(X < 16\%)$, $P_*(X > 45\%)$, $med_{P_*}(X)$, $\mathbb{E}(P_*)$, $\text{Var}(P_*)$.

Statistic	$p = 0.1$	$p = 0.5$	$p = 1$	Raw data
% underweight	$10.1 \pm 6.2\%$	$6.5 \pm 3.8\%$	$4.9 \pm 2.4\%$	4.4%
% morbidly obese	$6.3 \pm 3.6\%$	$3.6 \pm 1.9\%$	$2.6 \pm 1.4\%$	2.6%
Median	26.62 ± 1.26	26.62 ± 0.74	26.60 ± 0.92	26.50
Mean	27.58 ± 1.08	27.32 ± 0.36	27.31 ± 0.30	27.32
Variance	119.95 ± 60.04	74.36 ± 38.45	58.56 ± 18.36	59.08

4.3 California Workers Salary

As second real data experiment, we have analyzed the data-set of salaries of California workers already considered in [5]. In [5], it is discussed how problematic it is to recover the population median when data are privatized using additive Laplace noise. The authors derive a non-trivial specialized mechanism Q and estimator T for median estimation. As discussed in our Introduction, a problem is that, in the Non-interactive setting, the data curator cannot choose Q to target a specific task, because the released data-set will be used by different users interested in different statistical tasks. In most of the applications, Q will often consists in noise addition, so the specialized estimator T of [5] will not have the same theoretical guarantees derived for a different Q. Instead, our methodology works for every Q, and in this experiment we show how it can approximate well the true distribution P. Figure 3 a) shows how dramatically Laplace mechanism deteriorates the raw data in this example. However, in panel b) we can see that the posterior approximation obtained with our methodology, using only the sanitized data, approximates the true data distribution reasonably well. Table 3 also shows posterior summaries where it is shown that the true median, mean and variance always fall within the posterior credible regions.

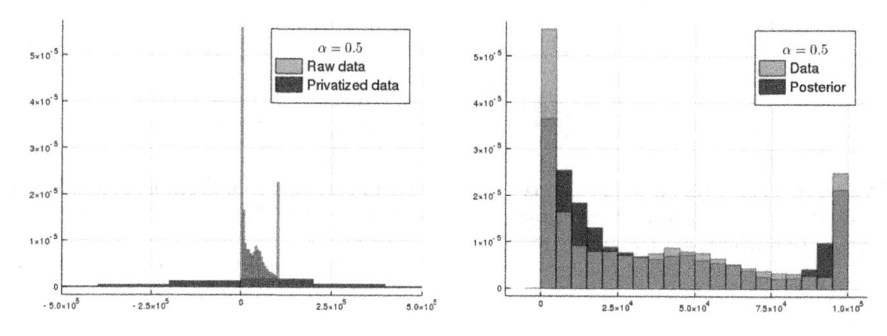

Fig. 3. California Salary data-set, size N, Gaussian noise $\alpha = 0.5$. From left to right: a) Raw vs Privatized data; b) Estimated and True density of P_*, sample size $n = N$.

Table 3. California Salary data-set. Laplace mechanism, $\alpha = 0.5$. Posterior estimates and credible intervals of $med_{P_*}(X)$, $\mathbb{E}(P_*)$, $\mathrm{Var}(P_*)$, computed with Dirichlet Process Mixture (DPM) and sample estimators (naive). True values in column 4 (raw data).

Statistic	DPM	Naive	Raw data
Median	24420 ± 2437	35912	26529
Mean	36364 ± 266	36720	36350
Variance	$1.10e9 \pm 5.45e7$	$1.51e9$	$1.19e9$

5 Conclusions

In this work, we have proposed the first Bayesian Nonparametric approach to Differential Private data. The proposed methodology reformulates the problem of learning the distribution of the raw data, which has been transformed using a mechanism satisfying Differential Privacy, as a latent variable mixture models. The proposed strategy is a "black box" procedure that can be applied to any mechanism Q and does not require any deep analytic study of the problem. Moreover, it is fully non-parametric and does not require any assumption on P. We have described computational algorithms to learn the distribution of the original data, and functionals of it, and provided theoretical guarantees in terms of posterior consistency. Finally, the proposed methodology is shown to have good empirical performance in simulated and real data experiments.

Appendix A: Algorithm for Laplace Mechanism

In this Section, we derive the posterior $\mathbb{P}(dX_k^*|Z_{j_1:j_{n_k}})$ in the case of Laplace Mechanism. Together with Algorithm 2 in the main text, this posterior offers an efficient MCMC algorithm to perform posterior estimation when the Laplace mechanism has been applied to the original data. We remark that, even though the posterior (4) might look complicated at first glance, it is actually just a mixture distribution. For most choices of P_0, it is very easy to compute the weights of this mixture and sample from it. After the proof of Proposition A, we will detail a specific example of (4) for P_0 being Gaussian, which will be used in the experiments. The parameters r and λ_α are chosen as in [5] so that the Laplace Mechanism satisfies Differential Privacy.

Proposition A (Posterior with Laplace Mech.). *Let $r > 0$ and $\Pi_{[-r,r]}$ denote the projection operator on $[-r, r]$, defined as $\Pi_{[-r,r]}(x) = (x)\min(|x|, r)$. Let $Z_i|X_i \sim$ Laplace$(\Pi_{[-r,r]}(X_i), \lambda_\alpha)$ $i = 1, \dots, n$ and let $Z_{j_1}, \dots, Z_{j_{n_k}}$ denote the n_k observations currently assigned to cluster k, i.e. with $c_{j_i} = k$, assumed w.l.o.g. to be ordered increasingly. Let also $i_- := \min\{i \mid Z_{j_i} \geq -r\}$ ($i_- = m+1$ if the set is empty) and $i_+ := \max\{i \mid Z_{j_i} \leq r\}$ ($i_+ = 0$ if the set is empty) and $\widetilde{Z}_{i_--1} = -r$, $\widetilde{Z}_{i_++1} = r$ and for $i \in [i_-, i_+]$, $\widetilde{Z}_i = Z_{j_i}$. Then, the posterior distribution $\mathbb{P}(dX_k^*|Z_{j_1:j_{n_k}})$ is proportional to*

$$\begin{aligned}
&\propto \mathbb{I}_{X_k^* < -r}\, C_{i_--1}\, e^{\frac{2i_- - n_k - 2}{\lambda_\alpha} r} P_0(dX_k^*) + \mathbb{I}_{X_k^* \geq -r}\, C_{i_+}\, e^{-\frac{2i_+ - n_k}{\lambda_\alpha} r} P_0(dX_k^*) \\
&\quad + \sum_{j=i_--1}^{i_+} \mathbb{I}_{X_k^* \in [\widetilde{Z}_j, \widetilde{Z}_{j+1})}\, C_j\, e^{-\frac{2j-n_k}{\lambda_\alpha} X_k^*} P_0(dX_k^*)
\end{aligned} \tag{4}$$

where $C_j = e^{\frac{1}{\lambda_\alpha}\left(\sum_{i=1}^{j} \widetilde{Z}_i - \sum_{i=j+1}^{n} \widetilde{Z}_i\right)}$ for $j = \{i_- - 1, \dots, i_+\}$.

Normal Base Measure: Let $P_0(dX) = \frac{1}{\sqrt{2\pi}\sigma} e^{-\frac{(X-\mu)^2}{2\sigma^2}} dX$ be a Normal distribution. Let us denote $\tilde{\mu}_j = \frac{(n-2j)\sigma^2}{\lambda_\alpha} + \mu$. Then the posterior (4) specializes into

$$\mathbb{P}(X_k^*|Z_{j_1:j_{n_k}}) \propto \mathbb{I}_{X_k^*<-r}\ C_{i_--1}\ e^{\frac{2i_- - n_k - 2}{\lambda_\alpha} r} \frac{1}{\sqrt{2\pi}\sigma} e^{-\frac{(X_k^*-\mu)^2}{2\sigma^2}}$$
$$+ \sum_{j=i_--1}^{i_+} \mathbb{I}_{X_k^* \in [\widetilde{Z}_j, \widetilde{Z}_{j+1})}\ C_j\ e^{\frac{\tilde{\mu}_j^2-\mu^2}{2\sigma^2}} \frac{1}{\sqrt{2\pi}\sigma} e^{-\frac{(X_k^*-\tilde{\mu}_j)^2}{2\sigma^2}}$$
$$+ \mathbb{I}_{X_k^* \geq r}\ C_{i_+}\ e^{-\frac{2i_+ - n_k}{\lambda_\alpha} r} \frac{1}{\sqrt{2\pi}\sigma} e^{-\frac{(X_k^*-\mu)^2}{2\sigma^2}},$$

where we have used the fact that

$$\frac{2j-n_k}{\lambda_\alpha} X_k^* + \frac{(X_k^*-\mu)^2}{2\sigma^2} = \frac{1}{2\sigma^2}\left(X_k^{*2} - 2\tilde{\mu}_j X_k^* + \tilde{\mu}_j^2\right) + \frac{\mu^2 - \tilde{\mu}_j^2}{2\sigma^2}.$$

Let us denote, for $j = i_- - 2, .., i_+ + 1$,

$$\Pi_{i_--2} = C_{i_--1}\ e^{\frac{2i_- - n_k - 2}{\lambda_\alpha} r}\left[1 + \text{erf}\left(\frac{-r-\mu}{\sqrt{2}\sigma}\right)\right]$$
$$\Pi_j = C_j e^{\frac{\tilde{\mu}_j^2-\mu^2}{2\sigma^2}} [\text{erf}\left(\frac{\widetilde{Z}_{j+1} - \tilde{\mu}_j}{\sqrt{2}\sigma}\right) - \text{erf}\left(\frac{\widetilde{Z}_j - \tilde{\mu}_j}{\sqrt{2}\sigma}\right)] \quad \text{for } j = i_- - 1, .., i_+;$$
$$\Pi_{i_++1} = C_{i_+}\ e^{-\frac{2i_+ - n_k - 2}{\lambda_\alpha} r}\left[1 - \text{erf}\left(\frac{r-\mu}{\sqrt{2}\sigma}\right)\right]$$

where erf denotes the Gauss error function. Let $(\pi_j)_j = (\Pi_j / \sum_k \Pi_k)_j$ the normalized weights. The posterior is then a mixture of truncated Normals with disjoint supports. In order to sample for it, we can proceed in 2 steps. First, we sample the a categorical variable J such that $\mathbb{P}(J = j) = \pi_j$. If $J = i_- - 2$, we sample X_k^* from a truncated Normal with mean and variance respectively μ and σ^2 restricted on $(-\infty, -r)$. If $J = i_+ + 1$, we sample X_k^* from a truncated Normal with same mean and variance on (r, ∞). Otherwise, sample X_k^* from a truncated Normal with mean and variance respectively $\tilde{\mu}_j$ and σ^2 restricted to $(\widetilde{Z}_j, \widetilde{Z}_{j+1})$.

Appendix B: Proof of Proposition 1

Denote first $M_P(Z_i) = \int Q(Z_i|X_i)P(dX_i)$, the marginal of the observations when the sensitive data is distributed according to P. Therefore, denoting P_* the true distribution of the sensitive data X_i, it comes that the true marginal distribution of Z_i is M_{P_*}. We will prove Proposition 2, following these steps,

1. Step 1: We show that

$$\forall \epsilon > 0,\ \Pi(h(M_P, M_{P^*}) > \epsilon \mid Z_{1:n}) \to 0 \text{ a.s.} \tag{5}$$

Here, Π denotes the Dirichlet process prior and $\Pi(\cdot|Z_{1:n})$ denotes the posterior under the DPM model and h the Hellinger distance.

2. Step 2: We will show that for any $\delta > 0$,

$$W_2(P, P_*)^2 \leq C_\delta h(M_P, M_{P^*})^{3/4} + C\delta^2 \tag{6}$$

where W_2 is the $\mathbb{L}_2$ Wasserstein distance.

3. Conclusion: Using step 1 and 2, we will show that for any $\epsilon > 0$,

$$\Pi(W_1(P, P_*) > \epsilon \mid Z_{1:n}) \to 0 \text{ a.s.} \tag{7}$$

Now, since W_1 is convex and uniformly bounded on the space of probability measures on $\mathcal{X} \subset [a, b]$, Theorem 6.8 of [13] gives that $\mathbb{E}(P|Z_{1:n})$ converges almost surely to P_* for the W_1 metric. Since $[a, b]$ is compact, this implies that it also converges for any W_k for $k \geq 1$.

To simplify the reading of the proof, in the following C will refer to constant quantities (in particular they do not depend on n), that can change from line to line.

Let us start with the easiest step, which is Eq. (7) of step 3. Let $\epsilon > 0$, from Eq. (6), we know that

$$\Pi(W_2(P, P_*)^2 \leq \epsilon \mid Z_{1:n}) \geq \Pi(C_\delta h(M_P, M_{P^*})^{3/4} + C\delta^2 \leq \epsilon \mid Z_{1:n}).$$

Take δ such that $C\delta^2 \leq \epsilon/2$. We can hence lower bound the left hand side of previous inequality by $\Pi(C_\delta h(M_P, M_{P^*})^{3/4} \leq \epsilon/2 \mid Z_{1:n})$, which we know from Eq. (5) converges almost surely to 1, proving convergence in W_2, which implies (7) since $\mathcal{X} \subset [a, b]$.

Now let us consider Step 1. The Dirichlet prior Π defines a prior on the marginals of Z_i, M_P (also denoted Π). Since $Z_i \overset{iid}{\sim} M_{P_*}$, Schwartz theorem guarantees that (5) holds as long as M_{P_*} is in the Kullback-Leibler support of Π. We will use Theorem 7.2 of [13] to prove it. Let

$$\underline{Q}(Z_i; \mathcal{X}) = \inf_{x \in \mathcal{X}} Q(Z_i|x).$$

Let $Z_i \in \mathcal{Z}$, for any $X_i \in \mathcal{X}$, the differential privacy condition gives

$$Q(Z_i|X_i) \leq e^\alpha \underline{Q}(Z_i; \mathcal{X}) < +\infty,$$

which corresponds to condition (A1) in the theorem of [13]. We only need to prove that (A2) holds, i.e.

$$\int \log\left(\frac{M_{P_*}(Z_i)}{\underline{Q}(Z_i; \mathcal{X})}\right) M_{P_*}(dZ_i) < +\infty,$$

for any probability measure P on $\mathcal{X}$. To see this we rewrite the expression in the log as follows

$$\frac{M_{P_*}(Z_i)}{\underline{Q}(Z_i; \mathcal{X})} = \int \frac{Q(Z_i|X_i)}{\underline{Q}(Z_i; \mathcal{X})} P(dX_i) \leq e^\alpha \int P(dX_i) = e^\alpha$$

where last inequality is due to the differential privacy property of Q. This proves Step 1.

It remains to prove Step 2. We remark first that since the noise is additive in our setting, $Q(Z_i|X_i) = C_Q e^{-\alpha\rho(X_i - Z_i)/\Delta}$ where C_Q is a constant (independent of X_i). Denote $f : t \mapsto C_Q e^{-\alpha\rho(t)/\Delta}$ and $\mathcal{L}(f)$ its Fourier transform. Denote $P * f$ the convolution of P and f. We also recall that

$$\mathcal{L}(M_P) = \mathcal{L}(P * f) = \mathcal{L}(P)\mathcal{L}(f)$$

This part follows the same strategy as the proof of Theorem 2 in [20], the main difference being that here we are not interested in rates and hence need weaker conditions on f. In a similar way, we define a symmetric density K on $\mathbb{R}$ whose Fourier transform $\mathcal{L}(K)$ is continuous, bounded and with support included in $[-1, 1]$. Let $\delta \in (0, 1)$ and $K_\delta(x) = \frac{1}{\delta}K(x/\delta)$. Following the lines of the proof of Theorem 2 in [20], we find that

$$W_2^2(P, P_*) \leq C(||P * K_\delta - P_* * K_\delta||_2^{3/4} + \delta^2), \tag{8}$$

where C is a constant (depending only on K), and that

$$||P * K_\delta - P_* * K_\delta||_2 \leq 2d_{TV}(M_P, M_{P_*})||g_\delta||_2$$

where g_δ is the inverse Fourier transform of $\frac{\mathcal{L}(K_\delta)}{\mathcal{L}(f)}$ and d_{TV} the total variation distance. Now, using Plancherel's identity it comes that

$$||g_\delta||_2^2 \leq C \int \left| \frac{\mathcal{L}(K_\delta)^2(t)}{\mathcal{L}(f)^2(t)} \right|^2 dt \leq C \int_{[-1/\delta, 1/\delta]} \left| \frac{\mathcal{L}(K_\delta)^2(t)}{\mathcal{L}(f)^2(t)} \right|^2 dt \leq C \sup_{[-1/\delta, 1/\delta]} |\mathcal{L}(f)|^{-2}$$

where second line comes from the fact that the support of $\mathcal{L}(K_\delta)$ is in $[-1/\delta, 1/\delta]$, and third line from the fact that it is bounded. Since $|\mathcal{L}(f)|$ is strictly positive (from assumptions) and continuous, it comes that $C_\delta^2 = C \sup_{[-1/\delta, 1/\delta]} |\mathcal{L}(f)|^{-2} < +\infty$. Using the bound $d_{TV} \leq \sqrt{2}\, h$, we can write

$$||P * K_\delta - P_* * K_\delta||_2 \leq C_\delta h(M_P, M_{P_*}), \tag{9}$$

which together with (8) gives

$$W_2^2(P, P_*) \leq C_\delta h(M_P, M_{P*})^{3/4} + C\delta^2$$

Convergence of moments follows directly from [22] (Theorems 6.7 and 6.8).

References

1. Antoniak, C.E.: Mixtures of Dirichlet processes with applications to Bayesian nonparametric problems (1974)
2. Barak, B., Chaudhuri, K., Dwork, C., Kale, S., McSherry, F., Talwar, K.: Privacy, accuracy, and consistency too: a holistic solution to contingency table release. In: Proceedings of the Twenty-Sixth ACM SIGMOD-SIGACT-SIGART Symposium on Principles of Database Systems, pp. 273–282. ACM (2007)

3. Borgs, C., Chayes, J., Smith, A.: Private graphon estimation for sparse graphs. In: Advances in Neural Information Processing Systems, pp. 1369–1377 (2015)
4. Borgs, C., Chayes, J., Smith, A., Zadik, I.: Revealing network structure, confidentially: improved rates for node-private graphon estimation. In: 2018 IEEE 59th Annual Symposium on Foundations of Computer Science (FOCS), pp. 533–543. IEEE (2018)
5. Duchi, J.C., Jordan, M.I., Wainwright, M.J.: Minimax optimal procedures for locally private estimation. J. Am. Stat. Assoc. **113**(521), 182–201 (2018)
6. Dwork, C., McSherry, F., Nissim, K., Smith, A.: Calibrating noise to sensitivity in private data analysis. In: Halevi, S., Rabin, T. (eds.) TCC 2006. LNCS, vol. 3876, pp. 265–284. Springer, Heidelberg (2006). https://doi.org/10.1007/11681878_14
7. Eland, A.: Tackling urban mobility with technology. Google Europe Blog, 18 November 2015
8. Erlingsson, Ú., Pihur, V., Korolova, A.: RAPPOR: randomized aggregatable privacy-preserving ordinal response. In: Proceedings of the 2014 ACM SIGSAC Conference on Computer and Communications Security, pp. 1054–1067. ACM (2014)
9. Fienberg, S.E., Rinaldo, A., Yang, X.: Differential privacy and the risk-utility tradeoff for multi-dimensional Contingency Tables. In: Domingo-Ferrer, J., Magkos, E. (eds.) PSD 2010. LNCS, vol. 6344, pp. 187–199. Springer, Heidelberg (2010). https://doi.org/10.1007/978-3-642-15838-4_17
10. Gaboardi, M., Lim, H.W., Rogers, R.M., Vadhan, S.P.: Differentially private chi-squared hypothesis testing: Goodness of fit and independence testing. In: ICML 2016 Proceedings of the 33rd International Conference on International Conference on Machine Learning-Volume 48. JMLR (2016)
11. Gaboardi, M., Rogers, R.: Local private hypothesis testing: chi-square tests. arXiv preprint arXiv:1709.07155 (2017)
12. Gao, F., van der Vaart, A., et al.: Posterior contraction rates for deconvolution of Dirichlet-Laplace mixtures. Electron. J. Stat. **10**(1), 608–627 (2016)
13. Ghosal, S., Van der Vaart, A.: Fundamentals of Nonparametric Bayesian Inference, vol. 44. Cambridge University Press, Cambridge (2017)
14. Karwa, V., Slavković, A., et al.: Inference using noisy degrees: differentially private β-model and synthetic graphs. Ann. Stat. **44**(1), 87–112 (2016)
15. Kasiviswanathan, S.P., Nissim, K., Raskhodnikova, S., Smith, A.: Analyzing graphs with node differential privacy. In: Sahai, A. (ed.) TCC 2013. LNCS, vol. 7785, pp. 457–476. Springer, Heidelberg (2013). https://doi.org/10.1007/978-3-642-36594-2_26
16. Lo, A.Y.: On a class of Bayesian nonparametric estimates: I. Density estimates. Ann. Stat. **12**, 351–357 (1984)
17. Machanavajjhala, A., Kifer, D., Abowd, J., Gehrke, J., Vilhuber, L.: Privacy: theory meets practice on the map. In: Proceedings of the 2008 IEEE 24th International Conference on Data Engineering, pp. 277–286. IEEE Computer Society (2008)
18. McSherry, F., Talwar, K.: Mechanism design via differential privacy. In: FOCS 2007, pp. 94–103 (2007)
19. Neal, R.M.: Markov chain sampling methods for Dirichlet process mixture models. J. Comput. Graph. Stat. **9**(2), 249–265 (2000)
20. Nguyen, X., et al.: Convergence of latent mixing measures in finite and infinite mixture models. Ann. Stat. **41**(1), 370–400 (2013)
21. Rinott, Y., O'Keefe, C.M., Shlomo, N., Skinner, C., et al.: Confidentiality and differential privacy in the dissemination of frequency tables. Stat. Sci. **33**(3), 358–385 (2018)

22. Villani, C.: Optimal Transport: Old and New, vol. 338. Springer, Heidelberg (2008). https://doi.org/10.1007/978-3-540-71050-9
23. Wang, Y., Lee, J., Kifer, D.: Revisiting differentially private hypothesis tests for categorical data. arXiv preprint arXiv:1511.03376 (2015)
24. Wasserman, L., Zhou, S.: A statistical framework for differential privacy. J. Am. Stat. Assoc. **105**(489), 375–389 (2010)

A Partitioned Recoding Scheme for Privacy Preserving Data Publishing

Chris Clifton[1], Eric J. Hanson[2], Keith Merrill[2], Shawn Merrill[1(✉)], and Amjad Zahraa[1]

[1] Department of Computer Science and CERIAS Purdue University, West Lafayette, IN 47907, USA
clifton@cs.purdue.edu, smerrill@purdue.edu
[2] Department of Mathematics, Brandeis University, Waltham, MA 02453, USA
{ehanson4,merrill2}@brandeis.edu

Abstract. There is growing interest in Differential Privacy as a disclosure limitation mechanism for statistical data. The increased attention has brought to light a number of subtleties in the definition and mechanisms. We explore an interesting dichotomy in *parallel composition*, where a subtle difference in the definition of a "neighboring database" leads to significantly different results. We show that by "pre-partitioning" the data randomly into disjoint subsets, then applying well-known anonymization schemes to those pieces, we can eliminate this dichotomy. This provides potential operational benefits, with some interesting implications that give further insight into existing privacy schemes. We explore the theoretical limits of the privacy impacts of pre-partitioning, in the process illuminating some subtle distinctions in privacy definitions. We also discuss the resulting utility, including empirical evaluation of the impact on released privatized statistics.

1 Introduction

The main approaches to sanitizing data in the computer science literature have been based on either generalization (i.e., replacing quasi-identifiers such as birthdate with less specific data such as year of birth), or randomization (i.e., changing the birthdate by some randomly chosen number of days). There are a variety of privacy definitions based on these methods; many generalization definitions build on k-anonymity [10, 11] and many randomization definitions are based on (ε-)differential privacy [1, 2].

Often we work on subsets of the overall population as a result of sampling, to address computational limitations, or for other reasons. This raises a question - should privatization be applied to the entire data, or can it just be done on the subset? This turns out to be a complex question, impacting the privacy guarantees in subtle ways. In the setting of differential privacy, *amplification* [5] can be used to reduce the noise needed to obtain differential privacy on a sample. This only works, however, when it is unknown whether any given individual is in the sample.

J. Domingo-Ferrer and K. Muralidhar (Eds.): PSD 2020, LNCS 12276, pp. 49–61, 2020.
https://doi.org/10.1007/978-3-030-57521-2_4

In the present paper we present a random partitioning approach that can be applied with many existing schemes. In many cases, this provides the same privacy guarantee as treating the entire dataset as a whole. We focus in particular on differential privacy. In this setting, depending on how the partitioning is done, there can be subtle differences in how privacy budgets from different partition elements are combined under differential privacy. We explore and clarify these issues, in particular emphasizing the difference between bounded and unbounded neighboring databases. We also show that when the differentially private results computed on each partition element are released to distinct, non-colluding data users, the secrecy of which individuals fall into which partition elements allows us to make use of amplification.

The layout of this paper is as follows: In Sect. 2, we give an overview of basic definitions and necessary background. In Sect. 3, we formally define partitioned preprocessing and show that under differential privacy, this provides equivalent privacy to treating the dataset as a whole. We give both analytical and computational examples using counting and proportion queries in Sect. 4. Finally, we extend the results of amplification to bounded differential privacy and discuss their significance to partitioned preprocessing in Sect. 5.

2 Background and Definitions

We begin with the definition of differential privacy, which has emerged as a baseline for formal privacy models.

Definition 1 (Differential Privacy). *Let $\varepsilon > 0$ be a real number. Let $\mathcal{A}$ be a randomized algorithm that takes as input a dataset D from some universe $\mathcal{D}$ and assigns a value in* Range($\mathcal{A}$). *We say that $\mathcal{A}$ is an ε-*differentially private mechanism *(or satisfies ε-DP) if for any two datasets $D, D' \in \mathcal{D}$ with $d(D, D') = 1$ (that is, there exist tuples $t \in D$ and $t' \in D'$ with $(D \setminus \{t\}) \cup \{t'\} = D'$) and for any (measurable) subset $S \subseteq$* Range($\mathcal{A}$), *we have*

$$P(\mathcal{A}(D) \in S) \leq e^{\varepsilon} \cdot P(\mathcal{A}(D') \in S).$$

We emphasize that we are taking as our definition the *bounded* version of differential privacy. That is, the datasets D and D' contain the same number of entries. This can be thought of as *changing* an entry to move from D to D'. The alternative is the *unbounded* version of differential privacy. In this case, the dataset D is the same as the dataset D' with one entry added, or vice versa. This can be thought of as *adding or deleting* an entry to move from D to D'. As we will see, this distinction leads to a subtle but important difference that is narrowed by this paper.

Bounded differential privacy is well suited for dealing with a dataset where the population size is known. An example would be a census where the total count is released, but characteristics of individuals should be protected.

We now recall a feature of certain sanitization schemes that plays a central role in what follows.

Definition 2 (Parallel Composition). *We say that a privacy guarantee $\mathcal{P}$ satisfies* parallel composition *if, for all sanitization schemes $\mathcal{A}_1, ..., \mathcal{A}_n$ meeting $\mathcal{P}$ and disjoint datasets $D_1, ..., D_n$ (i.e., any individual appearing in any of the D_i appears in only that dataset), the union $\bigcup_{i=1}^{n} \mathcal{A}_i(D_i)$ satisfies the privacy guarantee $\mathcal{P}$.*

ε-differential privacy is known to satisfy parallel composition [2]:

Lemma 1. *Let $D_1, \ldots, D_n$ be disjoint datasets, that is, datasets that can be assumed to have no individual in common. Let $\mathcal{A}$ satisfy ε-DP. Let $\mathbf{D} = (D_1, \ldots, D_n)$, considered as a vector of datasets, and let $\mathbf{D}'$ be any other vector with $\|\mathbf{D} - \mathbf{D}'\|_1 = 1$. Then for any outcome $\mathbf{S} \subseteq \text{Range}(\mathcal{A})^n$,*

$$P(\mathcal{A}(\mathbf{D}) \in \mathbf{S}) \leq e^{\varepsilon} \cdot P(\mathcal{A}(\mathbf{D}') \in \mathbf{S}).$$

There is a subtle caveat to the above parallel composition statement for differential privacy: The datasets must be *given* as disjoint datasets. Specifically, a change to one of the datasets is not allowed to affect the others. This is not quite the same as being able to partition a dataset along the input domain and then apply the anonymization technique. In the latter case, the following shows that the privacy parameter doubles.

Lemma 2. *Let $S_i, i = 1, \ldots, n$ be a partitioning of the input domain, and let $M_i, i = 1, ..., n$ satisfy ε-DP. Given a dataset D, the mechanism that returns the sequence $M_i(D \cap S_i)$ satisfies 2ε-DP.*

It is again important to note that the reason for the 2ε term is that we are using *bounded* differential privacy. This means two of the subdatasets S_i can change with a single change to the dataset D. For example, if we partition based on whether each tuple represents a minor, an adult under 65, or a senior citizen, we could impact two partition elements by replacing a minor with a senior citizen. On the other hand, with unbounded differential privacy, the addition or deletion of a tuple only impacts a single partition element, meaning ε-differential privacy is satisfied as in Lemma 1. This key distinction lead to two of the earlier papers on differential privacy giving two different parallel composition theorems corresponding to Lemmas 1 [7] and 2 [2] due to different definitions of neighboring databases.

3 Partitioned Preprocessing

The main contribution of this work is to establish a random partitioning scheme (partitioned preprocessing) that leads to ε-differential privacy in the *bounded* case (see Theorem 1). In other words, we show the result of Lemma 1 also applies to bounded differential privacy when the distinction between partition elements is not data-dependent.

We note that this is not the first work to look at partitioning in a mechanism for differential privacy. Ebadi, Antignac, and Sands propose partitioning-based approaches to deal with limitations in PINQ [3]. Their work assumes the unbounded case (where Lemma 1 applies); their approach could be a potential application of our work in the case where we *cannot* assume unbounded sensitivity.

We start with our definition.

Definition 3 (Partitioned preprocessing). *Let D be a dataset of size n. Choose a random decomposition $\mathbf{n} = \{n_i\}$ of n into positive integers from any distribution, and choose a permutation π on n letters uniformly at random. Let D_π denote the dataset whose ℓ-th entry is the $\pi^{-1}(\ell)$-th entry of D. Denote by $D_{\pi,1}$ the first n_1 elements of D_π, by $D_{\pi,2}$ the next n_2 elements, and so on. We refer to the collection of datasets $\{D_{\pi,i}\}_i$ as the* partitioned preprocessing *of D.*

We remark that this definition is consistent for any distribution on the decompositions of n. For example, the sizes of the partition elements, $\mathbf{n}$, can be fixed ahead of time. This corresponds to taking the distribution that selects $\mathbf{n}$ with probability 1.

The idea behind partitioned preprocessing is to apply a mechanism (or set of mechanisms) $\mathcal{A}$ that satisfies some privacy definition to each partition element separately. When the definition satisfies parallel composition, it is often the case that this preserves the original privacy guarantee.

3.1 Differential Privacy Under Partitioned Preprocessing

As we have seen in Lemma 2, under bounded differential privacy, naively partitioning a dataset and ensuring that the use of each partition element satisfies ε-differential privacy only guarantees 2ε-differential privacy. We now show that under partitioned preprocessing, if we satisfy ε-DP for each partition element, we still satisfy ε-DP overall.

Theorem 1 (Partitioned Preprocessing for ε-DP). *Let D be a data-set with n elements and let j be a positive integer. Choose a decomposition $\mathbf{n}$ of n with j elements based on any distribution and choose a permutation π on n elements uniformly at random. Consider the partitioned preprocessing of the dataset D_π into j pieces $\{D_{\pi,i}\}_{1\leq i\leq j}$. For $1 \leq i \leq j$, let $\mathcal{A}_i$ be a mechanism which satisfies ε_i-DP. Apply $\mathcal{A}_i$ to the piece $D_{\pi,i}$, and return the (ordered) list of results. Then the scheme $\mathcal{A}$ returning $(\mathcal{A}_1(D_{\pi,1}),\ldots,\mathcal{A}_j(D_{\pi,j}))$ satisfies ε-DP, where $\varepsilon := \max\limits_{1\leq i\leq j} \varepsilon_i$.*

We note that when $j = 1$, this reduces to applying the mechanism $\mathcal{A}_1$ to the dataset as a whole with (total) privacy budget ε_1.

Proof. Let $D = (x_1,\ldots,x_n)$ be a dataset with n elements. For convenience, set $t := x_1$ and let t' be another tuple that is not necessarily in D. Let $D' := (t', x_2, ..., x_n)$. For a fixed positive integer j, we denote by $P_{n,j}$ the set of all decompositions of n into j pieces, i.e., all choices $\{n_1,\ldots,n_j\}$ satisfying that $n_i \in \mathbb{Z}_{>0}$ and $\sum_{i=1}^{j} n_i = n$. Let μ denote a probability measure on $P_{n,j}$, and represent an arbitrary element by $\mathbf{n}$. Let S_n denote the collection of all permutations on n

elements, so that $|S_n| = n!$. Given $\mathbf{n} \in P_{n,j}$ and an index $1 \leq \ell \leq j$, let $S_{\mathbf{n}}^{\ell}$ denote the set of permutations π which place t in the partition elements $D_{\pi,\ell}$. We note that $S_{\mathbf{n}}^{\ell}$ is precisely the set of $\pi \in S_n$ for which $n_1 + n_2 + ... + n_{\ell-1} < \pi^{-1}(1) \leq n_1 + n_2 + ... + n_{\ell}$, that the collection $\{S_{\mathbf{n}}^{\ell}\}_{1 \leq \ell \leq n}$ gives a disjoint decomposition of S_n, and that $|S_{\mathbf{n}}^{\ell}| = (n-1)! n_{\ell}$.

Now fix intervals $T_1, \ldots, T_j$ in $\mathbb{R}$. Then

$$
\begin{aligned}
P(\mathcal{A}(D) \in T_1 \times \cdots \times T_j) &= \sum_{\mathbf{n} \in P_{n,j}} \frac{\mu(\mathbf{n})}{n!} \sum_{\pi \in S_n} P\left(\bigcap_{k=1}^{j} \mathcal{A}_k(D_{\pi,k}) \in T_k \right) \\
&= \sum_{\mathbf{n} \in P_{n,j}} \frac{\mu(\mathbf{n})}{n!} \sum_{\pi \in S_n} \prod_{k=1}^{j} P(\mathcal{A}_k(D_{\pi,k}) \in T_k) \\
&= \sum_{\mathbf{n} \in P_{n,j}} \frac{\mu(\mathbf{n})}{n!} \sum_{\ell=1}^{j} \sum_{\pi \in S_{\mathbf{n}}^{\ell}} \prod_{k=1}^{j} P(\mathcal{A}_k(D_{\pi,k}) \in T_k).
\end{aligned}
$$

Now for a fixed $\mathbf{n} \in P_{n,j}$, $\pi \in S_{\mathbf{n}}^{\ell}$, and a fixed index k, if $k \neq \ell$ we have $D_{\pi,k} = D'_{\pi,k}$, and hence

$$P(\mathcal{A}_k(D_{\pi,k}) \in T_k) = P(\mathcal{A}_k(D'_{\pi,k}) \in T_k).$$

On the other hand, if $k = \ell$, by the definition of ε-differential privacy, we have

$$P(\mathcal{A}_\ell(D_{\pi,\ell}) \in T_\ell) \leq e^{\varepsilon_\ell} P(\mathcal{A}_\ell(D'_{\pi,\ell}) \in T_\ell).$$

Therefore, returning to our formula,

$$
\begin{aligned}
P(\mathcal{A}(D) \in T_1 \times \cdots \times T_j) &\leq \sum_{\mathbf{n} \in P_{n,j}} \frac{\mu(\mathbf{n})}{n!} \sum_{\ell=1}^{j} \sum_{\pi \in S_{\mathbf{n}}^{\ell}} e^{\varepsilon_\ell} \prod_{k=1}^{j} P(\mathcal{A}_k(D'_{\pi,k}) \in T_k) \\
&\leq e^{\varepsilon} \sum_{\mathbf{n} \in P_{n,j}} \frac{\mu(\mathbf{n})}{n!} \sum_{\pi \in S_n} \prod_{k=1}^{j} P(\mathcal{A}_k(D'_{\pi,k}) \in T_k) \\
&= e^{\varepsilon} \sum_{\mathbf{n} \in P_{n,j}} \frac{\mu(\mathbf{n})}{n!} \sum_{\pi \in S_n} P\left(\bigcap_{k=1}^{j} \mathcal{A}_k(D'_{\pi,k}) \in T_k \right) \\
&= e^{\varepsilon} P(\mathcal{A}(D') \in T_1 \times \cdots \times T_j).
\end{aligned}
$$

The crucial difference between the above theorem and Lemma 2 is that the partitioning is done in a *data-independent* manner. This is what allows us to preserve the privacy parameter ε instead getting 2ε. The key is that the partitioning of the data is completely determined by the sizes of the partition elements $\mathbf{n}$ and the permutation π used to order the elements; once we condition on those choices, replacing $t \mapsto t'$ therefore only affects a single partition element, and hence introduces only a single factor of e^{ε}.

We note that because the preprocessed extension of any differentially private mechanism is again differentially private, all the usual results about post-processing, sequential and parallel composition, and the like still hold for this extension.

4 Detailed Examples

In this section, we provide analytical and computational examples of results obtained from using partitioned preprocessing and differential privacy. This aims to enlighten both the process of our method and the utility of the noisy query results it generates.

For differentially private mechanisms, the notion of utility is frequently measured by the amount of noise the mechanism needs to add in order to satisfy the definition. We will focus here on the variance of the noise, the idea being that the smaller the variance the more faithful the statistics, and hence the more useful for data analysis and data mining.

For specificity, fix a positive integer n. We recall that the (global) *sensitivity* of a query f at size n is

$$\Delta_n f := \max_{d(D,D')=1} |f(D) - f(D')|, \qquad |D| = |D'| = n.$$

In other words, $\Delta_n f$ represents the maximum impact that any one individual can have on the answer to the query f between an arbitrary dataset of size n and its neighbors. We note that for the purposes of computing sensitivities at two different sizes, the set of possible tuples of the datasets are taken to be the same, but the size of such allowed datasets has changed.

We will focus for the remainder of this section on the well-known Laplace mechanism [2], a differentially private mechanism that returns for a query f on a dataset D (with $|D| = n$) the answer $f(D)$ *plus* some noise, drawn from a Laplace (also called a double exponential) distribution with mean 0 and variance $2(\Delta_n f/\varepsilon)^2$, where ε is our privacy budget.

We recall that in differential privacy, one needs to set the number of queries allowed on the dataset ahead of time, since the amount of noise added to each query must compensate for the total number of queries (otherwise by asking the same query repeatedly, one could appeal to the (Weak) Law of Large Numbers to obtain an accurate estimate of $f(D)$, since the noise has mean 0 and would "cancel out" in the long run). Traditionally, if we are going to allow k queries on the dataset, we would add noise corresponding to ε/k.

One way to interpret our expansion is the following: Instead of immediately splitting the privacy budget ε, we first prepartition (at random) the dataset D into j pieces, $D_{\pi,1}, ..., D_{\pi,j}$, for some permutation $\pi \in S_n$ and sizes $\mathbf{n} = (n_1, \dots, n_j)$. We then ask some number of queries on each piece. The motivating idea is that each piece of the dataset can be used to answer the queries of a different data user. For example, suppose a data user wishes to know the answer to $k' < k$ queries. Then on that data user's piece of the data, we add only ε/k' noise to the query responses.

For example, suppose a data user has been 'assigned' the partion element $D_{\pi,i}$ and that they wish to run only some query f. The mechanism then returns $f(D_{\pi,i}) + \text{Lap}(\Delta_{n_i} f/\varepsilon)$ and the data user is free to use this information as they wish. This includes sharing this information with other data users, as the guarantee of Theorem 1 makes no assumption that a data user is given access to only the results of the queries asked on a single partition element. When this is the case, we can leverage amplification, as detailed in Sect. 5.

We observe that the noise added to satisfy differential privacy is smaller under partitioned preprocessing than when all queries are answered on the entire dataset. This is primarily because partitioning the data is effectively the same as sampling, which introduces noise that may outweigh the benefit of having larger ε. We will demonstrate this in Sect. 4.2. Another potential complication comes from the need to account for query responses being computed on a smaller sample than D. For example, counting queries will need to be scaled by a factor of $|D|/|D_{\pi,i}|$. This has the impact of scaling the sensitivity by the same factor. Since we are using bounded differential privacy, the value of $|D|$ and $|D'|$ are the same, and under partitioned processing, the size of the partitions $|D_{\pi,i}|$ do not change. As a result, these values are sensitivity 0 and can be released without compromising privacy.

The remainder of this section is aimed at understanding the variance (both analytically and empirically) of proportion queries answered under partitioned preprocessing.

4.1 Variance of Proportion Queries

Suppose that f is a proportion query, that is, $f(D)$ is the *proportion* of records that satisfy some attribute. Then we have $\Delta_m f = 1/m$ for all m. As in the previous section, we suppose $|D| = n$ and we have partitioned D into $D_{\pi,1}, \dots, D_{\pi,j}$ with $|D_{\pi,i}| = n_i$. For simplicity, we assume we are running a total of j queries.

If we run f on the entire dataset D, we return $f(D) + \text{Lap}(\Delta_n f/(\varepsilon/j))$, whereas under partitioned preprocessing, we would run f on only $D_{\pi,i}$, and return $f(D_{\pi,i}) + \text{Lap}(\Delta_{n_i} f/\varepsilon)$. Since probabilistically we expect that $f(D_{\pi,i}) \approx f(D)$, we compare the variances in each of these cases.

In the former, we have variance equal to $2(\Delta_n f/(\varepsilon/j))^2 = 2\frac{j^2}{|D|^2\varepsilon^2}$. In the latter case, we recall that we have two independent sources of noise: that coming from the partitioning (hypergeometric) and that coming from differential privacy (Laplacian). The total variance is the sum of the two:

$$\begin{aligned}\sigma^2 &= \frac{p(1-p)}{|D_{\pi,i}|}\frac{|D| - |D_{\pi,i}|}{|D| - 1} + 2(\Delta_{n_i} f/\varepsilon)^2 \\ &= \frac{p(1-p)}{|D_{\pi,i}|}\frac{|D| - |D_{\pi,i}|}{|D| - 1} + \frac{2}{|D_{\pi,i}|^2\varepsilon^2} \approx \frac{p(1-p)}{|D_{\pi,i}|}\frac{|D| - |D_{\pi,i}|}{|D| - 1} + \frac{2j^2}{|D|^2\varepsilon^2}.\end{aligned}$$

We see that in this case, the variance under partitioned preprocessing is always slightly larger than in the traditional scheme by an additional factor of

$$\frac{p(1-p)}{|D_{\pi,i}|}\frac{|D|-|D_{\pi,i}|}{|D|-1} \approx \frac{p(1-p)}{|D_{\pi,i}|}\frac{(j-1)|D_{\pi,i}|}{|D|-1} = \frac{p(1-p)(j-1)}{|D|-1} \leq \frac{j-1}{4(n-1)}.$$

Even though the partitioned preprocessing has greater variance than the original mechanism in the proportion case, we note that the noise coming from partitioning (sampling) is identical for our method and the traditional method of partitioning. Thus when a data curator has opted for the use of partitioning and bounded differential privacy, using partitioned preprocessing rather than the traditional method effectively doubles the privacy budget without changing the noise coming from partitioning.

4.2 Empirical Demonstration

Perhaps an easier way to understand the utility of random partitioning is through the impact on real queries. We give an example of a proportion query on a 1940 U.S. Census dataset released for disclosure avoidance tests [9]. This dataset consists of over 100M individuals, and avoids the complex use methods (and consequent difficulty of determining sensitivity) of many other public use microdata sets.

We use a typical proportion query, the proportion of adult males to all males. This is run on a 10% sample of the data, and then on an element of a partition of that sample into 10 pieces (essentially a 1% sample) with a correspondingly higher privacy budget (Fig. 1); the box plots show the distribution of results over 1000 runs.

The idea is that if we had other queries, they could be run on other partition elements without affecting this query – but if run on the full dataset, they would need shares of the total privacy budget to achieve the same privacy level, thus requiring a smaller value of epsilon.

We see that as expected, for small values of ε, the impact of partitioning is small relative to the noise added to protect privacy. The distribution of results for the partitioned data and the full 10% sample is basically the same for $\varepsilon \leq 0.001$. (Note that even at $\varepsilon = 0.0005$, the majority of the time the error is less than 0.5%.) For larger values of ε, little noise is required for differential privacy, and the sampling error dominates – but even so, at a 90% confidence interval, the error is well under 0.5% for the partitioned approach.

5 Amplification

We now give an overview of the amplification technique of [5] and discuss its relationship with partitioned preprocessing. We prove a version of amplification

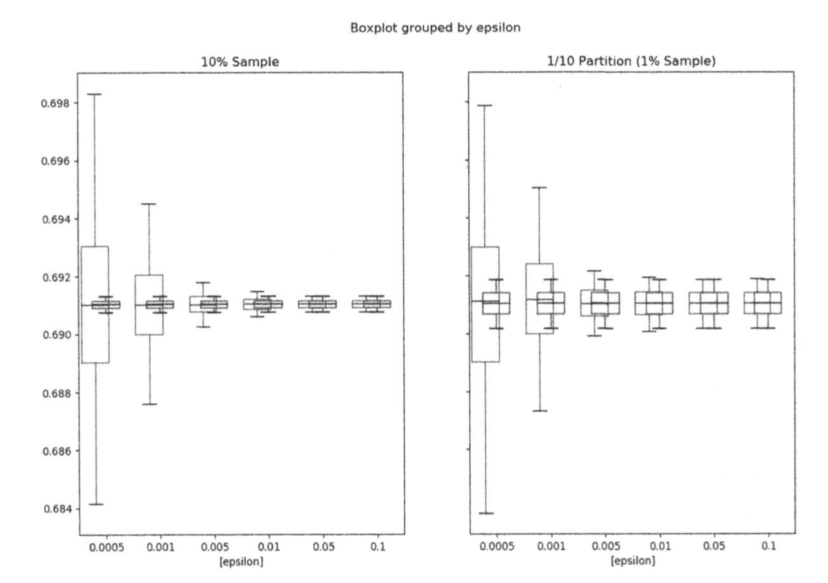

Fig. 1. Distribution of query results with and without differential privacy on a 10% sample database (left) vs. partition of the 10% sample into 10 pieces with correspondingly higher budget (right): Proportion of Adult Males to All Males in the 1940 Census. The outer/left box at each value of ε represents the result with differential privacy and the inner/right box at each value of ε represents the result without differential privacy (sampling error only).

for *bounded* differential privacy; the original paper is for unbounded differential privacy and the proof does not easily generalize.

The idea behind amplification is that sampling with known probability before answering queries greatly increases the privacy budget for queries answered on the sample. More precisely, we have the following, which we prove in the appendix.

Theorem 2 (Amplification for bounded ε-DP). *Let $\mathcal{A}$ satisfy ε-DP. Let D be a dataset with $|D| = n$ and choose an integer $n' < n$. We denote $\beta = n'/n$. Choose a subdataset $D' \subset D$ with $|D'| = n'$ uniformly at random. Then the mechanism which returns $\mathcal{A}(D')$ satisfies ε'-DP, where*

$$\varepsilon' = \ln\left(\frac{e^{\varepsilon}\beta + 1 - \beta}{1 - \beta}\right).$$

Our statement differs from that in [5] by the factor of $1 - \beta$ in the denominator, which comes from our use of bounded differential privacy. It also adds the assumption that the size n' of the subdataset is fixed for a given β and n, which will be subtly used in the proof. Moreover, it is possible that $\varepsilon' > \varepsilon$, in which case the mechanism still satisfies the original ε-DP guarantee. Table 1 shows some examples of this as well as examples where amplification significantly increases the privacy budget.

Table 1. Examples of maximum values of β for amplification to provide savings and values of ε' when $\beta = 1/10$ for various values of ε.

ε	.1	.5	1	2	5	10
max. β for $\epsilon' < \epsilon$	.087	.282	.387	.464	.498	.499
ε' at $\beta = 1/10$	N/A	.17	.26	.60	2.86	7.80

As an immediate consequence of Theorem 2, we have the following.

Corollary 1. *Let $\mathcal{A}$ satisfy ε-DP. Let D be a dataset with n elements and let j be a positive integer. Fix a decomposition $\mathbf{n}$ of n with j elements and choose a permutation π on n elements uniformly at random. Choose an index $1 \leq i \leq j$ and let $D_{\pi,i}$ be the i-th partition element of the partitioned preprocessing of D. Then the mechanism that returns $\mathcal{A}(D_{\pi,i})$ satisfies ε'-DP, where ε' is as defined in Theorem 2 with $\beta = n_i/n$.*

The caveat to this result is that the mechanism is *only* returning the result computed on $D_{\pi,i}$. Indeed, if we return $(\mathcal{A}(D_{\pi,i_1}), \ldots, \mathcal{A}(D_{\pi,i_k}))$, we have effectively changed our parameter β to $\frac{1}{n}(n_{i_1} + \cdots + n_{i_k})$. Amplification could still provide a benefit in this case, but only if this new sampling rate is still much smaller than 1. Overall, amplification with partitioned preprocessing is most appropriate when the results of queries run on different partition elements are released to distinct data users who do not share their results.

6 Conclusion

Independently sanitizing subsets of a dataset raises a number of challenges. By introducing partitioned preprocessing, we have provided an expansion of the tighter bound on the differential privacy parallel composition theorem for unbounded neighboring datasets to support bounded differential privacy in certain cases. Moreover, when the sizes of the partition elements are known and the results of the mechanisms run on different elements are given to distinct, non-colluding data users, we can leverage amplification to increase the privacy budget.

This does come at some cost in result quality, although not significant for small values of ε or high sensitivity queries. Even so, there are still logistical advantages to using random partitioning with differential privacy. One example is if we are only collecting a subset of the data to begin with (sampling). Randomly partitioning the population before sampling gives us the ability to collect further samples and treat the privacy budget independently. More structured partitioning (e.g., stratified sampling based on individual characteristics) can put us into the 2ε situation, where allowing queries to two partition elements requires they share a privacy budget. Another example is when a user would like to employ a differentially private mechanism that is superlinear in runtime;

in these cases, partitioning off a small portion of the data can enable a differentially private answer with substantial savings on time investment. One place such mechanisms show up is in the context of smooth sensitivity [8]. This is a concept that replaces global sensitivity with a weighted maximum of the impact of a single individual in datasets sufficiently close to the real dataset, which can often be substantially smaller. As a concrete example, the mechanism used in [12] to build a Naïve Bayes classifier using smooth sensitivity has a runtime of $\mathcal{O}(n^2)$.

The idea of random partitioning also applies to techniques based on generalization. For example, assuring k-anonymity, ℓ-diversity [6], or t-closeness [4] on each partition element provides the same guarantee (with the same parameter) on the dataset as a whole.[1] With techniques based on generalization, the possibility exists that queries may be answerable on some partition element, whereas the necessarily granularity could be generalized away globally. Further discussion is left for future work.

Acknowledgments. This work was supported by the United States Census Bureau under CRADA CB16ADR0160002. The views and opinions expressed in this writing are those of the authors and not the U.S. Census Bureau.

Appendix: Proof of Theorem 2

We first fix some notation. Let $\mathcal{A}$ be a ε-differentially private mechanism. Let D be a dataset with $|D| = n$ and choose an integer $n' < n$. Fix some tuple $t \in D$. We denote by Y_t the set of all subdatasets $D_s \subset D$ with $|D_s| = n'$ and $t \in D_s$ and by N_t the set of all subdatasets $D_s \subset D$ with $|D_s| = n'$ and $t \notin D_s$. We observe

$$|Y_t| = \binom{n-1}{n'-1} \qquad |N_t| = \binom{n-1}{n'}.$$

For $D' = D \setminus \{t\} \cup \{t'\}$ a neighbor of D, we define Y'_t and N'_t analogously. We observe that $N_t = N'_t$.

We will need the following lemma.

Lemma 3. *Let* $t \in D$ *and* $S \subset \mathrm{range}(\mathcal{A})$. *Then*

$$\sum_{D_s \in Y_t} \frac{P(\mathcal{A}(D_s) \in S)}{|Y_t|} \leq e^{\varepsilon} \sum_{D_s \in N_t} \frac{P(\mathcal{A}(D_s) \in S)}{|N_t|}.$$

Proof. For each $D_s \in Y_t$, we can replace the tuple t by any of the $n - n'$ tuples in $D \setminus D_s$ to create a dataset in N_t that is a neighbor of D_s. Similarly, given any $D_s \in N_t$, we can replace any of the n' tuples in D_s with t to create a dataset in Y_t that is a neighbor of D_s.

[1] There is some subtlety here, as k-anonymity under global recoding is not assured, even if each partition element satisfies it.

Now consider

$$(n-n')\sum_{D_s \in Y_t} P(\mathcal{A}(D_s) \in S)$$

as counting each $D_s \in Y_t$ with multiplicity $n-n'$. Thus we replace the $n-n'$ copies of $D_s \in Y_t$ in this sum with its $n-n'$ neighbors in N_t. By differential privacy, each such change causes the probability to grow by no more than e^ε. Moreover, each dataset in N_t will occur n' times in the new sum. Thus

$$(n-n')\sum_{D_t \in Y_t} P(\mathcal{A}(D_t) \in S) \leq e^\varepsilon n' \sum_{D_t \in N_t} P(\mathcal{A}(D_t) \in S).$$

The result now follows from the observation that $\frac{n'}{n-n'} = \frac{|Y_t|}{|N_t|}$.

This lemma captures the reason we have assumed the size of the subdataset to be fixed. In the unbounded case, if we delete a tuple t to pass from dataset D to D', then for each $D_s \subseteq D$ with $t \in D_s$, there is a unique $D'_s \subseteq D'$ with $d(D_s, D'_s) = 1$. Lemma 3 is our generalization of this fact to the unbounded case.

We are now ready to prove Theorem 2, which we restate here for convenience.

Theorem 2. *Let $\mathcal{A}$ satisfy ε-DP. Let D be a dataset with $|D| = n$ and choose an integer $n' < n$. We denote $\beta = n'/n$. Choose a subdataset $D' \subset D$ with $|D'| = n'$ uniformly at random. Then the mechanism which returns $\mathcal{A}(D')$ satisfies ε'-DP, where*

$$\varepsilon' = \ln\left(\frac{e^\varepsilon \beta + 1 - \beta}{1-\beta}\right).$$

Proof. Let $S \subset \text{range}(\mathcal{A})$ and let $D' = D \setminus \{t\} \cup \{t'\}$ be a neighbor of D. We will use the law of total probability twice, conditioning first on whether $t \in D_s$ (i.e. on whether $D_s \in Y_t$ or $D_s \in N_t$), then on the specific subdataset chosen as D_s. This gives

$$\begin{aligned}
P(\mathcal{A}(D_s) \in S) &= \beta \sum_{D_t \in Y_t} \frac{P(\mathcal{A}(D_t) \in S)}{|Y_t|} + (1-\beta) \sum_{D_t \in N_t} \frac{P(\mathcal{A}(D_t) \in S)}{|N_t|} \\
&\leq \beta e^\varepsilon \sum_{D_t \in N_t} \frac{P(\mathcal{A}(D_t) \in S)}{|N_t|} + (1-\beta) \sum_{D_t \in N_t} \frac{P(\mathcal{A}(D_t) \in S)}{|N_t|} \\
&= (\beta e^\varepsilon + 1 - \beta) \sum_{D_t \in N_t} \frac{P(\mathcal{A}(D_t) \in S)}{|N_t|} \\
&= (\beta e^\varepsilon + 1 - \beta) \sum_{D'_s \in N'_t} \frac{P(\mathcal{A}(D'_s) \in S)}{|N'_t|}.
\end{aligned}$$

by the lemma and the fact that $N_t = N'_t$. By analogous reasoning, we have

$$\begin{aligned}
P(\mathcal{A}(D'_s) \in S) &= \beta \sum_{D'_t \in Y'_t} \frac{P(\mathcal{A}(D'_t) \in S)}{|Y'_t|} + (1-\beta) \sum_{D'_t \in N'_t} \frac{P(\mathcal{A}(D'_t) \in S)}{|N'_t|} \\
&\geq (1-\beta) \sum_{D'_t \in N'_t} \frac{P(\mathcal{A}(D'_t) \in S)}{|N'_t|}.
\end{aligned}$$

Combining these two inequalities yields

$$P(\mathcal{A}(D_s) \in S) \leq \frac{\beta e^{\varepsilon} + 1 - \beta}{1 - \beta} P(\mathcal{A}(D'_s) \in S).$$

References

1. Chawla, S., Dwork, C., McSherry, F., Smith, A., Wee, H.: Toward privacy in public databases. In: Kilian, J. (ed.) TCC 2005. LNCS, vol. 3378, pp. 363–385. Springer, Heidelberg (2005). https://doi.org/10.1007/978-3-540-30576-7_20
2. Dwork, C., McSherry, F., Nissim, K., Smith, A.: Calibrating noise to sensitivity in private data analysis. In: Halevi, S., Rabin, T. (eds.) TCC 2006. LNCS, vol. 3876, pp. 265–284. Springer, Heidelberg (2006). https://doi.org/10.1007/11681878_14
3. Ebadi, H., Antignac, T., Sands, D.: Sampling and partitioning for differential privacy. In: 14th Annual Conference on Privacy, Security and Trust (PST), pp. 664–673, Auckland, NZ, 12–14 December 2016
4. Li, N., Li, T.: t-closeness: privacy beyond k-anonymity and l-diversity. In: Proceedings of the 23nd International Conference on Data Engineering (ICDE 2007), Istanbul, Turkey, 16–20 April 2007
5. Li, N., Qardaji, W., Su, D.: On sampling, anonymization, and differential privacy: or, k-anonymization meets differential privacy. In: 7th ACM Symposium on Information, Computer and Communications Security (ASIACCS'2012), pp. 32–33, Seoul, Korea, 2–4 May 2012
6. Machanavajjhala, A., Gehrke, J., Kifer, D., Venkitasubramaniam, M.: l-diversity: privacy beyond k-anonymity. ACM Trans. Knowl. Discov. Data (TKDD) **1**(1), 3-es (2007)
7. McSherry, F.: Privacy integrated queries: an extensible platform for privacy-preserving data analysis. In: Proceedings of the ACM SIGMOD International Conference on Management of Data, pp. 19–30, Providence, Rhode Island, 29 June - 2 July 2009
8. Nissim, K., Raskhodnikova, S., Smith, A.: Smooth sensitivity and sampling in private data analysis. In: STOC, pp. 75–84 (2007)
9. Ruggles, S., et al.: IPUMS USA: version 8.0 extract of 1940 Census for U.S. census bureau disclosure avoidance research [dataset] (2018). https://doi.org/10.18128/D010.V8.0.EXT1940USCB
10. Samarati, P.: Protecting respondent's privacy in microdata release. IEEE Trans. Knowl. Data Eng. **13**(6), 1010–1027 (2001)
11. Sweeney, L.: k-anonymity: a model for protecting privacy. Int. J. Uncertainty, Fuzziness Knowl. Based Syst. **10**(5), 557–570 (2002)
12. Zafarani, F., Clifton, C.: Differentially private naive bayes classifier using smooth sensitivity. Under review by The European Conference on Machine Learning and Principles and Practice of Knowledge Discovery in Databases, 14–18 September 2020

Explaining Recurrent Machine Learning Models: Integral Privacy Revisited

Vicenç Torra[1,2](✉), Guillermo Navarro-Arribas[3], and Edgar Galván[4]

[1] Department of Computing Science, Umeå University, Umeå, Sweden
vtorra@ieee.org
[2] School of Informatics, Skövde University, Skövde, Sweden
[3] Department Information and Communications Engineering – CYBERCAT, Universitat Autònoma de Barcelona, Bellaterra, Catalonia, Spain
guillermo.navarro@uab.cat
[4] Naturally Inspired Computation Research Group, Department of Computer Science, Maynooth University, Maynooth, Ireland
Edgar.Galvan@mu.ie

Abstract. We have recently introduced a privacy model for statistical and machine learning models called integral privacy. A model extracted from a database or, in general, the output of a function satisfies integral privacy when the number of generators of this model is sufficiently large and diverse. In this paper we show how the maximal c-consensus meets problem can be used to study the databases that generate an integrally private solution. We also introduce a definition of integral privacy based on minimal sets in terms of this maximal c-consensus meets problem.

Keywords: Integral privacy · Maximal c-consensus meets · Clustering · Parameter selection

1 Introduction

The output of any function computed from a database can be sensitive. It can contain traces of the data used. A simple example is the computation of the mean of a set of salaries. The presence in the database of a person with a high salary may affect the mean salary so significantly that this person presence can be singled out. When the data is sensitive, this type of disclosure can be problematic. This situation applies if we consider an intruder accessing the mean salary of patients in the psychiatric department of a hospital of a small town.

Data-driven models inherit the same problems. Membership attacks [6] are to infer that a particular record has been used to train a machine learning or a statistical model.

Data privacy [7] is to provide techniques and methodologies to ensure that disclosure does not take place. Privacy models are computational definitions of privacy.

Differential privacy [1] is one of the privacy models that focus on this type of attacks. Informally, we say that the output of a function satisfies differential privacy when this output value does not change significantly when we add or remove a record

J. Domingo-Ferrer and K. Muralidhar (Eds.): PSD 2020, LNCS 12276, pp. 62–73, 2020.
https://doi.org/10.1007/978-3-030-57521-2_5

from a database. This means in our case, that the presence or absence of a particular individual in the database cannot be singled out. That is, the presence or absence of the person with a high salary does not change much the *mean* salary. To make differential privacy possible, outputs are randomized so what needs to be similar is the probability distribution on the possible outcomes.

We introduced [8] integral privacy as an alternative privacy model that focuses on the privacy of the generators of the output of a function. Informally, we require that the set of databases that can produce this output is sufficiently large and diverse. We have developed solutions for this privacy model to compute statistics as the mean and deviations [3], decision trees [5] and linear regressions [4]. Our results show that integral privacy can provide in some scenarios solutions with good quality (i.e., good utility) and that solutions may be better with respect to utility than those of differential privacy.

A solution from an integral privacy point of view is one that can be produced by different databases. The more databases that can generate the solution the better, and the more diverse these databases are, the better. Integral privacy formalises this idea.

Our original definition requires databases to be different. In this paper we propose a formalization based on minimal sets. It permits to strengthen our privacy requirements on the generators of the model. This new definition is proposed in Definition 4.

1.1 Model Selection and Integral Privacy

Our definitions of integral privacy are proposed as alternatives to differential privacy. Our goal is to select machine and statistical learning modes that are good from a privacy point of view. Machine and statistical learning is, in short, a model selection problem from a set of candidate ones.

When we need to select a model from candidate solutions (a data-driven model) and we want the model to be integrally private and optimal with respect to other parameters (e.g., accuracy, fairness, bias-free) we need to navigate among sets of records. This is so because each candidate solution needs to be generated by at least a database, and integrally private solutions need to be generated by several alternative databases.

For the sake of explainability we consider that it is appropriate to provide tools to actually explore and navigate on these sets of databases. While from a formal point of view, this is not required, and we can just define a methodology to produce integrally private solutions, as we proceeded in [5], it is convenient to develop techniques to understand the space of databases and, in particular, the databases that generate an integrally private solution. In this paper we study how the maximal c-consensus meets [9] of a set of records can be used for this purpose. The new definition of integral privacy is based on the solution of maximal c-consensus meets. Solutions of this problem represent key records for a given model.

1.2 Structure of the Paper

The structure of the paper is as follows. In Sect. 2 we review integral privacy and in Sect. 3 the maximal c-consensus meets. Then, in Sect. 4, we introduce a new definition for integral privacy based on maximal c-consensus meets. This is the definition on minimal sets. In Sect. 5 we prove some results related to the maximal c-consensus meets and the related definition for integral privacy. These results explain integrally private solutions in terms of maximal c-consensus meet solutions. Section 6 describes how we compute the solutions. The paper finishes with some conclusions and directions for future research.

2 On Privacy Models: Integral Privacy

In this section we review two privacy models that focus on the privacy of individual records in a database when this database is used to compute a function. That is, we consider a data set X and we apply a function, a query, or an algorithm A to this database and generate an output y. So, $y = A(X)$. Then, we are interested in avoiding inferences on individual records x in X from the information that we have on y.

Differential privacy [1] is one of the privacy models for this type of scenario. Its focus is on the presence or an absence of a record in the database, and that this presence or absence cannot be inferred from y. The definition is based on a comparison of two probability distributions on the space of outputs of A. We assume that different executions of algorithm A may lead to different outcomes, and that the distributions obtained from two databases that differ in only one record, say x, are similar enough. When the distributions are similar enough we cannot infer from the distributions that this particular record x was used.

Definition 1. *A randomized algorithm A is said to be ε-differentially private, if for all neighbouring data sets X and X', and for all events $E \subset Range(A)$,*

$$\frac{Pr[A(X) \in E]}{Pr[A(X') \in E]} \leq e^{\varepsilon}.$$

In this definition we have that X and X' are neighboring data sets when they differ in one record. We represent that X and X' are neighboring data sets with $d(X, X') = 1$.

Integral privacy [8] is similar to differential privacy in that it focuses on the consequences of knowing the output of a function. I.e., on the inferences that can be obtained from this output. The original definition considered not a single database and a single output but two databases and their corresponding outputs. Then, inferences are not only on the databases but also on the modifications (transitions) that have been applied to one database to transform it into another one. Here we focus on a single database.

The cornerstone of the definition of integral privacy is the concept of generator of an output. That is, the set of databases that can generate the output. We formalize this concept as follows.

Let P be the population in a given domain $\mathscr{D}$. Let A be an algorithm that given a data set $S \subseteq P$ computes an output $A(S)$ that belongs to another domain $\mathscr{G}$. Then for any G in $\mathscr{G}$ and some previous knowledge S^* with $S^* \subset P$ on the generators, the set of possible generators of G is the set defined by $Gen(G, S^*) = \{S' | S^* \subseteq S' \subseteq P, A(S') = G\}$. We use $Gen^*(G, S^*) = \{S' \setminus S^* | S^* \subseteq S' \subseteq P, A(S') = G\}$. When no information is known on S^*, we use $S^* = \emptyset$. Note that previous knowledge is assumed to be a set of records present in the database used to generate G.

Then, integral privacy is about having a large and diverse set of generators. This is to avoid inferences on records in the set and, in particular, avoid membership attacks.

Naturally, it is not always possible to achieve integral privacy because if the only way to generate G from S^* is to have $S^* \cup \{x\}$, previous knowldge S^* implies that $Gen^*(G,S^*) = \{\{x\}\}$. This is the case if we know all patients in the psychiatric department except the rich one. We can infer from G (the mean) that the rich is also in the set.

Definition 2. *Let P represent the data of a population, A be an algorithm to compute functions from databases $S \subseteq P$ into $\mathcal{G}$. Let $G \in \mathcal{G}$, let $S \subseteq P$ be some background knowledge S^* on the data set used to compute G, let $Gen(G,S^*)$ represent the possible databases that can generate G and are consistent with the background knowledge S^*. Then, i-integral privacy is satisfied when the set $Gen(G,S^*)$ is* large *and*

$$\cap_{g \in Gen^*(G,S^*)} g = \emptyset.$$

The intersection is to avoid that all generators share a record. This would imply that there is a record that is necessary to construct G. Following [8] we can distinguish two definitions of *large* in the previous definition.

One follows k-anonymity and requires $Gen(G,S^*)$ to have at least k elements. This means that there are at least k different databases that can be used to build G.

The second definition considers minimal sets in $Gen(G,S^*)$. Let us consider that there are 10 databases that generate a model G. Then, 5 of them share the record r and the other 5 share a record r'. Then, the model G would satisfy at least k-anonymous integral privacy for $k = 2$. In this paper we formalize this second approach in Definition 4.

An important concept in privacy is plausible deniability. We can define it for integral privacy as follows.

Definition 3. *Let G, A, S^*, P, $Gen(G,S^*)$ and $Gen^*(G,S^*)$ as in Definition 2. Integral privacy satisfies plausible deniability if for any record r in P such that $r \notin S^*$ there is a set $\sigma \in Gen^*(G,S^*)$ such that $r \notin \sigma$.*

Naturally, integral privacy satisfies by definition plausible deniability for all records not in S^*. This is so because the intersection of data sets in $Gen^*(G,S^*)$ is the empty set.

Differential privacy and integral privacy have fundamental differences. They are due to the fact that the former requires a *smooth* function, as the addition of any record does not change much the function (i.e., $A(D) \sim A(D \oplus x)$ where $D \oplus x$ means to add the record x to D). In contrast, integral privacy does not require *smoothness* as we do not focus on neighbourhoods. We require that the output of the function for any database results always in what we call *recurrent* models. If $f^{-1}(G)$ is the set of all (real) databases that can generate the output G, we require $A^{-1}(G)$ to be a large set for G. Consider the following example of integrally private function.

Example 1. Let D be a database, let A be an algorithm that is 1 if the number of records in D is even, and 0 if the number of records in D is odd. That is, $f(D) = 1$ if and only if $|D|$ is even.

If this function is applied to an arbitrary subset of the population in Sweden, then the function is integrally private and, therefore, satisfies plausible deniability. The function is not differentially private.

Example 2. Let $P = \{r_1 = 1000, r_2 = 1200, r_3 = 1400, r_4 = 1200, r_5 = 1000, r_6 = 1000, r_7 = 1200, r_8 = 1400, r_9 = 1800, r_{10} = 800\}$ salaries of a population, let $G = 1200$ the mean salary of a database extracted from P. This mean salary will be k-anonymous integral privacy for at least $k = 4$ because the following databases $\{r_1 = 1000, r_2 = 1200, r_3 = 1400\}$, $\{r_4 = 1200\}$, $\{r_5 = 1000, r_8 = 1400\}$, and $\{r_9 = 1800, r_{10} = 800\}$ generate a mean of 1200, and these databases do not share records.

3 Maximal c-consensus Meets

In the previous section we have discussed that the same model can be obtained from different databases. From a privacy perspective, we are interested in these *recurrent* models. Nevertheless, we have also discussed that the recurrence of a model is not enough. When all databases that generate a model share a record, the model is vulnerable to membership attacks.

We have recently introduced [9] maximal c-consensus meets, which can be used to study sets of databases. We show in the next section that this definition permits to define integral privacy in terms of minimal sets.

Given a reference set, the maximal c-consensus meets problem is about finding a set of representatives for a collection of subsets of the reference set. Using notation from lattice theory, we are interested in finding a set of meets that are maximal, in the sense that they have a large number of elements. The problem has similarities (see [9] for details) with other combinatorial optimization problems. In particular, it is related to max-k-intersect, consensus/ensemble clustering, and the minimum set cover problem.

Maximal c-consensus meets is defined in terms of a parameter c, which is the number of representatives we are looking for. It is similar to the number of clusters in clustering algorithms. E.g., k in k-means, c in fuzzy c-means.

3.1 Formalization of the Problem

Let X be a reference set. Let $n = |X|$ be its cardinality, $x_1, \ldots, x_n$ be the elements in X and $\wp(X)$ the set of all subsets of X. The subsets of X define a partially ordered set. Let $A, B \subseteq X$, we use $A \leq B$ when $A \subseteq B$. Therefore, $(\wp(X), \leq)$ is a partially ordered set. I.e., this relationship satisfies reflexivity, antisymmetry and transitivity.

For a partially ordered set $(L, \leq)$, given a subset Y of L we say that $u \in L$ is an upper bound when for all $y \in Y$ we have $y \leq u$. Similarly, $l \in L$ is a lower bound when for all $y \in Y$ we have $l \leq y$. In lattice theory we have the concepts of least upper bound (or join or supremum) and greatest lower bound (or meet or infimum). Then, $(L, \leq)$ is a lattice when each $a, b \in L$ have a join and a meet. We use $\vee$ and $\wedge$ to represent, respectively, the join and the meet as usual. E.g., $a \vee b$ is the join of a and b, and $a \wedge b$ is the meet of a and b.

Given a finite reference set X, the partially ordered set $(\wp(X), \leq)$ is a lattice when the meet is the intersection and the join is the union. This is the lattice we consider in this paper.

Maximal c-consensus meets [9] is defined in terms of a collection S of η subsets of X. Let $S_i \subseteq X$ for $i = 1, \ldots, \eta$, where η is the number of these sets. Then, $S = \{S_1, \ldots, S_\eta\}$. The goal of the problem is to find c parts of the collection whose meets are maximal. Let π_j be a part of S, then, the size of the corresponding meet is $|\cap_{s \in \pi_j} S|$. Let Π be the partition of S with elements π_j for $j = 1, \ldots, c$.

Table 1 gives an example with $X = \{1,2,3,4,5,6,7,8,0\}$ and sets $S_i \subseteq X$ for $i = 1, \ldots, 36$.

When we consider that the total size of the meets of Π is $\sum_{j=1}^{c} |\cap_{s \in \pi_j} S|$ (i.e., the addition of all sizes), we can formalize the maximal c-consensus meets problem as the maximization of the total size of the meets as follows.

$$\begin{aligned} &\text{maximize} \sum_{j=1}^{c} |\cap_{s_i \in \pi_j} S_i| \\ &\text{subject to} \sum_{j=1}^{c} \mu_j(S_i) = 1 \quad \text{for all } i = 1 \ldots \eta \\ &\qquad \mu_j(S_i) \in \{0,1\} \quad \text{for all } i = 1 \ldots \eta \text{ and all } j = 1, \ldots, c \end{aligned} \tag{1}$$

In this formulation μ defines a partition of S. This is so because of the constraints on μ in the problem.

Solutions of the problem above do not require that all meets are large. A few large ones (or all but one large ones) can be enough to lead to a good optimal solution. Because of that, we introduced an alternative definition that we call well-balanced maximal c-consensus meets. In this case we consider the size of the meet with the smallest size. The size of this meet is the one that we want to maximize. The definition follows.

$$\begin{aligned} &\text{maximize} \min_{j=1}^{c} |\cap_{s_i \in \pi_j} S_i| \\ &\text{subject to} \sum_{j=1}^{c} \mu_j(S_i) = 1 \quad \text{for all } i = 1 \ldots \eta \\ &\qquad \mu_j(S_i) \in \{0,1\} \quad \text{for all } i = 1 \ldots \eta \text{ and all } j = 1, \ldots, c \end{aligned} \tag{2}$$

To solve this problem we proposed in [9] the use of a k-means like clustering algorithm and the use of genetic algorithms.

4 Using Maximal c-consensus Meets to Define Integral Privacy

Let P represent the data of a population, A be an algorithm to compute a model (a statistic or a function). Then, different subsets $S \subset P$ will produce models $A(S) \in \mathcal{G}$. Here $\mathcal{G}$ is the space of all possible models.

Let us focus on a particular model $G \in \mathcal{G}$, then $Gen(G, S^*)$ represents all databases that can generate G. From an integral privacy perspective, we are interested in obtaining information on the databases in $Gen(G, S^*)$ that can generate G. The maximal c-consensus meets provide information on this.

Table 1. Set of records corresponding to the problem BC4.

$\{1,2,3,4,5,6,8,0\}$, $\{1,2,3,4,5,6,8\}$, $\{1,2,3,4,5,6,0\}$, $\{1,2,3,5,6,8,0\}$, $\{1,2,4,5,6,8,0\}$, $\{2,3,4,5,6,8,0\}$, $\{1,2,3,4,5,6\}$, $\{1,2,3,5,6,8\}$, $\{1,2,3,5,6,0\}$, $\{1,2,4,5,6,8\}$, $\{1,2,4,5,6,0\}$, $\{1,2,5,6,8,0\}$, $\{1,3,4,5,8,0\}$, $\{2,3,4,5,6,8\}$, $\{2,3,4,5,6,0\}$, $\{2,3,5,6,8,0\}$, $\{2,4,5,6,8,0\}$, $\{1,2,4,5,6\}$, $\{1,2,5,6,8\}$, $\{1,2,5,6,0\}$, $\{1,3,5,8,0\}$, $\{1,4,5,8,0\}$, $\{2,3,4,5,6\}$, $\{2,3,5,6,8\}$, $\{2,3,5,6,0\}$, $\{2,4,5,6,8\}$, $\{2,4,5,6,0\}$, $\{2,5,6,8,0\}$, $\{3,4,5,8,0\}$, $\{1,5,8,0\}$, $\{2,4,5,6\}$, $\{2,5,6,8\}$, $\{2,5,6,0\}$, $\{3,5,8,0\}$, $\{4,5,8,0\}$, $\{5,8,0\}$

Observe that with respect to maximal c-consensus meets it is irrelevant whether we consider $Gen(G,S^*)$ or $Gen^*(G,S^*)$ as the difference of the corresponding two optimization problems will be the same and the objective functions only differ on a constant.

Observe that Table 1 can be seen from this perspective. Let us consider that the reference set $X = \{1,2,3,4,5,6,7,8,0\}$ represents the individuals of the whole population P and each set in Table 1 represents a database. For the sake of illustration we consider here that when we apply algorithm A to all these databases we obtain the same output.

Then, the maximal c-consensus meets permits us to find clusters of databases that share a large number of records. We will use this perspective to formalize the second definition of integral privacy sketched above. The one that is based on minimal sets in $Gen(G,S^*)$.

Observe that given a set of databases $Gen(G,S^*)$, when we find the optimal partition Π of these databases (in terms of the maximal c-consensus meets) for given a value c, the partition permits us to compute the set of common records $\cap_{s_i \in \pi_j} S_i$ for each $\pi_j \in \Pi$. Let m_j represent this set of common records. Then, from a privacy perspective, a good model G is the one that $m_i \cap m_j = \emptyset$. That is, any pair of meets m_i and m_j share no elements.

This permits to formalize meet-based integral privacy as follows. The definition is based on the parameter c. The larger the c, the larger the privacy. Naturally, if we require a very large c (say 10 or 100) this means that we need to be able to generate the same output with a large number of databases that do not share any record.

Definition 4. *Let P represent the data of a population, A be an algorithm that computes a function from databases $S \subseteq P$ in the set $\mathcal{G}$. Let $G \in \mathcal{G}$, let $S^* \subseteq P$ be some background knowledge on the data set used to compute G, let $Gen(G,S^*)$ represent the possible databases that can generate G and are consistent with the background knowledge S^*, and $Gen^*(G,S^*)$ the same set removing S^* (see definitions above).*

Then, G satisfies c-meets-based integral privacy if there is a solution Π of the maximal c-consensus meets for $Gen^*(G,S^*)$ according to Eq. 2 such that for all $\pi_i \neq \pi_j \in \Pi$ satisfies

$$m_i \cap m_j = \emptyset$$

with $m_i = \cap_{S \in \pi_i} S$ and $m_j = \cap_{S \in \pi_j} S$.

Note that there may be several solutions Π of the optimization problem with the same objective function. We require only that one of them satisfies $m_i \cap m_j = \emptyset$ for all $\pi_i \neq \pi_j \in \Pi$.

This definition implies that a solution G is c-meets based integral privacy if for each $x \neq S^*$ there are at least $c-1$ databases in $Gen^*(G, S^*)$ such that x is not there.

We illustrate this definition with the following example.

Example 3. Note that for $c = 4$, the 4 generators of Example 2 above will satisfy the constraint $m_i \cap m_j = \emptyset$ as we have $m_1 = S_1$, $m_2 = S_2$, $m_3 = S_3$, and $m_4 = S_4$.

5 On the Effects of the Parameter c

Both the maximal c-consensus meets and the definition of integral privacy based on these meets depend on the parameter c. We can study how different parameters c influence the solutions of the optimization problem and the effects on the definition of integral privacy. We first prove results on the objective functions of both optimization problems.

Proposition 1. *For the problem based on addition (Eq. 1), the objective function (OF) is strictly monotonic (increasing) with respect to increasing c. We have* $OF_1 = |\cap_{i=1}^{\eta} S_i|$ *for* $c = 1$, *and* $OF_\eta = \sum_{i=1}^{\eta} |S_i|$ *for* $c = \eta$.

Proof. When $c = 1$, there is a single π_1, and therefore all sets are assigned to it (i.e., $\pi_1 = S$). Therefore, the corresponding meet will be the intersection of all $S_1, \ldots, S_\eta$ and, thus, $OF = |\cap_{i=1}^{\eta} S_i|$.

When $c = \eta$, the optimal assignment is to assign each S_i to a different part. I.e., $\pi_i = \{S_i\}$. In this case, $OF = \sum_{i=1}^{\eta} |S_i|$.

Then, to prove that it is strictly monotonic consider a given c and a given partition $\Pi = \{\pi_1, \ldots, \pi_c\}$ with $c < \eta$ with its corresponding objective function OF_i. Let us consider a part π_i with at least two S_j and S_k assigned to it. As $c < \eta$ such part exists. Then, let define π_i' as π_i without S_j and π_i'' as just S_j (i.e., $\pi_i' = \pi_i \setminus \{S_j\}$ and $\pi_i'' = \{S_j\}$). Finally define a new partition with $c+1$ parts as the previous one replacing π_i by the two new sets π_i' and π_i''. That is, $\Pi' = \{\pi_1, \ldots, \pi_c\} \setminus \{\pi_i\} \cup \{\pi_i', \pi_i''\}$. The cardinality of the meets of π_i' and π_i'' is at least as the same as the cardinality of π_i. Therefore as we add these numbers, the objective function will be larger. □

Proposition 2. *For the problem based on the minimum (Eq. 1), the objective function (OF) is monotonic (increasing) with respect to increasing c. We have* $OF_1 = |\cap_{i=1}^{\eta} S_i|$ *for* $c = 1$, *and* $OF_\eta = \min_{i=1}^{\eta} |S_i|$ *for* $c = \eta$.

Proof. The proof of this proposition is similar to the previous one. We can prove the monotonicity of the objective function using the same sets. Nevertheless, as when we build π_i' and π_i'' from π_i and we include them in the objective function, this objective function just takes the min of the cardinality, the objective function may not strictly increase. E.g., if we have $\pi_i = \{\{1,2,3\}, \{1,2,3,4\}, \{1,2,3,5\}\}$ and we define $\pi_i'' = \{1,2,3\}$ and $\pi_i' = \{\{1,2,3,4\}, \{1,2,3,5\}\}$, the objective function will not increase. □

These two results show that the larger the number of parameters, we have, in general, a larger value of the objective function.

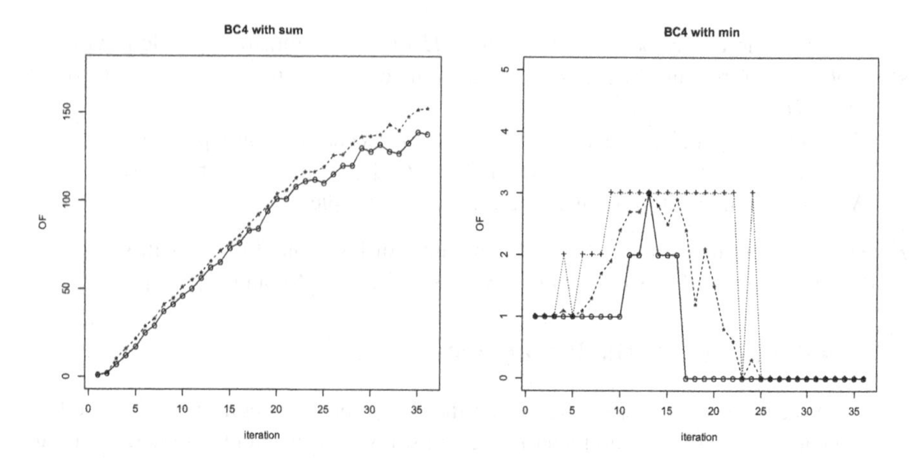

Fig. 1. Values of the objective function for the problem BC4. The figure on left corresponds to the maximal c-consensus meets and the one on the right corresponds to the well-balanced maximal c-consensus meets. Minimum, mean and maximum values of the objective function in each iteration are displayed.

5.1 Open Research Questions

With respect to the definition for integral privacy, it is clear that the larger the c, the more difficult will be to have a model that is compliant with the definition. Nevertheless, an open question is whether a model G that is integrally private for c is also integrally private for any c' such that $c' < c$.

Another open question is whether when a model G with generators $Gen(G,S^*)$ is integrally private for a parameter c, the model is also integrally private when another database is added into the set. That is, if we have two sets of generators $Gen(G,S^*)$ and $Gen(G',S'^*)$ such that $Gen(G,S^*) \subseteq Gen(G',S'^*)$, if integral privacy for $Gen(G,S^*)$ ensures integral privacy for $Gen(G',S'^*)$. We can show with an example that the objective function can decrease when a database is added. It is left as an open problem if this can cause that there is no integrally private solution.

Example 4. Let $A_1 = \{a_1,a_2,a_3\}$, $A_2 = \{a_1,a_2,a_3,a_4\}$, $B_1 = \{b_1,b_2,b_3\}$, $B_2 = \{b_1, b_2, b_3, b_4\}$, $C_1 = \{c_1,c_2,c_3,b_1,b_2\}$, and $C_2 = \{c_1,c_2,c_3,a_1,a_2\}$. An optimal solution for this problem with $c = 3$ is $\pi_1 = \{A_1,A_2\}$, $\pi_2 = \{B_1,B_2\}$, $\pi_3 = \{C_1,C_2\}$. Therefore, $|\pi_1| = |\pi_2| = |\pi_3| = 3$ and the objective function is 3.

If we consider $S' = \{a_1,a_2,b_1,b_2,c_1,c_2\}$ we have that we cannot reach an objective function equal to 3. The assignment $\pi_1 = \{A_1,A_2,C_2\}$, $\pi_2 = \{B_1,B_2,C_1\}$, $\pi_3 = \{S'\}$ results into an objective function equal to 2.

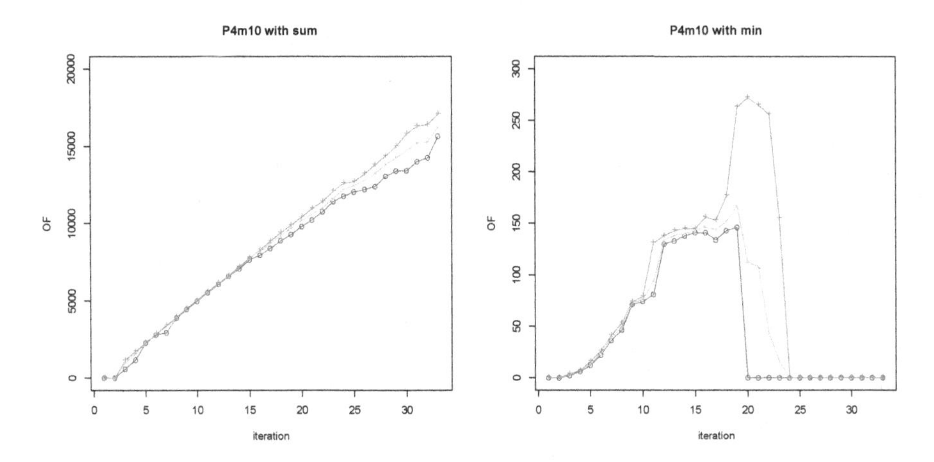

Fig. 2. Values of the objective function for another problem consisting on computing an integrally private *mean* of a file with 1080 records by means of rounding and sampling (following [3]).

6 Solutions Using Genetic Algorithms

In [9] we proposed the use of genetic algorithms to find solutions for the optimization problem described above. We illustrate here the solutions obtained for different values of c. We show the performance of our solution using genetic algorithms when compared with the theoretical results obtained in the previous section. We show that when the objective function uses addition, we obtain results that are consistent with the theoretical ones, but that this is not the case when the objective function uses minimum.

We have considered the solutions obtained for different values of c for the same problems BC1-BC9 considered in [9]. We describe in detail the results of problem *BC*4. This problem consists of the sets of records listed in Table 1. There are 36 sets with at most 9 elements.

We have used 60 iterations with 20 chromosomes each, with a probability of 0.4 for structural mutation and 0.4 for structural crossover. We have used a c that ranges from 1 to η, where η is the number of sets. For *BC*4 we have $\eta = 36$. Each problem is solved 10 times and the minimum, mean, and maximum values of the objective function are recorded. These results for BC4 are displayed in Fig. 1.

It is important to note that while the results above apply for the optimal solutions, solutions found by genetics algorithms are not necessarily optimal. Therefore they do not necessarily satisfy monotonicity. In particular, it is possible that due to structural mutation and structural crossover some of the parts of S are empty, which affects drastically the value of the objective function. Figure 1 gives (left) the results for the first formulation (i.e., Eq. 1) and (right) the results for the second formulation (i.e., Eq. 2).

It can be seen the genetic algorithm is able to obtain results that follow the results in Proposition 1 for the first formulation. That is, the objective function is monotonic with respect to the number of clusters c (except for a few cases). In contrast, our implementation with genetic algorithms does not lead to solutions fully consistent with Proposition 2 for the second formulation. For $c > 15$ the genetic algorithms are not always able to find the optimal, and for $c > 25$ the genetic algorithms are never able to find the optimal. The second objective function is quite sensitive to parts that are empty and they lead to an objective function with a value equal to zero. Recall that the objective function is the size of the smallest set. In fact, when the number of parts is relatively large there are several parts with no sets associated to them. In addition, the meets of the other parts are still rather small. See e.g., for $c = 23$ we have that the best solution has objective function because there are three parts with zero sets assigned to it, and in addition, even removing these parts the minimum meet is of size only 2. Such *optimal solution* is given in Table 2.

Figure 2 shows another example based on the results explained in [3]. There are 33 sets each consisting of up to 1080 records. These sets are obtained from the computation of an integrally private *mean* of the file by means of rounding and sampling. The details on the file and the computation are found in [3]. We can observe that the results obtained by the algorithms are similar for both problems. In the case of using the objective function with the minimum, the optimal is not achieved for c larger than 20.

Table 2. Optimal solution found for $c = 23$ with the problem BC4.

{1 2 3 5 6} {} {2 5 6 8} {0 4 5 8} {0 3 5 8}
{} {2 5 6 8} {0 1 2 4 5 6} {2 4 5 6} {0 2 5 6}
{0 2 3 4 5 6 8} {0 1 3 5 8} {0 5} {1 2 4 5 6 8}
{5 8} {} {0 4 5}{2 5 6}{0 1 2 5 6 8}{2 3 5 6}
{1 5}{1 2 4 5 6}{2 3 4 5 6}

7 Conclusions and Future Work

In this paper we have shown how the maximal c-consensus meets can be used in the context of integral privacy to find the common records of sets of databases that can produce the same solution. We have proven some results related to the monotonicity of the optimal value of the objective function with respect to the number of parts. We have also seen that our approach based on genetic algorithms for solving the optimization problem is not successful for large values of c.

For understanding sets of databases, a smaller c is preferable. How to select a small c with a high objective function is an open problem. We plan to use multiobjetive optimization for this problem.

The maximal c-consensus meets have also been used to formalize a definition for integral privacy. We plan to develop methods to find integrally private models (e.g., decision trees) and statistics (e.g., *means* and *variances*) using the new definition. These solutions need to be evaluated with respect to their utility. Our definition is based on intruder's background knowledge, represented by means of S^*. Further work is needed to analyse what kind of background knowledge can be available.

Acknowledgments. Partial support of the project Swedish Research Council (grant number VR 2016-03346) is acknowledged.

References

1. Dwork, C.: Differential privacy. In: Bugliesi, M., Preneel, B., Sassone, V., Wegener, I. (eds.) ICALP 2006. LNCS, vol. 4052, pp. 1–12. Springer, Heidelberg (2006). https://doi.org/10.1007/11787006_1
2. Rahman, M.A., Rahman, T., Laganière, R., Mohammed, N.: Membership inference attack against differentially private deep learning model. Trans. Data Priv. **11**(1), 61–79 (2018)
3. Senavirathne, N., Torra, V.: Integral privacy compliant statistics computation. In: Pérez-Solà, C., Navarro-Arribas, G., Biryukov, A., Garcia-Alfaro, J. (eds.) DPM/CBT -2019. LNCS, vol. 11737, pp. 22–38. Springer, Cham (2019). https://doi.org/10.1007/978-3-030-31500-9_2
4. Senavirathne, N., Torra, V.: Approximating robust linear regression with an integral privacy guarantee. Proc. PST **2018**, 1–10 (2018)
5. Senavirathne, N., Torra, V.: Integrally private model selection for decision trees. Comput. Secur. **83**, 167–181 (2019)
6. Shokri, R., Stronati, M., Song, C., Shmatikov, V.: Membership inference attacks against machine learning models. In: Proceedings IEEE Symposium on Security and Privacy (2017). arXiv:1610.05820
7. Torra, V.: Data Privacy: Foundations, New Developments and the Big Data Challenge. SBD, vol. 28. Springer, Cham (2017). https://doi.org/10.1007/978-3-319-57358-8
8. Torra, V., Navarro-Arribas, G.: Integral privacy. In: Foresti, S., Persiano, G. (eds.) CANS 2016. LNCS, vol. 10052, pp. 661–669. Springer, Cham (2016). https://doi.org/10.1007/978-3-319-48965-0_44
9. Torra, V., Senavirathne, N.: Maximal *c*-consensus meets. Inf. Fusion **51**, 58–66 (2019)

Utility-Enhancing Flexible Mechanisms for Differential Privacy

Vaikkunth Mugunthan(✉), Wanyi Xiao, and Lalana Kagal

Massachusetts Institute of Technology, Cambridge, MA 02139, USA
vaik@mit.edu

Abstract. Differential privacy is a mathematical technique that provides strong theoretical privacy guarantees by ensuring the statistical indistinguishability of individuals in a dataset. It has become the de facto framework for providing privacy-preserving data analysis over statistical datasets. Differential privacy has garnered significant attention from researchers and privacy experts due to its strong privacy guarantees. However, the accuracy loss caused by the noise added has been an issue. First, we propose a new noise adding mechanism that preserves (ϵ, δ)-differential privacy. The distribution pertaining to this mechanism can be observed as a generalized truncated Laplacian distribution. We show that the proposed mechanism adds optimal noise in a global context, conditional upon technical lemmas. We also show that the generalized truncated Laplacian mechanism performs better than the optimal Gaussian mechanism. In addition, we also propose an (ϵ)-differentially private mechanism to improve the utility of differential privacy by fusing multiple Laplace distributions. We also derive the closed-form expressions for absolute expectation and variance of noise for the proposed mechanisms. Finally, we empirically evaluate the performance of the proposed mechanisms and show an increase in all utility measures considered, while preserving privacy.

Keywords: Differential privacy · Generalized truncated Laplacian distribution · Merging of Laplacian distributions

1 Introduction

In the modern era, there has been a rapid increase in the amount of digital information collected by governments, social media, hospitals, etc. Though this data holds great utility for business and research purposes, inappropriate use of data can lead to a myriad of issues pertaining to privacy. For example, Target inferred that a teen girl was pregnant before her family knew and started sending her coupons related to baby products [8]. Few years ago, Uber's poor privacy practices caught news' attention: their employees misused customer data to track their customers, including politicians and celebrities, in real time, and blogged about "Rides of Glory", where Uber was able to track one night stands [13]. [11]

J. Domingo-Ferrer and K. Muralidhar (Eds.): PSD 2020, LNCS 12276, pp. 74–90, 2020.
https://doi.org/10.1007/978-3-030-57521-2_6

used the Internet Movie Database as a source of prior knowledge to re-identify the anonymized Netflix records of users, unveiling their alleged political preferences and other sensitive information. Due to these and other similar incidents, governments and policymakers start to recognize the importance of protecting personal data. The European Union (EU) recently proposed the General Data Protection Regulation (GDPR) to protect all EU citizens from privacy breaches in today's data-driven world [17] and other countries are contemplating similar regulation meanwhile. Unfortunately, the gaps in current privacy preserving techniques make it difficult for data collectors to support this kind of privacy regulation. However, differential privacy helps organizations to comply with these regulations. The key idea of differential privacy is to reduce the privacy impact on individuals whose information is available in the dataset. Hence, it is not possible to identify individual records and sensitive information pertaining to a particular user.

Many possible approaches can be taken to preserve the privacy of datasets. Early techniques included simple mechanisms for anonymizing datasets by redacting or removing certain fields from datasets and operating on them normally. However, it quickly became apparent that an adversary with auxiliary information could learn significant information from these anonymized datasets. This led to the development of k-anonymity, which generalizes quasi-identifiers (pieces of data that by themselves are not unique identifiers but can be combined with others to act like one) and ensures that a particular user's data is indistinguishable from that of at least $(k-1)$ other users [16]. Though k-anonymity can protect against identity disclosure, it is susceptible against homogeneity and background-knowledge based attacks [14]. l-diversity overcomes this problem and protects against inference-based attacks [10]. However, the semantic relationship between the sensitive attributes makes l-diversity prone to skewness and similarity-based attacks as it is inadequate to avoid attribute exposure [14]. Differential privacy, which provides strong theoretical privacy guarantees, was proposed to provide statistical indistinguishability of datasets.

Differentially private mechanisms are used to release statistics of a dataset as a whole while protecting the sensitive information of individuals in the dataset. Basically, differential privacy guarantees that the released results reveal little or no new information about an individual in the dataset. As an individual sample cannot affect the output significantly, the attackers thus cannot infer the private information corresponding to a particular individual.

Though there has been a myriad of significant contributions in the field of differential privacy, the reasons that it has not yet been adopted by many in the industry are: first, lack of flexibility in the existing mechanisms due to dearth of configurable parameters, second, concerns over reduced utility and privacy. In this paper, we address these issues and offer solutions. Our contributions are as follows:

1. First, we propose the generalized truncated Laplacian mechanism. We also derive the upper bounds on noise amplitude and noise power for the proposed mechanism. We also show that the generalized truncated Laplacian

mechanism offers better flexibility than existing (ϵ, δ)-differentially private mechanisms and performs better than the optimal Gaussian mechanism by reducing the noise amplitude and noise power in all valid privacy regimes [1].
2. Second, we propose how different Laplacian distributions can be merged based on different breakpoints and we also prove that the resulting distribution is differentially private. We also show how it can enhance the utility while guaranteeing privacy.

The proposed mechanisms enable data controllers to fine-tune the perturbation that is necessary to protect privacy for use case specific distortion requirements. This also mitigates the problems pertaining to inaccuracy and provides better utility in bounding noise.

The paper is organized as follows. Section 2 compares and contrasts our work with related work in the field of differential privacy. Section 3 provides background on differential privacy. In Sect. 4 and Sect. 5, we present the generalized truncated Laplacian mechanism and the merging of Laplacian distribution mechanism, respectively. In Sect. 6 we conclude with a summary and a discussion of our future work.

2 Related Work

For numeric queries, ϵ-differential privacy [3] is achieved by adding Laplacian noise to the query result. It has been the de facto approach in a number of works pertaining to differential privacy [4,9] and [5]. [2] proposed (ϵ, δ)-differential privacy, which can be interpreted as ϵ-differential privacy "with probability $1-\delta$". In spite of its near-ubiquitous use, the Laplacian mechanism has no single substantiation of its optimality. [6] proposes a truncated Laplacian mechanism which draw noises from truncated Laplacian distribution. They have shown that the mechanism is more optimal than the optimal Gaussian mechanism as it significantly reduces the noise amplitude and noise power in a myriad of privacy regimes. [6] offers approximate differential privacy and is defined for the symmetric truncated region, that is, $[-A, A]$. [15] propose piecewise mixture distributions that preserve differential privacy and elucidate the importance of flexibility. Most mechanisms and algorithms in differential privacy uses probability distribution with density functions where ϵ is the only variable, a predefined and fixed sensitivity, and minimal amount of additional flexibility for the query-mechanism designer. In this paper, we propose other mechanisms that offer greater flexibility and provide better privacy guarantees than the existing mechanisms. In order to make use of the perturbed query outputs, we have to understand the trade-off between accuracy and privacy. ϵ plays a significant role in determining this trade-off. It is inversely proportional to the scale parameter in the Laplacian distribution. If the value of ϵ is close to zero, the response to two queries made on neighboring datasets is virtually indistinguishable. However, this makes the queries useless as a large amount of noise would have been added to the result and make it futile. In prior literature pertaining to the accuracy and privacy of differentially private mechanisms, the metric of accuracy is in terms of the amount of noise added to the output of a query or in terms of variance. [7] studied the trade-off between privacy and error for answering a group of linear queries in a differentially private manner, where the error is defined as the lowest expectation of the ℓ^2-norm of the noise among the query outputs. They also derived the boundary conditions on the error given the differential privacy constraint. [12] were able to extend the result on the trade-off between privacy and error to the case of (ϵ, δ)-differential privacy.

3 Background

In this section, we will provide an overview of differential privacy, describe the privacy-accuracy trade-off under (ϵ)-differential privacy and (ϵ, δ)-differential privacy, and provide the cost functions that are commonly used in evaluating the utility and privacy trade-off of mechanisms that satisfy differential privacy.

3.1 Differential Privacy

Consider a query function,

$$q : \mathcal{D} \rightarrow \mathbb{R},$$

where $\mathcal{D}$ denotes the set of all possible datasets. The query function q is applied to a dataset or subsets of datasets and returns a real number. Any two datasets $\mathcal{D}_1 \in \mathcal{D}$ and $\mathcal{D}_2 \in \mathcal{D}$ are called *neighboring datasets* if they differ by at most one element. In other words, one dataset is a subset of the other and $|\mathcal{D}_1 - \mathcal{D}_2| \leq 1$. We denote two neighboring datasets $\mathcal{D}_1$, $\mathcal{D}_2$ as $\mathcal{D}_1 \sim \mathcal{D}_2$. A randomized query-answering mechanism $\mathcal{A}$ is a function of the query function q, and will randomly output a real number with certain probability distribution $\mathcal{P}$ depending on $q(\mathcal{D})$, where $\mathcal{D}$ is the dataset.

A more relaxed notion of ϵ-differential privacy is (ϵ, δ)-differential privacy, which can be interpreted as the algorithm that is mostly ϵ-differentially private with the factor δ denoting the probability that it fails to be. Formally, we have the following definition.

Definition 1 *((ϵ, δ)-Differential Privacy). A randomized mechanism $\mathcal{A} : \mathcal{D} \rightarrow \mathcal{O}$ preserves (ϵ, δ)-differential privacy ((ϵ, δ)-DP) when there exists $\epsilon > 0$, $\delta > 0$ such that,*

$$Pr\ [\mathcal{A}(\mathcal{D}_1) \in \mathcal{T}] \leq e^{\epsilon} Pr\ [\mathcal{A}(\mathcal{D}_2) \in \mathcal{T}] + \delta$$

holds for every subset $\mathcal{T} \subseteq \mathcal{O}$ and for any two neighboring datasets $\mathcal{D}_1 \sim \mathcal{D}_2$.

Definition 2 *(Global Sensitivity). For a real-valued query function $q : \mathcal{D} \rightarrow \mathbb{R}$, where $\mathcal{D}$ denotes the set of all possible datasets, the global sensitivity of q, denoted by Δ, is defined as*

$$\Delta = \max_{\mathcal{D}_1 \sim \mathcal{D}_2} |q(\mathcal{D}_1) - q(\mathcal{D}_2)|,$$

for all $\mathcal{D}_1 \in \mathcal{D}$ and $\mathcal{D}_2 \in \mathcal{D}$.

Note when the query function q is a counting query or a histogram query, the global sensitivity $\Delta = 1$ because removing one user from the dataset $\mathcal{D}$ only affects the output of the query by at most 1.

3.2 Utility Model

In this section, we discuss the way that we will be using to evaluate the utility and privacy of a differentially private mechanism. Consider a cost function $\mathcal{L} : \mathbb{R} \rightarrow \mathbb{R}$, which is a function of the random additive noise in the mechanism $\mathcal{A}$. Given a random additive noise x, the cost function for it is defined as $\mathcal{L}(x)$. Therefore, we can derive the expectation of the cost over the probability distribution $\mathcal{P}$ by solving:

$$\int_{x \in \mathbb{R}} \mathcal{L}(x)\mathcal{P}(x)dx$$

Upper bounds on the minimum noise amplitude and noise power, correspond to the l^1 cost function $\mathcal{L}(x) = |x|$ and l^2 cost function $\mathcal{L}(x) = x^2$, respectively. Our objective is to minimize such expectation of the cost over the probability distribution for preserving differential privacy.

3.3 Differentially Private Mechanisms

For the case of real output, introducing noise in an additive manner is a standard technique to preserve differential privacy. Thus, we will be discussing mechanisms $\mathcal{A}$ that preserves ϵ or (ϵ, δ)- differential privacy by adding a random noise X drawn from a probability distribution $\mathcal{P}$. So we will reserve the notation $\mathcal{A}$ for mechanisms that take the standard formula:

$$\mathcal{A}(\mathcal{D}) = q(\mathcal{D}) + X.$$

We will also reserve the variable X for the additive random noise drawn from the probability distribution $\mathcal{P}$ from now on unless stated otherwise.

One of the most well-known differentially private mechanism is the Laplacian mechanism, which uses random noise X drawn from the symmetric Laplacian distribution. The zero-mean Laplacian distribution has a symmetric probability density function $f(x)$ with a scale parameter λ defined as:

$$f(x) = \frac{1}{2\lambda} e^{-\frac{|x|}{\lambda}}.$$

Given the global sensitivity, Δ, of the query function q, and the privacy parameter ϵ, the *Laplacian mechanism* $\mathcal{A}$ uses random noise X drawn from the Laplacian distribution with scale $\lambda = \frac{\Delta}{\epsilon}$. The Laplacian mechanism preserves ϵ-differential privacy [2].

A variant of the symmetric Laplacian mechanism is the truncated Laplacian mechanism, which uses a random noise generated from the truncated Laplace distribution. The zero-mean truncated Laplace distribution has a symmetric-bounded probability density function $f(x)$ with scale λ defined as:

$$f(x) = \begin{cases} Be^{-\frac{|x|}{\lambda}}, \text{ for } x \in [-A, A] \\ 0, \text{ otherwise} \end{cases}$$

where

$$A = \frac{\Delta}{\epsilon}\ln(1 + \frac{e^{\epsilon} - 1}{2\delta}) \text{ and } B = \frac{1}{2\frac{\Delta}{\epsilon}(1 - \frac{1}{1 + \frac{e^{\epsilon}-1}{2\delta}})}.$$

Given the global sensitivity Δ of the query function q, and the privacy parameters ϵ, δ, the *truncated Laplacian mechanism* $\mathcal{A}$ uses random noise X drawn from the truncated Laplacian distribution with scale $\lambda = \frac{\Delta}{\epsilon}$. It has been proven to be (ϵ, δ)-differentially private for $\delta < \frac{1}{2}$ [6].

Remark 1. Note that an ϵ or (ϵ, δ)-differential private mechanism $\mathcal{A}$ with additive noise X drawn from probability distribution $\mathcal{P}$ will still be ϵ or (ϵ, δ)-differential private when the mean μ of $\mathcal{P}$ is any finite real number instead of 0. Therefore, we will just be discussing and proving the $\mu = 0$ case in this paper. However, the proof for any real number μ is similar.

4 Generalized Truncated Laplacian Mechanism

In this section, we propose an (ϵ, δ)-differentially private mechanism that offers better flexibility than the symmetrically bounded truncated Laplacian mechanism [6] and better accuracy than the optimal Gaussian mechanism [1]. First, we state the probability density function and the cumulative distribution function of the generalized truncated Laplacian distribution. Then, we elucidate the (ϵ, δ)-differentially private mechanism. Finally, we evaluate the upper bound on noise amplitude and noise power.

4.1 Generalized Truncated Laplace Distribution

The probability distribution can be viewed as a generalized truncated Laplace distribution. Such a probability distribution is motivated by the symmetrically bounded Laplace distribution proposed by [6]. The proposed distribution in this paper is a more general version as it is asymmetrically bounded.

To construct such a distribution, we set the privacy parameter ϵ and δ. In contrast to most of the existing (ϵ, δ)-differential private mechanisms, where ϵ and δ are the only two variables in the algorithm design, the generalized truncated Laplacian distribution allows another parameter to specify the upper or lower bound of the probability density function. Therefore, with the additional bounding parameter, not depending on the value of ϵ or δ, the proposed generalized truncated Laplace distribution provides more flexibility.

Definition 3. *The zero-mean generalized truncated Laplace distribution has a probability density function $f(x)$ with scale λ, and is asymmetrically bounded by A and B where $A < 0 < B$, defined as:*

$$f(x) = \begin{cases} Me^{-\frac{|x|}{\lambda}} & \text{for } x \in [A, B] \\ 0 & \text{otherwise} \end{cases} \quad \text{where, } M = \frac{1}{\lambda(2 - e^{\frac{A}{\lambda}} - e^{-\frac{B}{\lambda}})}$$

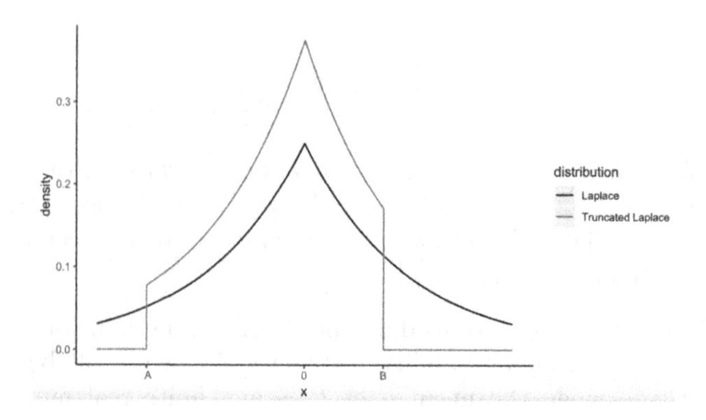

Fig. 1. Laplacian mechanism vs Generalized truncated Laplacian mechanism

Figure 1 depicts a zero-mean Laplace distribution and generalized truncated Laplacian distribution with a scale factor of 2.

The proposed probability distribution is valid, as its probability density function $f(x)$ is greater than 0 for x in the sample space and $\int_{\infty}^{\infty} f(x)dx = 1$

Then we present the closed form of the cumulative distribution function, $F(x)$, for the generalized truncated Laplacian distribution.

The cumulative distribution function is defined as,

$$F(x) = \begin{cases} \frac{e^{\frac{x}{\lambda}} - e^{\frac{A}{\lambda}}}{2 - e^{\frac{A}{\lambda}} - e^{-\frac{B}{\lambda}}} & \text{if } x < 0 \\ \frac{2 - e^{\frac{A}{\lambda}} - e^{-\frac{x}{\lambda}}}{2 - e^{\frac{A}{\lambda}} - e^{-\frac{B}{\lambda}}} & \text{if } x \geq 0 \end{cases}$$

4.2 Mechanism

Given the global sensitivity Δ of the query function q, and the privacy parameters ϵ, δ, the *Generalized Truncated Laplacian mechanism* $\mathcal{A}$ uses random noise X drawn from the generalized truncated Laplacian distribution in Definition 3 with the following parameters:

$$\lambda = \frac{\Delta}{\epsilon}, A + \Delta \leq 0 \leq B - \Delta$$

If $|A| \geq |B|$,

$$\begin{cases} A = \lambda \ln \left[2 + (\frac{1-\delta}{\delta})e^{-\frac{B}{\lambda}} - (\frac{1}{\delta})e^{-\frac{B-\Delta}{\lambda}} \right] \\ B = \text{any positive real number satisfying} |A| \geq |B| \end{cases}$$

If $|A| < |B|$,

$$\begin{cases} A = \text{any negative real number satisfying } |A| < |B| \\ B = -\lambda \ln \left[2 + (\frac{1-\delta}{\delta})e^{\frac{A}{\lambda}} - (\frac{1}{\delta})e^{\frac{A+\Delta}{\lambda}} \right] \end{cases}$$

Theorem 1. *The generalized truncated Laplacian mechanism preserves (ϵ, δ)-differential privacy.*

Proof. The proof for Theorem 1 relies on the following two lemmas, and the proof for those lemmas can be found in Appendix A.

Lemma 1.

$$\max\Big(\int_A^{A+\Delta} f(x)dx, \int_{B-\Delta}^{B} f(x)dx\Big) = \delta$$

for the probability density function $f(x)$, λ, A and B of the generalized truncated Laplace distribution.

Lemma 2. *A mechanism $\mathcal{A}(\mathcal{D}) = q(\mathcal{D}) + X$ that adds a random noise X drawn from probability distribution $\mathcal{P}$ with probability density function $f(x)$, satisfies (ϵ, δ)-differential privacy when*

$$\mathcal{P}(\mathcal{S}) - e^{\epsilon}\mathcal{P}(\mathcal{S} + d) < \delta$$

holds for any $|d| \leq \Delta$, and any measurable set $\mathcal{S} \subseteq \mathbb{R}$, where Δ is the global sensitivity for the query function q.

Using Lemma 2, in order to prove that our mechanism is (ϵ, δ)-differential private, we need to show that for the global sensitivity, Δ, of our query function q,

$$\mathcal{P}(\mathcal{S}) - e^{\epsilon}\mathcal{P}(\mathcal{S} + d) \leq \delta$$

holds for any $|d| \leq \Delta$, and any measurable set $\mathcal{S} \subseteq \mathbb{R}$. Equivalently, it is sufficient to show that

$$\max(\mathcal{P}(\mathcal{S}) - e^{\epsilon}\mathcal{P}(\mathcal{S} + d)) \leq \delta$$

If $x \in (A + \Delta, B - \Delta)$ then,

$$\frac{f(x)}{f(x+d)} = \frac{Me^{-\frac{|x|}{\lambda}}}{Me^{-\frac{|x+d|}{\lambda}}} = e^{\frac{|x+d|-|x|}{\lambda}} \leq e^{\frac{|d|}{\lambda}} \leq e^{\epsilon},$$

which implies $\forall |d| \leq \Delta, f(x) - e^{\epsilon} f(x + d) \leq 0$ when $x \in (A + \Delta, B - \Delta)$.

Thus, for measurable set $\mathcal{S}$,

$$\mathcal{P}(\mathcal{S}) - e^{\epsilon}\mathcal{P}(\mathcal{S} + d) \leq \mathcal{P}(\mathcal{S}') - e^{\epsilon}\mathcal{P}(\mathcal{S}' + d)$$

for $\mathcal{S} \subseteq \mathbb{R}$ and $\mathcal{S}' = \mathcal{S} \backslash (A + \Delta, B - \Delta)$. Therefore, $\mathcal{P}(\mathcal{S}) - e^{\epsilon}\mathcal{P}(\mathcal{S} + d)$ is maximized for some set $\mathcal{S} \subseteq (-\infty, A + \Delta]$ or $\mathcal{S} \subseteq [B - \Delta, \infty)$. Since the distribution changes exponentially with rate $\frac{1}{\lambda} = \frac{\epsilon}{\Delta}$, multiplying the probability distribution $\mathcal{P}$ by e^{ϵ} will result in shifting the probability distribution by Δ. Therefore,

$$\sup_{\mathcal{S} \subseteq \mathbb{R}} \mathcal{P}(\mathcal{S}) - e^{\epsilon}\mathcal{P}(\mathcal{S} + d) \leq \max\Big(\int_{-\infty}^{A+\Delta} f(x)dx, \int_{B-\Delta}^{\infty} f(x)dx\Big) = \max\Big(\int_{A}^{A+\Delta} f(x)dx, \int_{B-\Delta}^{B} f(x)dx\Big)$$

From Lemma 1, we have the desired inequality

$$\mathcal{P}(\mathcal{S}) - e^{\epsilon}\mathcal{P}(\mathcal{S} + d) \leq \delta.$$

Remark 2. We claim that

$$0 < \delta \leq \min\left(\int_A^0 f(x)dx, \int_0^B f(x)dx\right).$$

Proof. If $|A| \geq |B|$, then

$$\int_A^0 f(x)dx > \int_0^B f(x)dx$$
$$\Rightarrow \min\left(\int_A^0 f(x)dx, \int_0^B f(x)dx\right) = \int_0^B f(x)dx$$

Additionally,

$$\max\left(\int_A^{A+\Delta} f(x)dx, \int_{B-\Delta}^B f(x)dx\right) = \int_{B-\Delta}^B f(x)dx = \delta$$

Since $0 \leq B - \Delta$,

$$\delta = \int_{B-\Delta}^B f(x)dx \leq \int_0^B f(x)dx.$$

If $|A| < |B|$, the proof is similar.

4.3 Upper Bound on Noise Amplitude and Noise Power

We apply the generalized truncated Laplacian mechanism to derive upper bounds on the minimum noise amplitude and noise power, corresponding to the l^1 cost function $\mathcal{L}(x) = |x|$ and l^2 cost function $\mathcal{L}(x) = x^2$, respectively.

When $\mathcal{L}(x) = |x|$, the upper bound on minimum noise amplitude is

$$\frac{2\lambda - (\lambda - A)e^{\frac{A}{\lambda}} - (\lambda + B)e^{-\frac{B}{\lambda}}}{2 - e^{\frac{A}{\lambda}} - e^{-\frac{B}{\lambda}}}, \text{where } \lambda = \frac{\Delta}{\epsilon}, \text{ A and B are specified in Theorem 1.}$$

This result is obtained by evaluating

$$inf_{\mathcal{P}\in\mathcal{P}_{\epsilon,\delta}} \int_{x\in\mathbb{R}} |x|\mathcal{P}(dx) = \int_A^B |x| f(x)dx = M\left(\int_A^0 -xe^{\frac{x}{\lambda}}dx + \int_0^B xe^{-\frac{x}{\lambda}}dx\right)$$

As the noise with probability density function $f(x)$ satisfies (ϵ, δ)-differential privacy, this provides an upper bound on $\inf_{\mathcal{P}\in P_{\epsilon,\delta}} \int_{x\in\mathbb{R}} |x|\mathcal{P}(dx)$.

Similarly, we derive the upper bound on the minimum noise power by having $\mathcal{L}(x) = x^2$, and we get

$$\frac{4\lambda^2 - (2\lambda^2 - 2\lambda A + A^2)e^{\frac{A}{\lambda}} - (2\lambda^2 + 2\lambda B + B^2)e^{-\frac{B}{\lambda}}}{2 - e^{\frac{A}{\lambda}} - e^{-\frac{B}{\lambda}}}$$

where $\lambda = \frac{\Delta}{\epsilon}$, and A and B are specified in Sect. 4.2.
As the noise with probability density function $f(x)$ satisfies (ϵ, δ)-differential privacy, this provides an upper bound on $\inf_{\mathcal{P}\in P_{\epsilon,\delta}} \int_{x\in\mathbb{R}} x^2 \mathcal{P}(dx)$.

We performed experiments to compare the performance of the generalized truncated Laplacian mechanism with the optimal Gaussian mechanism [1]. [1] calculate the variance of the optimal Gaussian mechanism using the cumulative density function instead of a tail bound approximation. The ratio of the noise amplitude and noise power of generalized truncated Laplacian mechanism and the optimal Gaussian mechanism is always less than 1 for appropriate values of δ, A and B as shown in Appendix B. Compared to the optimal Gaussian mechanism, the generalized truncated Laplacian mechanism reduces the noise power and noise amplitude across all privacy regimes.

5 Merged Laplacian Mechanism

In this section, we propose an ϵ-differentially private mechanism that merges different Laplacian distributions on different breakpoints. We also evaluate the l^1 cost function $\mathcal{L}(x) = |x|$ and l^2 cost function $\mathcal{L}(x) = x^2$ for the proposed mechanism, compare it with the Laplacian mechanism and show that our proposed mechanism achieves better utility.

Definition 4. *The zero-mean merged Laplacian distribution has a probability density function $f(x)$ with n break points $0 < c_1 < c_2 < \cdots < c_n = \infty$ and n scale parameters $\lambda_1, \lambda_2, ..., \lambda_n$ defined as:*

$$f(x) = f_m(x) \text{ for } x \in (-c_m, -c_{m-1}] \cup [c_{m-1}, c_m)$$
$$\text{Let } c_0 = 0, \forall m \in \{1, 2, ..., n\} \text{ where } f_m(x) = a_m e^{-\frac{|x|}{\lambda_m}},$$

and all $a_m > 0$ computed by

$$\sum_{m=1}^{n} \int_{c_{m-1}}^{c_m} a_m e^{-\frac{|x|}{\lambda_m}} dx = \frac{1}{2}, \tag{1}$$

$$\text{and } f_m(c_m) = f_{m+1}(c_m). \tag{2}$$

Remark 3. Note that (1) and (2) gives sufficient inputs to calculate a_m for $m \in \{1, 2, .., n\}$ as we can write a_m's in terms of $\lambda_1, \lambda_2, ..., \lambda_n$ and a_1 and by inductively applying

$$f_m(c_m) = f_{m+1}(c_m) \implies a_m e^{-\frac{c_m}{\lambda_m}} = a_{m+1} e^{-\frac{c_m}{\lambda_{m+1}}} \implies a_{m+1} = a_m e^{\frac{c_m}{\lambda_{m+1}} - \frac{c_m}{\lambda_m}}.$$

Then, we can rewrite (1) with a_1 as the only variable to solve for the value of a_1. Hence, we can get the values for the rest of the a_m's.

Now we will prove that the probability distributions that we proposed is a valid one. To do this, we need to show that its probability density function $f(x)$ is continuous and greater than 0 for x in the domain and the cumulative probability from $-\infty$ to ∞ is 1.

Proof. First, it is easy to see that $f(x) > 0$ for x in the domain as $e^{-\frac{|x|}{\lambda_m}} > 0$ for $m \in \{1, 2, ..., n\}$, and all $a_m > 0$. Thus, $f_m(x) > 0 \Rightarrow f(x) > 0$.

Additionally, $f(x)$ is continuous on $\cup_{m=0}^{n}(c_{m-1}, c_m)$ as $e^{-\frac{|x|}{\lambda_m}}$ is continuous. At each break point c_m, the continuity is ensured by $f_m(c_m) = f_{m+1}(c_m)$. Now we will show that the cumulative probability from $-\infty$ to ∞ is 1.

$$\int_{-\infty}^{\infty} f(x)dx = \sum_{m=1}^{n}\left(\int_{-c_m}^{-c_{m-1}} f_m(x)dx + \int_{c_{m-1}}^{c_m} f_m(x)dx\right) = 2\sum_{m=1}^{n}\int_{c_{m-1}}^{c_m} f_m(x)dx = 2 \cdot \frac{1}{2} = 1$$

Now we propose a differentially private mechanism which adds noise drawn from the Merged Laplacian distribution as defined in Definition 4.

Theorem 2. *Given the global sensitivity, Δ, of the query function q, and the privacy parameter $\epsilon = \epsilon_1$, the Merged Laplacian mechanism $\mathcal{A}$ uses random noise X drawn from the merged Laplacian distribution with scale parameter $\lambda_m = \frac{\epsilon_m}{\Delta}$ where $\lambda_1 > \lambda_2 > \cdots > \lambda_n$ and preserves ϵ - differential privacy.*

Proof. To prove that our mechanism preserves ϵ - differential privacy, we need to show that for $\mathcal{D}_1 \sim \mathcal{D}_2$,

$$Pr[\mathcal{A}(\mathcal{D}_1) \in \mathcal{T}] \leq e^{\epsilon} Pr[\mathcal{A}(\mathcal{D}_2) \in \mathcal{T}]$$

for any subset $\mathcal{T} \subseteq \mathcal{O}$, where $\mathcal{O}$ is the set of all outputs of the mechanism. And the above inequality is equivalent to

$$\frac{Pr[\mathcal{A}(\mathcal{D}_1) = t]}{Pr[\mathcal{A}(\mathcal{D}_2) = t]} \leq e^{k\epsilon}, \forall t \in \mathcal{T} \Leftrightarrow \qquad \frac{Pr[X = t - q(\mathcal{D}_1)]}{Pr[X = t - q(\mathcal{D}_2)]} \leq e^{k\epsilon}.$$

We will prove this inductively. Our base case is when $n = 1$, then the mechanism becomes the well-known Laplacian mechanism, which is ϵ - differentially private as $\epsilon = \max(\epsilon_1)$. Now, notice that since $\lambda_m = \frac{\epsilon_m}{\Delta}$ and $\lambda_1 > \lambda_2 > \cdots > \lambda_n$, then $\max(\epsilon_1, \epsilon_2, ..., \epsilon_n) = \epsilon_1 = \epsilon$.

Now, assume with the same break points $c_1, c_2, ..., c_{k-1}$ where $0 < c_1 < c_2 < \cdots < c_k = \infty$, the merged Laplacian mechanism is $\epsilon = \epsilon_1$ - differentially private. We want to prove that adding one more break point $c_k < \infty$ to the new merged mechanism satisfies ϵ - differential privacy. We will prove the case where $t - q(\mathcal{D}_1)$ and $t - q(\mathcal{D}_2)$ are negative, as the other cases follows the similar proof with a few sign changes. For $m \in \{1, 2, ..., k-1\}$, we have

$$\frac{Pr[X = t - q(\mathcal{D}_1)]}{Pr[X = t - q(\mathcal{D}_2)]} = \frac{a_m e^{\frac{t-q(\mathcal{D}_1)}{\lambda_m}}}{a_k e^{\frac{t-q(\mathcal{D}_2)}{\lambda_k}}} = \frac{a_m}{a_k} \cdot e^{\frac{t-q(\mathcal{D}_1)}{\lambda_m} - \frac{t-q(\mathcal{D}_2)}{\lambda_k}}$$

We also know that,

$$a_k = a_{k-1} e^{\frac{c_{k-1}}{\lambda_k} - \frac{c_{k-1}}{\lambda_{k-1}}} = a_{k-2} e^{\frac{c_{k-2}}{\lambda_{k-1}} - \frac{c_{k-2}}{\lambda_{k-2}} + \frac{c_{k-1}}{\lambda_k} - \frac{c_{k-1}}{\lambda_{k-1}}} = a_m e^{\sum_{i=m-1}^{k-1} \left(\frac{c_i}{\lambda_{i+1}} - \frac{c_i}{\lambda_i} \right)}$$

Hence,

$$\frac{a_m}{a_k} = e^{\sum_{i=m-1}^{k-1} \left(\frac{c_i}{\lambda_i} - \frac{c_i}{\lambda_{i+1}} \right)}.$$

Notice that

$$\sum_{i=m-1}^{k-1} \left(\frac{c_i}{\lambda_i} - \frac{c_i}{\lambda_{i+1}} \right) = \sum_{i=m-1}^{k-1} \left(\frac{c_i \left(\lambda_{i+1} - \lambda_i \right)}{\lambda_i \lambda_{i+1}} \right) < 0 \text{ since } \lambda_1 > \lambda_2 > \cdots > \lambda_n.$$

Thus,

$$\frac{Pr[X = t - q(\mathcal{D}_1)]}{Pr[X = t - q(\mathcal{D}_2)]} = \frac{a_m}{a_k} \cdot e^{\frac{t-q(\mathcal{D}_1)}{\lambda_m} - \frac{t-q(\mathcal{D}_2)}{\lambda_k}} < e^{\frac{t-q(\mathcal{D}_1)}{\lambda_m} - \frac{t-q(\mathcal{D}_2)}{\lambda_k}} = e^{\frac{\lambda_k(t-q(\mathcal{D}_1)) - \lambda_m(t-q(\mathcal{D}_2))}{\lambda_m \lambda_k}}$$
$$< e^{\frac{\lambda_m(t-q(\mathcal{D}_1)) - \lambda_m(t-q(\mathcal{D}_2))}{\lambda_m \lambda_k}} = e^{\frac{t-q(\mathcal{D}_1)-t+q(\mathcal{D}_2)}{\lambda_k}} = e^{\frac{q(\mathcal{D}_2)-q(\mathcal{D}_1)}{\lambda_k}} = e^{\epsilon_k} < e^{\epsilon}.$$

Hence, we have proved that the proposed mechanism is ϵ - deferentially private.

We evaluate the l^1 cost function $\mathcal{L}(x) = |x|$ and l^2 cost function $\mathcal{L}(x) = x^2$, for the Laplacian, Merged Laplacian with 1 break point and Merged Laplacian with 2 break points as shown in Appendix C. We show that the cost for the Merged Laplacian with 2 break points is lower than that of the Laplacian mechanism and hence we achieve better utility for the same privacy loss.

6 Conclusion and Future Work

In this paper, we presented two novel differentially private mechanisms that provide better accuracy guarantees compared to existing mechanisms. Firstly, we presented a new noise adding mechanism that preserves (ϵ, δ)-differential privacy. The proposed mechanisms provide more scope for customization as they have more parameters to tune. Due to this customizable and flexible nature, appropriate values for different parameters in the mechanisms can be set. We also show that the generalized truncated Laplacian mechanism performs better than the optimal Gaussian mechanism. Next, we show that the proposed merging of Laplacian mechanisms demonstrates better in performance in terms of various metrics for l^1 and l^2 loss without sacrificing additional privacy. As a part of future work, we plan to perform an in-depth comparison of all (ϵ, δ)-differentially private and ϵ-differentially private mechanisms and highlight the pros and cons of every mechanism.

A Proof for Lemma 1 and Lemma 2

Here, we present the proof for Lemma 1 and Lemma 2 used in Sect. 4.2.

Proof (Proof for Lemma 1*).* Since the probability density function $f(x)$ is monotonically increasing when $x \geq 0$ and is monotonically decreasing when $x < 0$,

$$\max\Big(\int_{A}^{A+\Delta} f(x)dx, \int_{B-\Delta}^{B} f(x)dx\Big) = \begin{cases} \int_{B-\Delta}^{B} f(x)dx \text{ when } |A| \geq |B| \\ \int_{A}^{A+\Delta} f(x)dx \text{ when } |A| < |B|. \end{cases}$$

We will first discuss the case when $|A| \geq |B|$,

$$\max\Big(\int_{A}^{A+\Delta} f(x)dx, \int_{B-\Delta}^{B} f(x)dx\Big) = \int_{B-\Delta}^{B} f(x)dx = \int_{B-\Delta}^{B} Me^{-\frac{x}{\lambda}}dx$$
$$= M\lambda\Big(e^{-\frac{B-\Delta}{\lambda}} - e^{-\frac{B}{\lambda}})\Big)$$

Plugging in $M = \frac{1}{\lambda(2-e^{\frac{A}{\lambda}}-e^{-\frac{B}{\lambda}})}$ as in our definition for the generalized truncated Laplacian distribution, $A = \lambda \ln\Big[2 + (\frac{1-\delta}{\delta})e^{-\frac{B}{\lambda}} - (\frac{1}{\delta})e^{-\frac{B-\Delta}{\lambda}}\Big]$ as specified in Theorem 1, we have

$$\int_{B-\Delta}^{B} f(x)dx = \frac{e^{-\frac{B-\Delta}{\lambda}} - e^{-\frac{B}{\lambda}}}{2 - e^{\frac{A}{\lambda}} - e^{-\frac{B}{\lambda}}} = \frac{e^{-\frac{B-\Delta}{\lambda}} - e^{-\frac{B}{\lambda}}}{(\frac{1}{\delta})\Big(e^{-\frac{B-\Delta}{\lambda}} - e^{-\frac{B}{\lambda}}\Big)} = \delta$$

We omit showing the computation for the case when $|A| \leq |B|$ as the derivation is very similar to that of the above mentioned case.

Now, we will proceed to prove Lemma 2.

Proof (Proof for Lemma 2*).* Given two neighboring datasets $\mathcal{D}_1 \sim \mathcal{D}_2$, we know that $|q(\mathcal{D}_1) - q(\mathcal{D}_2)| \leq \Delta$, thus the condition $\mathcal{P}(\mathcal{S}) - e^{\epsilon}\mathcal{P}(\mathcal{S}+d) \leq \delta$ for any $|d| \leq \Delta$ is equivalent to

$$\mathcal{P}(\mathcal{S}) - e^{\epsilon}\mathcal{P}(\mathcal{S} + q(\mathcal{D}_1) - q(\mathcal{D}_2)) \leq \delta \Leftrightarrow \mathcal{P}(\mathcal{S} - q(\mathcal{D}_1)) \leq e^{\epsilon}\mathcal{P}(\mathcal{S} - q(\mathcal{D}_2)) + \delta$$

Hence, for any $t \in \mathcal{S}$, the condition is equivalent to

$$\Pr(X = t - q(\mathcal{D}_1)) \leq e^{\epsilon}\Pr(X = t - q(\mathcal{D}_2)) + \delta$$
$$\Leftrightarrow \Pr(q(\mathcal{D}_1) + X = t) \leq e^{\epsilon}\Pr(q(\mathcal{D}_2) + X = t) + \delta$$
$$\Leftrightarrow \Pr\left[\mathcal{A}(\mathcal{D}_1) \in \mathcal{T}\right] \leq e^{\epsilon}\Pr\left[\mathcal{A}(\mathcal{D}_2) \in \mathcal{T}\right] + \delta,$$

which is the necessary condition for mechanism $\mathcal{A}$ to preserve (ϵ, δ)-differential privacy.

B Generalized Truncated Laplacian - Evaluation

We empirically show that the ratio of the noise amplitude L_1^* and noise power L_2^* of generalized truncated Laplacian mechanism and the optimal Gaussian mechanism is always less than 1 for appropriate values of δ, A and B as described in Sect. 4. Compared to the optimal Gaussian mechanism, the generalized truncated Laplacian mechanism reduces the noise power and noise amplitude across all privacy regimes. The implementation can be found in https://github.com/vaikkunth/DPMechanisms.

ϵ	δ	A	L_1^*	L_2^*
0.7	2.5e−06	−17.46	0.25	0.13
0.4	4.0e−06	−27.57	0.27	0.15
0.4	2.5e−06	−28.74	0.27	0.14
0.4	9.5e−06	−25.4	0.29	0.17
0.4	3.0e−06	−28.29	0.27	0.14
0.7	6.0e−06	−16.21	0.27	0.14
0.4	8.5e−06	−25.68	0.29	0.16
0.7	3.5e−06	−16.98	0.26	0.13
0.7	4.0e−06	−16.79	0.26	0.14
0.4	2.0e−06	−29.3	0.26	0.14
0.1	4.5e−06	−93.66	0.31	0.19
0.7	3.0e−06	−17.2	0.26	0.13
0.4	5.5e−06	−26.77	0.28	0.15
0.7	9.5e−06	−15.55	0.28	0.15
0.4	6.0e−06	−26.55	0.28	0.16
0.7	1.0e−06	−18.77	0.24	0.12
0.4	6.5e−06	−26.35	0.28	0.16
0.4	7.0e−06	−26.17	0.28	0.16
0.4	4.5e−06	−27.27	0.27	0.15
0.4	9.0e−06	−25.54	0.29	0.17
0.4	3.5e−06	−27.9	0.27	0.15
0.1	9.5e−06	−86.19	0.32	0.21
0.1	5.5e−06	−91.66	0.31	0.19
0.1	5.0e−06	−92.61	0.31	0.19
0.4	1.5e−06	−30.02	0.26	0.13
0.4	8.0e−06	−25.83	0.29	0.16

ϵ	δ	A	L_1^*	L_2^*
0.7	7.0e−06	−15.99	0.27	0.15
0.7	7.5e−06	−15.89	0.27	0.15
0.4	1.0e−06	−31.03	0.25	0.13
0.1	7.0e−06	−89.24	0.32	0.2
0.4	5.0e−06	−27.01	0.28	0.15

C Merging Laplacian Distributions - Evaluation

We evaluate the l^1 and l^2 cost for the Laplacian, Merged Laplacian with 1 break point and Merged Laplacian with 2 break points. We show that the cost for the Merged Laplacian with 2 break points is lower than that of the Laplacian mechanism and hence we achieve better utility for the same privacy loss. The implementation can be found in https://github.com/vaikkunth/DPMechanisms.

$(\epsilon_1, \epsilon_2, \epsilon_3)$	(c_1, c_2)	L_1^*	L_2^*
(0.2, 0.25, 0.33)	(1, 3)	(3.0, 3.03, 1.12)	(18.0, 18.22, 3.94)
(0.17, 0.25, 0.33)	(1, 3)	(3.0, 3.03, 1.12)	(18.0, 18.22, 3.96)
(0.17, 0.2, 0.33)	(1, 3)	(3.0, 3.05, 1.13)	(18.0, 18.35, 4.07)
(0.14, 0.25, 0.33)	(1, 3)	(3.0, 3.03, 1.12)	(18.0, 18.22, 3.97)
(0.14, 0.2, 0.33)	(1, 3)	(3.0, 3.05, 1.13)	(18.0, 18.35, 4.08)
(0.14, 0.17, 0.33)	(1, 3)	(3.0, 3.06, 1.13)	(18.0, 18.43, 4.15)
(0.14, 0.17, 0.2)	(1, 3)	(5.0, 5.01, 1.96)	(50.0, 50.15, 16.26)
(0.12, 0.25, 0.33)	(1, 3)	(3.0, 3.03, 1.13)	(18.0, 18.22, 3.98)
(0.12, 0.2, 0.33)	(1, 3)	(3.0, 3.05, 1.13)	(18.0, 18.35, 4.08)
(0.12, 0.17, 0.33)	(1, 3)	(3.0, 3.06, 1.13)	(18.0, 18.43, 4.16)
(0.12, 0.17, 0.2)	(1, 3)	(5.0, 5.01, 1.96)	(50.0, 50.15, 16.28)
(0.12, 0.14, 0.33)	(1, 3)	(3.0, 3.07, 1.14)	(18.0, 18.49, 4.21)
(0.12, 0.14, 0.2)	(1, 3)	(5.0, 5.02, 1.98)	(50.0, 50.26, 16.5)
(0.11, 0.25, 0.33)	(1, 3)	(3.0, 3.03, 1.13)	(18.0, 18.22, 3.98)
(0.11, 0.2, 0.33)	(1, 3)	(3.0, 3.05, 1.13)	(18.0, 18.35, 4.09)
(0.11, 0.17, 0.33)	(1, 3)	(3.0, 3.06, 1.14)	(18.0, 18.43, 4.16)
(0.11, 0.17, 0.2)	(1, 3)	(5.0, 5.01, 1.96)	(50.0, 50.15, 16.3)
(0.11, 0.14, 0.33)	(1, 3)	(3.0, 3.07, 1.14)	(18.0, 18.49, 4.21)
(0.11, 0.14, 0.2)	(1, 3)	(5.0, 5.02, 1.98)	(50.0, 50.26, 16.52)
(0.11, 0.12, 0.33)	(1, 3)	(3.0, 3.08, 1.14)	(18.0, 18.53, 4.25)
(0.11, 0.12, 0.2)	(1, 3)	(5.0, 5.03, 1.99)	(50.0, 50.34, 16.68)
(0.11, 0.12, 0.14)	(1, 3)	(7.0, 7.01, 3.28)	(98.0, 98.12, 42.57)
(0.2, 0.25, 0.33)	(1, 5)	(3.0, 3.03, 1.61)	(18.0, 18.22, 5.17)

$(\epsilon_1, \epsilon_2, \epsilon_3)$	(c_1, c_2)	L_1^*	L_2^*
(0.17, 0.25, 0.33)	(1, 5)	(3.0, 3.03, 1.61)	(18.0, 18.22, 5.19)
(0.17, 0.2, 0.33)	(1, 5)	(3.0, 3.05, 1.63)	(18.0, 18.35, 5.4)
(0.14, 0.25, 0.33)	(1, 5)	(3.0, 3.03, 1.61)	(18.0, 18.22, 5.2)
(0.14, 0.2, 0.33)	(1, 5)	(3.0, 3.05, 1.63)	(18.0, 18.35, 5.41)
(0.14, 0.17, 0.33)	(1, 5)	(3.0, 3.06, 1.64)	(18.0, 18.43, 5.55)
(0.14, 0.17, 0.2)	(1, 5)	(5.0, 5.01, 1.85)	(50.0, 50.15, 10.69)
(0.12, 0.25, 0.33)	(1, 5)	(3.0, 3.03, 1.62)	(18.0, 18.22, 5.21)
(0.12, 0.2, 0.33)	(1, 5)	(3.0, 3.05, 1.63)	(18.0, 18.35, 5.42)
(0.12, 0.17, 0.33)	(1, 5)	(3.0, 3.06, 1.64)	(18.0, 18.43, 5.56)
(0.12, 0.17, 0.2)	(1, 5)	(5.0, 5.01, 1.85)	(50.0, 50.15, 10.7)
(0.12, 0.14, 0.33)	(1, 5)	(3.0, 3.07, 1.65)	(18.0, 18.49, 5.65)
(0.12, 0.14, 0.2)	(1, 5)	(5.0, 5.02, 1.86)	(50.0, 50.26, 10.98)

References

1. Balle, B., Wang, Y.: Improving the Gaussian mechanism for differential privacy: analytical calibration and optimal denoising. CoRR abs/1805.06530 (2018). http://arxiv.org/abs/1805.06530
2. Dwork, C., Kenthapadi, K., McSherry, F., Mironov, I., Naor, M.: Our data, ourselves: privacy via distributed noise generation. In: Vaudenay, S. (ed.) EUROCRYPT 2006. LNCS, vol. 4004, pp. 486–503. Springer, Heidelberg (2006). https://doi.org/10.1007/11761679_29
3. Dwork, C., McSherry, F., Nissim, K., Smith, A.: Calibrating noise to sensitivity in private data analysis. In: Halevi, S., Rabin, T. (eds.) TCC 2006. LNCS, vol. 3876, pp. 265–284. Springer, Heidelberg (2006). https://doi.org/10.1007/11681878_14
4. Dwork, C., Roth, A.: The algorithmic foundations of differential privacy. Found. Trends Theor. Comput. Sci. **9**(3–4), 211–407 (2014). https://doi.org/10.1561/0400000042
5. Fan, L., Xiong, L.: Real-time aggregate monitoring with differential privacy. In: 21st ACM International Conference on Information and Knowledge Management, CIKM 2012, Maui, HI, USA, 29 October – 02 November 2012, pp. 2169–2173 (2012). https://doi.org/10.1145/2396761.2398595
6. Geng, Q., Ding, W., Guo, R., Kumar, S.: Truncated Laplacian mechanism for approximate differential privacy. arXiv preprint arXiv:1810.00877 (2018)
7. Hardt, M., Talwar, K.: On the geometry of differential privacy. In: Proceedings of the Forty-Second ACM Symposium on Theory of Computing, pp. 705–714. ACM (2010)
8. Hill, K.: How Target Figured Out a Teen Girl was Pregnant Before Her Father Did. Forbes, Inc., Jersey City (2012)
9. Li, C., Hay, M., Rastogi, V., Miklau, G., McGregor, A.: Optimizing linear counting queries under differential privacy. In: Proceedings of the Twenty-Ninth ACM SIGMOD-SIGACT-SIGART Symposium on Principles of Database Systems, PODS 2010, 6–11 June 2010, Indianapolis, Indiana, USA, pp. 123–134 (2010). https://doi.org/10.1145/1807085.1807104

10. Machanavajjhala, A., Gehrke, J., Kifer, D., Venkitasubramaniam, M.: l-diversity: privacy beyond k-anonymity. In: 22nd International Conference on Data Engineering (ICDE 2006), pp. 24–24. IEEE (2006)
11. Narayanan, A., Shmatikov, V.: How to break anonymity of the Netflix prize dataset. CoRR abs/cs/0610105 (2006). http://arxiv.org/abs/cs/0610105
12. Nikolov, A., Talwar, K., Zhang, L.: The geometry of differential privacy: the sparse and approximate cases. In: Proceedings of the Forty-Fifth Annual ACM Symposium on Theory of Computing, pp. 351–360. ACM (2013)
13. Perry, D.: Sex and Uber's "rides of glory": the company tracks your one-night stands–and much more'. Comput. Law Secur. Rev. (2014)
14. Rajendran, K., Jayabalan, M., Rana, M.E.: A study on k-anonymity, l-diversity, and t-closeness techniques. IJCSNS **17**(12), 172 (2017)
15. Smith, D.B., Thilakarathna, K., Kâafar, M.A.: More flexible differential privacy: the application of piecewise mixture distributions in query release. CoRR abs/1707.01189 (2017). http://arxiv.org/abs/1707.01189
16. Sweeney, L.: k-anonymity: a model for protecting privacy. Int. J. Uncertainty Fuzziness Knowl.-Based Syst. **10**(05), 557–570 (2002)
17. Voigt, P., Von dem Bussche, A.: The EU General Data Protection Regulation (GDPR). A Practical Guide, 1st edn. Springer, Cham (2017). https://doi.org/10.1007/978-3-319-57959-7

Plausible Deniability

David Sidi(✉) and Jane Bambauer

The University of Arizona, Tucson, AZ 85712, USA
{dsidi,janebambauer}@email.arizona.edu

Abstract. From the perspective of responsible data release, simulation is a useful tool for estimating risk from adversaries with an unknown amount of identified auxiliary information. We present a simple approach to simulation of attack on sampled datasets, along with an implementation, and demonstrate how a data steward might make use of it to evaluate the privacy risk of release for data gathered about students in the University of California system.

Keywords: Statistical disclosure control · Simulation · Education data

1 Introduction

Data stewards are eager to have a trustworthy and useful measure of privacy risk as they prepare the release of a publicly accessible research database. This paper advances a theory of plausible deniability as a measure of database privacy with several distinctive strengths:

- Ease of interpretability
- Ease of use, with released code[1]
- Scalability to large datasets
- Suitability for evaluation of sampling techniques

The basic idea of plausible deniability is that, when a released dataset that is known to be a random sample of the relevant population is attacked, some of the apparent matches that are found between identified auxiliary data and the research data will be false matches.[2]

We show for a particular educational dataset that plausible deniability remains quite high even when heroic assumptions are made about the attacker—viz., that they have quite complete and accurate data about many of the variables in identified form, and are able to impute some of the values they lack.

We conclude with discussion of our results in the light of recent work that has argued sampling does a poor job of protecting plausible deniability.

[1] The project repository is https://codeberg.org/bavajadas.de.benadam/PrivacySim.

[2] We use the terms 'attacker' and 'intruder' interchangeably throughout.

Electronic supplementary material The online version of this chapter (https://doi.org/10.1007/978-3-030-57521-2_7) contains supplementary material, which is available to authorized users.

J. Domingo-Ferrer and K. Muralidhar (Eds.): PSD 2020, LNCS 12276, pp. 91–105, 2020.
https://doi.org/10.1007/978-3-030-57521-2_7

$$\Pr(\text{attempt}) \times \Pr(\text{apparent match} \mid \text{attempt}) \times \Pr(\text{correct match} \mid \text{apparent match}) \times \Pr(\text{harm} \mid \text{correct match})$$

Fig. 1. Model for risk of correct matching. Beyond the estimation of the likelihood that an apparent match correctly identifies an individual, which is separately modeled by Rocher et al. (2019), the model includes factors for the likelihood of an attempt with particular auxiliary information resulting in an apparent match, and the probability of the attempt with those resources happening at all.

2 What Is Plausible Deniability?

Records that are unique in a sample may not be unique in the broader population, and attributes that apply to some number in a sample may apply to a greater number in the population. A consequence of these simple facts is that without auxiliary information about the entire population, an attacker can only do so well against sampled data: even if a match or an attribute disclosure is correct, the attacker cannot be certain that it is, in fact, correct.[3]

Consider a paradigmatic example: an attacker knows his target Alice is represented in a database, and the attacker is able to uniquely match several attributes to Alice in a released subsample. A consequent revelation of a sensitive attribute can be denied on the grounds that the match to Alice is false, explained by the fact that only a random sample of the full dataset has been released. More concretely, suppose Alice is a 67-year-old black female who lives in the US in zip code 85719, but that there is more than one person with those same demographics. The intruder can't be sure that Alice is the person who is apparently matched; perhaps a different 67-year-old black female from the same zip code is the one described in the matching record.

3 Plausible Deniability Is Useful and Underdeveloped

Plausible deniability can be used prospectively to evaluate a data steward's options for deidentification. The risk (or estimated costs) borne by research subjects are a product of several factors: the probability that an attack will be attempted, the probability that an attempted attack will yield a unique match to the data subject,[4] the probability that the match is accurate, and the harm that the research subject would suffer from successful reidentification (see Fig. 1).[5]

[3] In communications anonymity, a similar point is made by pointing out that "usability is a security property;" that is, that increased adoption of an anonymity system increases the total set of individuals that an attacker must individuate (as well as those individuals' diversity, but that is a separate point). See Serjantov et al. (2003), and discussion of "degree of anonymity" in Berthold et al. (2001).

[4] The insights in this paper can also be extended to attacks that seek out attribute disclosure, without unique reidentification to a single research subject.

[5] We could add to this model the harm that could arise from *incorrect* matches that are presumed by the attacker and others to be accurate, but with increased attention on the likelihood of false matches, this problem should be marginalized. (Put differently, every form of deidentification runs the risk that a careless or fraudulent intruder might claim that they have reidentified a research subject when the chance that they have actually done so is very low.)

Plausible deniability provides significant privacy protection to research subjects. Plausible deniability creates two layers of protection—first and most directly by driving down the probability that an apparent match is correct; and second and more indirectly by making the research dataset a less desirable target for intruders to attack in the first place.

The variety of database privacy represented by Plausible Deniability has been lurking in the background of many technical papers and real-world practices (indeed, it is one of the main reasons that data stewards release subsamples instead of full datasets), but it has not been adequately theorized and recognized as a rigorous way to substantially reduce reidentification risk without the complexity of other popular options (Federal Committee on Statistical Methodology 2005).

Consider, for example, an internal study produced by the U.S. Census Bureau in which a group of researchers attempted to attack the data from an individual-level public use dataset on over two million data subjects. The subjects were selected from three counties that were specifically chosen because of their vulnerability (residents in these counties are less transient, and therefore less likely to have noisy or stale data.) Next, the researchers purchased identified data on 700,000 people in the selected counties from a data aggregator and used all available overlapping key variables such as age, ethnicity, gender, and income. Out of the more than 2 million records in the research data files, the researchers' matching algorithm accurately linked 87 of the 700,000 identified records to records in the Census Bureau dataset. The study's preferred method of measuring reidentification risk was to report the proportion of records accurately matched: 87/700,000 – or 0.012%. But much more significant from the standpoint of this paper is that the algorithm had made *apparent matches* on 389 individuals. Of those 389 apparent matches, *only 87 were actually correct*—an accuracy rate of 22% (Ramachandran et al. 2012). The vast majority of apparent matches were wrong. These false matches would taint the value of the accurate matches for an intruder who lacks access to the Census Bureau's dataset. A more recent internal study found that outside commercial datasets allowed for a greater number of matches to Census Bureau records but that these matches were again predominantly erroneous—only 39% of the putative matches were in fact correct (Abowd 2020). Since the intruder would not know which of their matches are correct and which are not, all of the apparent matches are highly suspect and of low value.

The efficacy of sampling for preserving privacy has come under acute attack from Rocher et al. (2019), which showed that an intruder could accurately estimate an individual's uniqueness likelihood from sample data alone—that is, that an intruder could estimate the likelihood that a record unique in the sample was also unique in the broader population. The approach develops a probabilistic generative model by estimating marginal distributions for the variables separately from the covariance, then combining the estimates using a Gaussian copula. The model is applied to several large datasets.

Rocher et al. (2019) claims that its "results reject [that] ... sampling or releasing partial datasets provide plausible deniability." The authors do not give 'plausible deniability' a precise definition, but a straightforward interpretation of their study suggests that plausible deniability for an individual requires an individual's uniqueness likelihood to be below a threshold. Plausible deniability for a sample would require that some proportion of the individuals in the sample have plausible deniability, on this view.[6] The thresholds for individual uniqueness likelihood and sample proportion would, presumably, be left to the judgment of a data steward in a particular context.

Our work supplements individual likelihood estimation by representing it in the context of uncertainty about an attack: in our simple graphical model, individual uniqueness likelihood is represented by the likelihood of correct match given an apparent one, but its ultimate impact on privacy is also dependent on both the likelihood of an apparent match given an attack, and the probability of the attack itself (Fig. 1). The individual likelihood estimation of uniqueness is a good measure of plausible deniability only under the assumption that the intruder has auxiliary information about all of the variables used to render the data subject unique. Indeed, the results from the illustrative attack in Rocher et al. reinforces the power of sampling: to reach the target threshold of 90% chance of population uniqueness, Rocher et al. (2019) found that on average, an attacker needs data on 10 quasi-identifiers for each target (though this number will increase or decrease depending on the structure and distribution of the variables).

Our approach is compatible with the recent trend toward assuming that all information could be used as auxiliary information to attack individuals in a dataset, but does not require that assumption. Instead, we define Plausible Deniability to allow different predictions about the availability of auxiliary information to be explicitly represented in simulation. A data steward can test a range of assumptions about the auxiliary information that might be available to an attacker and express the plausible deniability across the range.[7] Moreover, the choices about the assumptions can be transparent and easy to interpret.

4 How Plausible Deniability Overlaps with and Differs from Other Measures of Database Privacy

If we consider some of the most common measures for reidentification risk through the lens of false matches and plausible deniability, we see that they fail to account for an attacker's lack of confidence. At first blush, this might seem to suggest that they are all too risk-averse, but this is not accurate, as other measures of privacy inevitably require certain assumptions or relaxations in order to accommodate the release of useful research data. A more precise summary is that plausible deniability is differently conservative and lax when compared to other measures of risk.

4.1 k-Anonymity

k-Anonymity ensures that any combination of values for quasi-identifier variables that are represented in the released dataset must be shared with at least $k - 1$ other records

[6] This interpretation might be generalized further to include cases in which groups of one kind or another share low plausible deniability.

[7] The assumptions need not be uniform, either. A data steward can assign greater likelihood of accessible auxiliary information for a data subject who is likely to be targeted, such as a Governor, or to the members of a vulnerable group who face special harms from successful attack.

(Samarati 2001; Sweeney 2002). The measure of reidentification risk is "k" itself—ensuring that a matching attack would do no better than link an identity to k different data subjects. Thus, the larger "k" is, the more privacy is afforded. The limitations of "k" as a measure of privacy are now well understood. Some, like attribute disclosure and overlapping releases, are amenable to technical refinement, but the greatest concern is that k-Anonymity requires a data steward to make assumptions about which variables an attacker could have (quasi-identifiers) and which he is expected to not have (sensitive attributes). For these reasons, k-Anonymity is understood to be an overestimate of privacy. Less attention has been paid to the ways in which it is an underestimate. But when sampling is involved, it is. Even if an intruder has information on the sensitive attributes (the ones that were not considered quasi-identifiers) that allows him to break through the k-Anonymity protections and uniquely link an identity to a released database, the match could be wrong. There could be other people in the general population but not represented in the sample who share the tuple of values.

For example, even the famous reidentification of Gov. William Weld in pre-HIPAA hospital records was more error-prone than Latanya Sweeney's original study suggests. Sweeney claimed to have reidentified Gov. Weld by finding his zip code and date of birth in the publicly available voter registration rolls, and then identifying the hospital record that had his unique combination of gender-birthdate-zip code (Ohm 2010; Barth-Jones 2012a, b). But these three quasi-identifiers alone would not be sufficient to make a confident match. In fact, there was more than a 35% chance that at least one other person would have shared Gov. Weld's gender, birthdate, and zip code, and if such a doppelgänger existed, there would be a 45% chance that he was not registered to vote (Barth-Jones 2012a, b). Thus, Rocher et al. (2019) points out that Latanya Sweeney's reidentification of William Weld, using only sex, zip code, and date of birth, was actually quite likely to be wrong.[8]

The false match errors from sampling are not typically incorporated into the measure of privacy protection in k-anonymity. In fact, under the traditional understanding of k-anonymity, even though using the technique on a small random sample of the data would intuitively and obviously reduce reidentification risk, the sampling would cause the measure of "k" to drop, implying that privacy has been lost. A data steward who starts by first extracting a small random sample would have to introduce more clustering, suppression, and generalization in order to meet a target k-value than he would if he released the entire dataset.

A less common approach to k-Anonymity (sometimes called "k-mapping") ensures that the k-threshold is reached for the population of potential targets, and not for the research database itself. This version of k-Anonymity is more analytically sound because it avoids exaggerating the privacy threat and allows a research release to have fewer than k subjects sharing a set of values of quasi-identifiers as long as the data steward has confidence that the intruder would not be able to disambiguate those research subjects

[8] Rocher et al. (2019) estimated that there was a 23% chance the match was wrong. Ironically, this study by Rocher et al. was the same study that was described in the New York Times' article with the misleading headline "Your Data Were 'Anonymized'? These Scientists Can Still Identify You."

from the k − 1 other individuals in the general population who share their values. k-mapping thus has a lot in common with plausible deniability as we have defined it. It ensures that any intruder who has access to identified records that contain the quasi-identifiers would have at best a 1/k chance of correctly attributing the right identity to a line of data.

Plausible deniability is much more flexible in its assumptions, however. It allows data stewards to make an unlimited range of assumptions about the quantity and quality of information that an intruder might access rather than restricting the data steward to a strict set of quasi-identifiers. For example, a data steward could assume that an intruder could have identified records that contain all variables, but that have wrong or missing values for a proportion of the cells. The data steward could even make a range of assumptions about the distribution of missing values. In other words, plausible deniability allows a data steward to model an attack any way they think is remotely feasible and then study the impact of various deidentification techniques on false match rates.

4.2 Unicity

Unicity is another measure of privacy in popular use today, particularly in studies that report results from demonstration attacks. Researchers decide on a set number of "tuples" and then report what proportion of data subjects are unique in a dataset when the specified number of tuples are randomly drawn from their data. For example, the "Unique in a Crowd" study found that 50% of data subjects were unique based on just two randomly chosen time-location tuples. The famous credit card and Netflix prize reidentification studies used uniqueness or unicity to measure privacy risk, too.

Unicity is closely related to k-anonymity. In fact, it is k-anonymity with k set to two and all tuple variables treated as quasi-identifiers. While the k-value is relatively weak (k is usually set to at least five), this is made up for by the fact that the researchers treat variables that are not typically thought of as indirect identifiers as quasi-identifiers—therefore making conservative (as in heroic) assumptions about the type of information an intruder might have about an identified target. For this reason, unicity is more appealing to disclosure control experts who want to avoid making assumptions about the types of auxiliary information that might be available for an attack in the future.

However, just as with k-Anonymity, unicity measurements do not take sampling into account. And, as with k-anonymity, even though sampling would make accurate matching attacks more difficult, sampling will make the dataset look worse for privacy.

For example, de Montjoye et al. (2015) found that 90% of the people in an anonymized credit card dataset were unique based on the date and place of four purchases. A reader is likely to think that this means a person really is unique based on four purchases, and therefore vulnerable to attack if an intruder happens to have this information in identified form. But the credit card data was incomplete. It only described a sample of people (cardholders for the particular credit card company), and it only described a subset of those cardholders' purchases (the transactions that involved that particular credit card). Treating sample uniqueness as a measure of reidentification risk is perverse because the reidentification risk will appear to rise when a data producer uses smaller and smaller samples. At the extreme, a dataset containing just one observation of one transaction by one unidentified person will produce a "reidentification" risk of 100% using unicity as its measure even though it will in fact be resistant to attack.

4.3 Differential Privacy

Like the "unicity" measure for privacy, Differential Privacy makes heroic assumptions about what type of data might be available to a data intruder, now or in the future. In fact, Differential Privacy goes further than the studies that measure unicity. While studies relying on unicity often, if implicitly, make some assumptions about what amount and type of data an attacker is most likely to use, the differential privacy model treats all attacks as equally plausible, including an attack by an intruder who knows everything about every person except for one last piece of information about one person (Dwork and Smith 2009; Bhaskar et al. 2011). In some respects, assuming that attackers are virtually omniscient makes life easier for the data stewards, as it relieves them from having to make educated (and sometimes difficult) guesses about which types of threats are plausible and which are not.

The problem with this approach, though, is that its privacy guarantees are deceptive. Because it makes such strong assumptions about attackers, it must allow for some information leakage (through the selection of ε) in order to provide any useful information from the research dataset. In fact, even the ε guarantee has been too constraining in practice, leading to further relaxations of Differential Privacy (for example, the introduction of δ). Yet these parameters, ε and δ, do not and cannot ensure that the relaxations in privacy promises are well-aligned with real world risk. They are just as likely to relax the guarantees that relate to a matching attack using common indirect identifiers like age and zip code as they are to relax the guarantees relating to attacks that use uncommon information like body temperature or GPS coordinates at a specific moment in time. Thus, in some contexts, the use of Differential Privacy is an abdication of responsibility, allowing data custodians to claim quantifiable "guarantees" and avoid doing the hard but necessary work of differentiating between realistic and unrealistic risks (Domingo-Ferrer and Muralidhar 2016; Christensen and Miguel 2016).[9]

Plausible Deniability, as we define it, makes assumptions about intruder attacks that would violate the premise of Differential Privacy. The theoretical intruder who has all information about everybody except for one last detail about one person will be able to attack a dataset with high plausible deniability because that intruder will be able to rule out every other person who is similar to the target and might have been left out

[9] Nayek et al. (2016) assesses the problem with formal privacy measures, like "differential privacy," concluding "... for developing practical disclosure control goals, it is essential for the agency to consider intruders with limited prior information about their target units.". Elliot et al. note "many authors have commented that this environment is inherently difficult—if not impossible—to understand and therefore directly assessing risk is itself impossible. This in turn has led to bad decision-making about data sharing (a strange mixture of over-caution and imprudence which is driven more often than not by the personality of the decision-maker rather than by rational processes."

of the sample.[10] By contrast, Plausible Deniability assumes that an attacker has limited auxiliary information and could not confidently rule out the possibility of a false match. However, as we show, data stewards can still make very generous assumptions about the type and quality of auxiliary data that an attacker will have and still preserve data utility through sampling.

That said, the power of sampling is not lost on the Differential Privacy research community. Li et al. (2012) have studied its value for meeting and preserving Differential Privacy guarantees. Like us, these authors sought to incorporate the routine practice of sampling into quantified notions of privacy: "[w]e observe that in many data publishing scenarios, random sampling is an inherent step. For example, the census bureau publishes a 1-percent microdata sample. In many research settings (such as when Netflix wants to publishing movie ratings), it is sufficient to publish a random sample of the dataset. Many times, even when the dataset is not the result of explicit sampling, one can view it as result of implicit sampling, because the process of selecting respondents involves randomness. The natural question is how one can benefit from such explicit or implicit sampling."

Li et al. (2012) goes on to show that sampling can be combined with k-anonymity to satisfy (ε, δ)- Differential Privacy in a query system. The authors explore (β, ε, δ)-Differential Privacy where β represents the sampling proportion. To be clear, the authors do not use sampling the way we do—as a means to make even a successful attack futile because it cannot be distinguished from a false match. Instead, they use sampling within the structure and goals of Differential Privacy, thereby embedding the benefits and drawbacks of Differential Privacy that we describe above.

We take a different approach. Data stewards can use Plausible Deniability metrics to report the privacy value of a random sample of microdata under a range of generous assumptions. And unlike Differential Privacy, the choices involved in the use of Plausible Deniability will be easy for smaller, less sophisticated agencies to use and for the general public to understand.

5 How to Measure Plausible Deniability

Plausible deniability measures the chance that a data intruder, who is presumed to have access to a certain range of data for a certain proportion of the population, will falsely attribute sensitive data to the wrong person. The higher plausible deniability is, the more privacy is preserved. Although there are a number of statistical disclosure control techniques that can achieve plausible deniability (including noise-adding, data-swapping, and generalization), we focus primarily on sampling-based approaches in this study.[11]

Plausible deniability can be measured using a stochastic process. Consider D0, a dataset controlled by a data steward who wishes to create and release a research database, R0. In order to reduce the risk of reidentification, the data steward plans to randomly

[10] More technically, sampling alone could never meet Differential Privacy standards because any microdata release that does not involve perturbation or the creation of synthetic data will violate the Differential Privacy guarantee.

[11] We allow simulation of noise levels in both the released data and the auxiliary data, though this is not explicit in the simulation steps below. See the code repository, above at note 1.

select data on only 1/t of the research subjects. To measure plausible deniability, the data steward can assign random IDs to each research subject (which will be used to identify false matches) and run a simulation:

1. Randomly select $1/t$ records, and potentially remove some variables from the original data, to create $R_i \subset D_0$. This will represent a potential RESEARCH database prepared for release.
2. Randomly select $1/s$ records to create $A_i \subset D_0$, and match the variables in the released data. This will represent a potential AUXILIARY set of identified records that an intruder will use for attack.
3. Randomly select p non-empty values to be suppressed in each individual record in A_i where $p \geq 1$. This represents the "payload"—the potentially sensitive information about an individual that the intruder does not yet possess. (Alternatively, the data steward can select one or more variables that the steward believes are not accessible to an intruder and suppress all values in that set of variables.[12])
4. Match unique records in A_i to unique records in R_i. Each matched pair represents an apparent reidentification.[13] Let M_i be the count of matched pairs.
5. (Optional). Use the methodology from Rocher et al. (2019) to estimate the confidence of each matched pair, and drop all matched pairs where the confidence level is lower than $c*$.
6. Use the unique IDs to identify which apparent matches are false. Let F_i be the count of false matched pairs.
7. Use the unique IDs to accurately match the records in A_i to the records in R_i. Let P_i be the count of potential matches.
8. Log M_i, F_i, and the unique IDs of data subjects included in matched pairs and in false matched pairs.
9. Repeat N times.
10. Report the average false discovery rate ($\frac{\sum Fi}{\sum Mi}$). This is the Plausible Deniability measure for R_0 under the assumptions supporting the settings for s and p in the auxiliary data.

Data stewards could also look at the range of risk by identifying the minimum and maximum false discovery rates for each individual data subject.

We anticipate that data stewards would assess the plausible deniability for a research sample under a range of different assumptions for parameters s and p so that, rather than reporting one single number, the steward would be able to report several Plausible Deniability metrics under a range of reasonable or conservative assumptions about present and future attacks.

Moreover, within the assumptions that are built into the simulation, the estimates of Plausible Deniability are conservative estimates because the simulation uses the same

[12] This decision would be similar to the judgments that must be made when differentiating between quasi-identifiers and non-identifiers when implementing k-anonymity.

[13] A slightly more sophisticated version of our methodology would include all matches, and sampling uniformly to decide which records from the released data to match with which from the auxiliary data.

source of data for both the simulated research databases and the simulated auxiliary data. In the real world, an intruder would be using data sourced from another origin, and therefore containing different time-dependent data or different errors from the data used in the research sample. Data stewards could incorporate this error into the simulations if they would like.

Conversely, a data steward could start with a target level of Plausible Deniability under a set of attack assumptions (s and p) and use simulations to find the right sampling proportion.

For some databases, a data steward could use the distributions and covariance of the variables to derive Plausible Deniability metrics analytically, but for large datasets, it will often be simpler to make use of the law of large numbers through repeated simulations.

6 Proof of Concept Illustration: Education Data

As an application of the foregoing ideas, attacks were simulated on a dataset released in 2008 by the University of California (UC) system, which includes data on undergraduate students from UC campuses. A graphical representation of a simulated attack on the UC data is presented in Fig. 2.

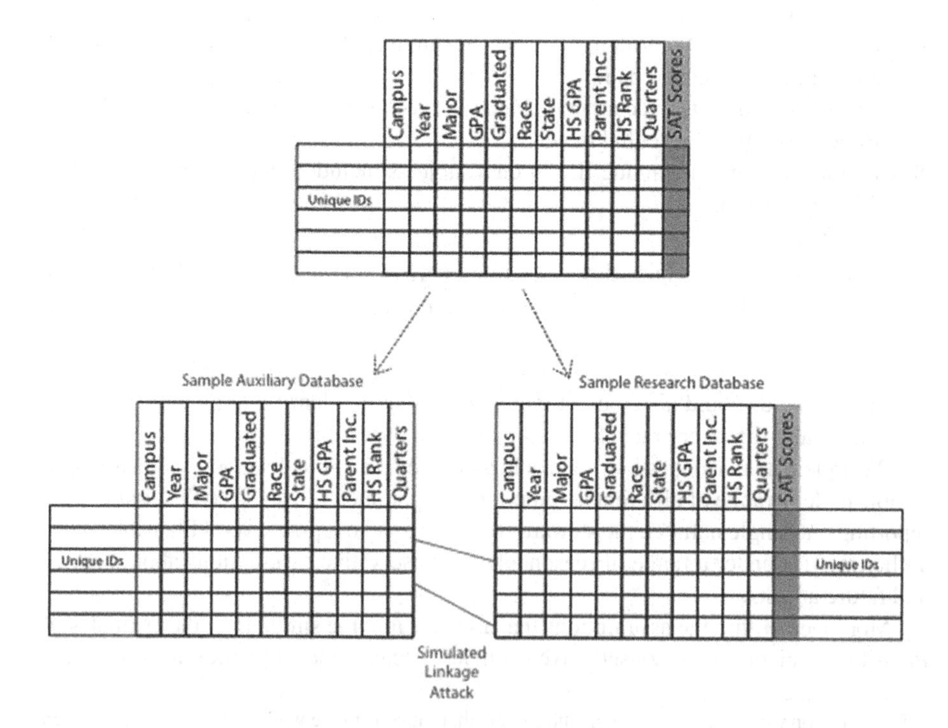

Fig. 2. Attacks against a dataset are performed by sampling once to construct a research database for release, and once to build the auxiliary data available to the attacker.

The UC data contains data for a range of years from 1992 to 2006, including race, whether the student applied as resident in California or from another state, SAT I and SAT II scores, raw and UC adjusted GPA scores, number of AP courses taken, parent income, the California API rank of the student's high school, whether the student enrolled in the UC system, whether a bachelor's degree was achieved, and the time to achieve the bachelor's degree.[14]

The simulation itself has a simple structure: first, a "munger" handles the particularities of preprocessing the data, providing a unified interface to different underlying datasets. So far, the munger is to be implemented for three datasets: the UC educational data discussed above, the 2010 1% and 5% PUMS data released by the US Census, and the MIMIC III hospital dataset, which includes health diagnosis and treatment information.[15]

Our simulations ran all combinations of settings for the parameters for 10 and 100 iterations.[16] Table 1 reports, for one set of parameters, the average number of possible matches, apparent matches, and true matches.

Table 1. Average Results for a Simulation Attack on UC Berkeley Enrollee Data Experiment 1. Attacker presumed to know Year, Race, Cumulative GPA, and Graduation Status on 20% of the Entire population. N = 54,618; auxiliary dataset contains 10,924 records; iterations = 100.

Research sample	20% Auxiliary Dataset (10,924 records)		
	Average # attempted matches	Average # correct attempted matches	False positive rate of error
50%	27.52	5.28	80.81%
25%	42.61	2.45	94.23%

Even though the attacker's auxiliary data had significant overlap with the research sample, he was able to make only a handful of matches, and many of those matches were actually wrong.

More concretely, consider the second line of Table 1. In an average attack simulation, the attacker had auxiliary information on over 10,000 UC enrollees. Of these, over two thousand were in the research sample and could have potentially been reidentified. However, on average, the attacker would have been able to attach an identity to only 43 records in the research sample, and almost all of those matches would have been false.

[14] The data dictionary can be found in the data directory of our repository. See above at note 1.

[15] See https:/www.census.gov/programs-surveys/acs/data/pums.html for the PUMS data, and https://mimic.physionet.org/ for the MIMIC III data.

[16] There were 1,620 parameter settings. For each iteration at a given setting, steps 1–9 mentioned above are performed. The full set of simulation runs is computationally intensive, so there are two implementations of the simulation code. One is designed to run serially, and is suitable for small, slow runs on a single laptop; the other is designed to run in parallel on a high-performance computing cluster (HPC). The cluster we used had some specific features, such as use of the PBS Scheduler, but minor modifications should allow the code to be used on a variety of HPC setups. See the experimental_actors branch of the repository referenced above at note 1.

The reason is that despite the large amount of information available to the attacker, most of his records would not be able to be linked to a unique line of data because other records in the research sample share the same values. When the attacker did make a match to a record that appeared to be unique in the research sample, it was actually not unique. Other records in the portion of the data that was not included in the research sample shared the same values. While the attacker occasionally got lucky and managed to link to the right (non-unique) record, more often the matches were false.

Note that an intruder is able to make more apparent matches when the UC releases a smaller sample of records. This may seem counterintuitive, but it makes sense. With fewer records in the research sample, there will be more records that look unique, and therefore more linkages between a unique record in the research sample and a unique record in the auxiliary sample. But this is an artifact of sampling; in fact, the unique-to-unique matches will be more likely to be wrong.

Next, consider a similar simulation exercise using a different set of variables. In this simulation, the attacker has access to Year, Race, and undergraduate GPA (UGPA) information. Given the low prevalence of undergraduate GPA available in identified form, the experiment uses both the 20% auxiliary sample and a still-conservative-but-more-realistic 1% auxiliary sample. Again, very few matches are able to be made to sampled UC data, and the vast majority of matches are wrong (Table 2).

Table 2. Results for a Simulation Attack on UC Berkeley Enrollee Data. The attacker is presumed to know Year, Race, and UGPA. N = 54,618; Iterations = 100.

Research sample	20% Auxiliary Dataset (10,924 records)			1% Auxiliary Dataset (546 records)		
	Average # attempted matches	Average # correct attempted matches	False positive rate of error	Average # attempted matches	Average # correct attempted matches	False positive rate of error
50%	21.1	8.8	58%	2.47	0.27	89%
25%	42.61	2.45	80%	5.91	0.15	97%

The simulation results show the power of sampling if the UC has already clustered the data (as it had with the 2008 data), even when the attacker is presumed to have access to extensive and unrealistic amounts of identifying information.

For the final two simulations, a still larger set of variables are assumed to be available to the attacker: in addition to everything the intruder in the first set of experiments had, the new attacker also has state of residence, major, high school GPA, parental income, high school rank, and quarters to degree. By using these additional variables, the attack will be able to make more matches because much more of the research data will have unique combinations of values for the full set of ten variables.

With more variables at the intruder's disposal, the attacks are able to make a lot more matches. Nevertheless, even under these conditions, the intruder would have very little confidence in his matches to a research dataset that contained only a 50% or 25% sample of the data subjects. Even if an intruder matches an identity to a unique record using all ten variables, each apparent match has a 40% probability of being wrong.

The final set of simulations repeated the simulation just described, but added the additional precaution of suppression (Table 3). In addition to releasing only a sample of the data, in this exercise, the UC suppresses a randomly selected 10% of the values of the key variables. This precaution reduces both the number of apparent matches and the percentage of apparent matches that are correct.

Table 3. Results for a Simulation Attack on UC Berkeley Enrollee Data Experiment 4. The attacker is presumed to know Year, Race, State, Major, Graduation Status, Undergraduate GPA, High School GPA, High School Rank, Parent Income, and Time to Degree on 20% of the entire population, and 10% of values are suppressed in the Research Sample. N = 54,618; Iterations = 1,000.

Research sample	20% Auxiliary Dataset (10,924 records)		
	Average # attempted matches	Average # correct attempted matches	False positive rate of error
50%	1,152	779	32%
25%	799	389	51%

7 Discussion

Plausible deniability measures the risk that an attack against a sampled dataset will produce too many matches that genuinely identify an individual. A data steward seeking to estimate the plausible deniability of a sample must therefore consider the likelihood of a genuine match given an apparent one—or more simply, the likelihood that an apparent match is correct.

The central message of Rocher et al. (2019) is that in estimating the likelihood that an apparent match is correct, the data steward and attacker are on almost even footing, since they have provided a method that allows accurate estimation of the individual likelihood from a sample, and the attacker is in possession of the sample. Accordingly, one casualty of Rocher et al. (2019) is an interpretation of the protections afforded by plausible deniability that is encouraged by the term itself: plausible deniability does not always protect by allowing an individual faced with a disclosure to say "The match to me is merely apparent. I am outside the sample, and you have not learned anything about the sensitive variables for me." For some data, protections of this sort may be important, and in those cases more weight should be placed upon individual uniqueness likelihoods, but for other data—equally possessed of plausible deniability—protection may be provided by the low probability of a person being matched in the first place.

Therefore the council of despair offered by Rocher et al (2019) over the privacy value of sampling is overbroad: attackers themselves must be modeled. The likelihood of a true match depends on the likelihood of an apparent match given an attack with particular resources, as well as the probability of those particular resources being assembled in the first place. Both of the latter aspects of plausible deniability are included in the approach we present, providing useful flexibility to data stewards with differing data that present differing threats. Our approach complements the methods of Rocher et al (2019), while tempering the gloomy conclusions drawn from it.

Finally, specifying the parameters of our approach provides a valuable source of interpretability: the size and quality of an attackers' dataset, the payload of variables that the attacker seeks, as well as the size and variables released, are all easily understood, and subject to justification by appeal to the particularities of an individual dataset release. In addition to flexibility and ease of use, our approach offers transparency.

References

Abowd, J.: Formal Privacy Methods for the 2020 Census. 2020 Census Program Memorandum Series: 2020.07 (2020)

Barth-Jones, D.: The Debate Over 'Re-identification' of Health Information: What Do We Risk? Health Affairs (2012a)

Barth-Jones, D.: The 'Re-identification' of Governor William Weld's Medical Information: A Critical Re-examination of Health Data Identification Risks and Privacy Protections, Then and Now. Draft (2012b). https://fpf.org/wp-content/uploads/The-Re-identification-of-Governor-Welds-Medical-Information-Daniel-Barth-Jones.pdf

Berthold, O., Pfitzmann, A., Standtke, R.: The disadvantages of free MIX routes and how to overcome them. In: Federrath, H. (ed.) Designing Privacy Enhancing Technologies. LNCS, vol. 2009, pp. 30–45. Springer, Heidelberg (2001). https://doi.org/10.1007/3-540-44702-4_3

Bhaskar, R., Bhowmick, A., Goyal, V., Laxman, S., Thakurta, A.: Noiseless database privacy. In: Lee, D.H., Wang, X. (eds.) ASIACRYPT 2011. LNCS, vol. 7073, pp. 215–232. Springer, Heidelberg (2011). https://doi.org/10.1007/978-3-642-25385-0_12

Christensen, G., Miguel, E.: Transparency, Reproducibility, and the Credibility of Economics Research. NBER Working Paper No. 22989 (2016)

de Montjoye, Y.-A., Radaelli, L., Singh, V.K.: Unique in the shopping mall: on the reidentifiability of credit card metadata. Science **347**, 536–539 (2015)

Domingo-Ferrer, F., Muralidhar, K.: New directions in anonymization: permutation paradigm, verifiability by subjects and intruders, transparency to users. Inf. Sci. **337**, 11–24 (2016)

Dwork, C., Smith, A.: Differential privacy for statistics: what we know and what we want to learn. J. Priv. Confidentiality **1**, 135–139 (2009)

Elliot, M., Domingo-Ferrer, J.: The future of statistical disclosure control. arXiv preprint arXiv: 1812.09204 (2018)

Federal Committee on Statistical Methodology: Statistical Policy Working Paper 22: Report on Statistical Disclosure Limitation Methodology (2nd version). Office of Management and Budget, Executive Office of the President (2005)

Li, N., Qardaji, W., Su, D.: On sampling, anonymization, and differential privacy or, k-anonymization meets differential privacy. In: Proceedings of the 7th ACM Symposium on Information, Computer and Communications Security (2012)

Nayak, T., Zhang, C., You, J.: Measuring Identification Risk in microdata Release and Its Control by Post-Randomization. Center for Disclosure Avoidance Research, U.S. Census Bureau Research Report Series #2016-02 (2016)

Ohm, P.: Broken Promises of Privacy, 57 UCLA L. Rev. 1701, 1719 (2010)

Ramachandran, A., Singh, L., Porter, E., Nagle, F.: Exploring re-identification risks in public domains. In: Tenth Annual International Conference on Privacy, Security and Trust, pp. 35–42 (2012)

Rocher, L., Hendrickx, J.M., De Montjoye, Y.-A.: Estimating the success of re-identifications in incomplete datasets using generative models. Nat. Commun. **10**, 1–9 (2019)

Samarati, P.: Protecting respondents' identities in microdata release. IEEE Trans. Knowl. Data Eng. **13**, 1010–1027 (2001)

Serjantov, A., Dingledine, R., Syverson, P.: From a trickle to a flood: active attacks on several mix types. In: Petitcolas, F.A.P. (ed.) IH 2002. LNCS, vol. 2578, pp. 36–52. Springer, Heidelberg (2003). https://doi.org/10.1007/3-540-36415-3_3

Sweeney, L.: Achieving k-anonymity privacy protection using generalization and suppression. Int. J. Uncertain. Fuzziness Knowl.-Based Syst. **10**(5), 571–588 (2002)

Microdata Protection

Analysis of Differentially-Private Microdata Using SIMEX

Anne-Sophie Charest(✉) and Leila Nombo

Université Laval, Quebec City, Canada
anne-sophie.charest@mat.ulaval.ca

Abstract. We are concerned here with the publication of microdata which preserve the confidentiality of the respondents, as measured by differential privacy. We borrow the SIMEX methodology from the measurement error literature to analyze microdata perturbed with the Laplace mechanism, hoping to recover unbiased estimates of the parameters of interest. Our simulations show that the strategy works in theory, but only for values of ϵ too large of be of used in practice. It might however be of use in the case of normal noise addition, when one does not require differential privacy.

Keywords: Differential privacy · Microdata · SIMEX

1 Introduction

We are concerned here with the publication of microdata which preserve the confidentiality of the respondents. This has of course been a topic of interest for decades, and various methods have already been proposed to achieve this goal, from recoding and suppression, through microaggregation, all the way to the generation of completely synthetic datasets. (See for example [12,24] or [6] for details). However, we are aiming here specifically at microdata which satisfy differential privacy.

Differential privacy [8] is a rigorous measure of the privacy offered by the process with which data or statistics computed from data are produced. It basically promises respondents that accepting to appear in the dataset will not allow an intruder to learn any more personal information about them than they could without their data being in the dataset. Several methods have been proposed to produce differentially-private discrete datasets, known as contingency tables in the statistics literature, and often referred to as histograms by computer scientists (see for e.g. [1,4,11]). More recently, a few methods have been proposed to create differentially-private synthetic microdata (e.g. [14,22,23]). Still, these methods remain rarely studied or used in practice for statistical disclosure control, probably for two main reasons.

First, differential privacy is often seen as excessive, or inappropriate, in comparison to privacy measures and constraints which have historically been used

J. Domingo-Ferrer and K. Muralidhar (Eds.): PSD 2020, LNCS 12276, pp. 109–120, 2020.
https://doi.org/10.1007/978-3-030-57521-2_8

for statistical disclosure control. Since this added guarantee comes with a need for more perturbation of the original data, it is not a very attractive criterion for practical use.

Second, the algorithms proposed to produce differentially-private data are often quite complicated, involving for example Fourier transforms or deep learning models. Consequently, they are hard to implement, and it is difficult to understand their impact on the quality of the data, and in particular to know how one should take into account the added variability in the data analysis. This contrasts with the methodology developed over the years to analyse synthetic data generated for privacy protection with methodology borrowed from the missing data literature [7,17–19].

In this paper, we propose to borrow from the literature on methodology for data with measurement errors to try to alleviate these two difficulties with differentially private microdata. More precisely, the idea is to use the very simple method of Laplace noise addition to generate the microdata, and combine it with the SIMEX methodology in order to take into account the noise mechanism and obtain better estimates.

To be fair, it has never been suggested to generate differentially private synthetic datasets by simply adding Laplace noise to continuous variables, and this is surely not the optimal method to do so. However, it is easily implemented and fits perfectly the framework of measurement error models, in particular because detailed information about the algorithm, such as the noise variance, can be published without impacting the privacy guarantee. Hence, while this strategy would most probably not be used in practice, it provides a good testing ground for the potential of SIMEX in analyzing differentially private datasets.

As for SIMEX, it is one of several techniques coming from the very large literature on statistics for measurement errors (see for e.g. [3]). The idea is simple: one gradually adds more noise to the observed data, hoping to observe a trend in how the estimator behaves with noise addition, which can be used to infer the estimate with the original dataset. SIMEX is applicable to a large class of models, and relatively easy to implement, in comparison to other methods sometimes used to take into account noise from statistical disclosure limitation methods.

The rest of the paper is divided as follows. Section 2 gives some background information on differential privacy and the Laplace mechanism, Sect. 3 presents the SIMEX methodology, Sect. 4 gives various simulation results and Sect. 5 offers a discussion and ideas for future research.

2 Differential Privacy

We review here the concept of differential privacy and the Laplace mechanism. We refer readers to [9] for more detailed presentations.

2.1 Definition

The formal definition of differential privacy may be given as follows:

Definition 1 (ϵ-Differential Privacy). *Let $\mathcal{D}$ be a set of all possible datasets. A randomized function κ satisfies ϵ-differential privacy if and only if for all neighboring datasets D_1 and D_2 both elements of $\mathcal{D}$, and for all $S \subseteq range(\kappa)$,*

$$e^{-\epsilon} \leq \frac{Pr[\kappa(D_1) \in S]}{Pr[\kappa(D_2) \in S]} \leq e^{\epsilon}. \tag{1}$$

For matrix datasets where rows correspond to respondents and columns to variables, we will say that two datasets are neighbors if their entries are identical in all but one of the rows. Note that to specify $\mathcal{D}$ the range of possible values for each variable must be known ahead of time.

The value of ϵ controls the level of privacy guaranteed by the randomized function κ and must be specified by the user. The smaller the value of ϵ, the greater the privacy protection. Note that Eq. (1) must be valid for any possible pair (D_1, D_2). Hence, for the extreme choice of $\epsilon = 0$, the output of the randomized function κ would have to be completely independent from the observed dataset. Moreover, for ϵ to be finite, it is necessary that the support of κ be identical for all possible input datasets.

Note also that the definition does not constrain the form of κ. Hence, the definition may be applied to functions which simply output one numerical statistic from the dataset, as well as to complicated processes which output synthetic datasets.

Interpretation. We first emphasize that differential privacy is not a measure of privacy of a specific output or dataset. Rather, it is a measure on the method used to generate such output or data. Hence, the phrase *differentially-private microdata*, which we use in this paper, should be understood as *microdata obtained using a differentially-private process.*

One may think of the privacy offered to respondents by differential privacy in terms of deniability. Recall that Eq. (1) requires that any output be *almost* as probable for any neighbouring datasets, where the quantitative meaning of this almost depends on the value of ϵ. Thus, seeing a specific output from the randomized function should not allow an intruder to identify which of two neighbouring datasets was indeed observed. So, even in the extreme scenario where the intruder knew the data of all but one of the respondents she would not be able to infer the data for that last respondent from the published output. No matter what the intruder claims their data to be, the respondent can deny that it is the case, and the intruder will not be able to prove otherwise. In that sense, we say that differential privacy provides deniability to the respondents.

Microdata. In order to produce differentially-private microdata, one must create a randomized function κ which generates synthetic data and satisfies Eq. (1). In general, this will be difficult because we must be able to study and constrain the distribution of this randomized function over the space of synthetic datasets. One may also decide to estimate parameters of a pre-determined statistical model under the constraint of differential privacy, and sample new data from it, but this still requires very technical calculations. A simpler way to produce differentially-private microdata, and which is directly amenable to treatment with measurement error methodology, is with the addition of Laplace noise.

2.2 Laplace Mechanism

Consider a dataset $D \in \mathcal{D}$, and a function $f : \mathcal{D} \to \mathbb{R}^k$. We want to publish $f(D)$. We first define the l_1-sensitivity of function f.

Definition 2 (l_1-sensitivity). *The l_1-sensitivity of a function $f : \mathcal{D} \to \mathbb{R}^k$ is*

$$\Delta f = \max_{D_1, D_2 \text{ neighbours}} \|f(D_1) - f(D_2)\|_1 \tag{2}$$

where $\|x\|_1 = \sum_{i=1}^{n} |x_i|$, the l_1-norm.

Then, one can prove that the following randomized function will satisfy ϵ-differential-privacy:

Definition 3 (Laplace mechanism). *Given any function $f : \mathcal{D} \to \mathbb{R}^k$, the Laplace mechanism is defined as*

$$\kappa(D, f(\cdot), \epsilon) = f(D) + (Y_1, \dots, Y_k) \tag{3}$$

where $Y_i \overset{iid}{\sim} \text{Laplace}(0, \Delta f / \epsilon)$.

Note that the Laplace(μ, b) distribution has probability density function given as $f(x|\mu, b) = \frac{1}{2b} \exp\left(-\frac{|x-\mu|}{b}\right)$, mean 0 and variance $2b^2$.

Microdata. To produce differentially-private microdata with the Laplace mechanism, we take f to be the identity function, since without any privacy constraints we would simply publish the observed dataset. Mathematically, f simply outputs a vector of length np where each entry is the value of a variable for an individual in the dataset. Since neighbouring datasets differ only by one individual, only p of these values will differ for two neighbouring datasets. Hence, the sensitivity of f is $\sum_{j=1}^{p} \Delta_j$ where Δ_j is the range of possible values for variable j.

3 SIMEX

The SIMEX (SIMulation EXtrapolation) method was originally proposed by [5] in order to correct the attenuation bias identified in regression models with measurement error. As the name indicates, the SIMEX algorithm consists of a simulation step followed by an extrapolation step. In the simulation step, we create new observations by gradually adding more noise to the data and observe the impact on the estimator. This experiment is repeated many times to minimize the variability of the results. The extrapolation step then fits a model to the estimate as a function of the amount of added noise, and extrapolates to the case of no noise.

Note that the method is quite general, and can be applied to various statistical models. It does however require that we know, or can estimate, the variance of the measurement error. Since differential privacy does not rely on anything being kept secret about the process, it is not a constraint for our work.

3.1 Original Simex

We present the method for a single perturbed variable; its multivariate extension is straightforward. See [2] for a detailed exposition.

Let (W, V) denote a dataset of n respondents with V a set of variables observed without measurement error, and W a noisy version of a variable X. The original SIMEX assumes that for $i = 1, \ldots, n$

$$W_i = X_i + \delta_i \tag{4}$$

where $\delta \sim N(0, \sigma_\delta^2)$ independent of X and V, with σ_δ^2 a known parameter. An extension to deal with Laplace noise addition was proposed in [13] and will be presented in the next section.

Let $\theta \in \Theta$ be the parameter of interest and T the function which associates a dataset to an estimate $\hat{\theta} \in \Theta$, so that $\hat{\theta}_{True} = T(\{X_i, V_i\}_1^n)$ and $\hat{\theta}_{Naive} = T(\{W_i, V_i\}_1^n)$.

Simulation Step. To do this, we first generate

$$\delta_{bi}^* \sim N(0, \sigma_\delta^2) \tag{5}$$

for observations $i = 1, \ldots, n$ and replications $b = 1, \ldots, B$. Here σ_δ^2 is assumed to be known in advance or else very well estimated independently [5,15,21]. The value of B is usually taken to be a few hundreds; increasing B reduces the variability of the estimates.

Then, for fixed values $0 \leq \lambda_1 < \lambda_2 < ... < \lambda_K$ we calculate new observations

$$W_{bi}(\lambda_k) = W_i + \sqrt{\lambda_k}\delta_{bi}^* \tag{6}$$

The values $\sqrt{\lambda_k}\delta_{bi}^*$ are called pseudo-errors. Note that [5] suggests $\lambda_1 = 0$ and $\lambda_K = 2$, and uses $K = 8$ in examples. For each λ_k, we obtain a set of B estimators $\{\hat{\theta}_b(\lambda_k)\}_{b=1}^B$ with

$$\hat{\theta}_b(\lambda_k) = T(\{W_{bi}(\lambda_k), V_i\}_1^n) \tag{7}$$

from which we compute

$$\hat{\theta}(\lambda_k) = \frac{1}{B}\sum_{b=1}^{B} \hat{\theta}_b(\lambda_k) \tag{8}$$

which is an estimate of the expected value of the estimator $\hat{\theta}$ for the noise level defined by λ_k.

Extrapolation Step. In the extrapolation step, we adjust a model for $\hat{\theta}(\lambda_k)$ as a function λ_k. [5] proposes a non-linear parametric model but [3] recommends a quadratic form. The graph of $\hat{\theta}(\lambda_k)$ versus λ_k can be used to choose an appropriate form. The SIMEX estimate is then the extrapolation of this model to $\lambda_k = -1$.

To see why the extrapolation is made at $\lambda_k = -1$, consider Eq. (5) which describes how the new observations $W_{bi}(\lambda_k), i = 1, ..., n$ are computed. The variance of $W_{bi}(\lambda_k)$ is given by:

$$\text{Var}(W_{bi}(\lambda_k)) = \sigma_x^2 + (1 + \lambda_k)\sigma_\delta^2 \tag{9}$$

where $\sigma_x^2 = \text{Var}(X)$. Since the goal is to return to the case where the measurement error is nonexistent, we want $(1+\lambda_k)\sigma_\delta^2 = 0$. Since $\sigma_\delta^2 \neq 0$, as the variance of a random variable cannot be zero, we must have $1 + \lambda_k = 0$ and therefore $\lambda_k = -1$.

3.2 Extension to Laplace Noise Addition

While the original SIMEX method is concerned with the addition of normal noise, we generate differentially-private microdata with the addition of Laplace noise. The SIMEX methodology extends to this noise distribution, but not as simply as one might expect.

Indeed, [13] shows not only that the original SIMEX produces an estimator which is no longer consistent for the parameter of interest if the true noise follows a Laplace distribution, but also that generating the pseudo-errors from a Laplace distribution does not solve the problem. In fact, the paper shows that pseudo-errors must be generated as the difference between two gamma random variables.

More precisely, if we suppose that

$$W_i = X_i + \delta_i \tag{10}$$

where δ_i are independently generated from a Laplace distribution with variance σ_δ^2, then the Laplace SIMEX requires to generate

$$W_{bi}(\lambda_k) = W_i + V_1 - V_2 \tag{11}$$

where V_1 and V_2 are generated independently from a Gamma $(p_k, \sigma_\delta^2/\sqrt{2})$ distribution. The p_k here is a scaling parameter which controls the level of noise, as did λ_k in the original method. The values $p_1 < p_2 < ... < p_K$ are fixed in advance, and also taken to be $p_1 = 0$ and $p_K = 2$ in [13], where by convention $p = 0$ means that all observations are equal to zero.

The extrapolation step models $\theta(\hat{\lambda}_k)$ as a function of p_k and extrapolates to $p = -1$. While a linear or quadratic model may be used, [13] recommends an extrapolant of the form $(a+b)/(c+p)$ for small values of σ_δ^2, or $(a+b)/(c+dp+p^2)$ otherwise. Non-linear least-squares estimates can be obtained using various methods; we use the `nls` function of the core R software [16], with the default Gauss-Newton algorithm.

4 Simulation Results

The objective of this work is to study if and how SIMEX can be used to provide accurate estimates of model parameters from microdata generated with Laplace noise addition in order to satisfy the rigorous differential privacy guarantee. We report below on simulation results for various linear regression models. A discussion will follow in the next section.

4.1 Simple Linear Regression

We first consider the simple linear regression model. Given a predictor X, we generate $Y_i = \beta_0 + \beta_1 X_i + \varepsilon_i$ where $\varepsilon_i \overset{iid}{\sim} N(0, \sigma_\varepsilon^2)$. The dataset (X, Y) is then perturbed with Laplace noise addition, with variance σ_ϕ^2 chosen to produce an ϵ-differentially private dataset (W, Z).

In this simple setup, one can show (see [3]) that the OLS estimator obtained naively from the regression of Z on W is consistent not for β_1 but for $\lambda\beta_1$ where $\lambda = \frac{\sigma_x^2}{\sigma_x^2+\sigma_\phi^2}$ which is necessarily smaller than 1. This is why we talk of attenuation bias. Note that the bias only depends on the variance of the predictor and the variance of the noise on that predictor, not on the variance of the regression residuals, or the size of the dataset. In our setup, the bias will depend on the variance of the added noise through the value of ϵ and the sensitivity of f, which itself depends on the range of the predictors and response.

We now want to see if SIMEX can correct this bias. Note that in this restricted case, SIMEX is not necessary. One could directly estimate λ and correct the slope estimate accordingly. This strategy was also implemented. However, it is problematic for small values of ϵ, where σ_ϕ^2 is so large that we often obtain negative estimates of σ_x^2. Since SIMEX is more versatile anyway, we only present results for this method.

In the simulation presented, we set $n = 500$, $X_i \overset{iid}{\sim} U(0,1), i = 1, \ldots, n$, and parameters $\beta_0 = 2, \beta_1 = 2$ and $\sigma_\varepsilon^2 = 1$. We generate differentially-private microdata for various values of ϵ. Since this requires that we know the range of possible values for Y, we first transform Y into $[0,1]$ by computing $Y_i^* = (Y_i - \min(Y))/(\max(Y) - \min(Y))$. The values of $\min(Y)$ and $\max(Y)$ would technically need to be computed in a differentially-private way, but we omit this step here. In fact, in real life, the same process would have to be applied to the predictors, but here we do not since we know that they take values in $[0,1]$.

Note that while traditional applications of SIMEX usually only involve noise in the predictors, we also perturb the response in order to achieve differential privacy for the whole dataset. This should not be a problem since adding random noise on the response simply increases the variance of the error in the linear model and does not bias the coefficient estimates.

Now, once the synthetic data has been generated, we compute the following quantities: the OLS estimate of β_1 with the true data (X, Y), the naive estimate of β_1 obtained with the dataset (W, Z) and SIMEX estimates of the same parameter obtained from the dataset (W, Z). For SIMEX, we use $B = 200$,

$p_0 = 0$, $p_K = K$ with $K = 20$, and we consider linear, quadratic and non-linear extrapolation of the form $(a+b)/(c+p)$ as proposed in [13].

Figure 1 shows the average relative bias on the slope over 500 replications of this experiment, for different values of ϵ. Bias is calculated in comparison to the estimated slope from the true dataset (X, Y). Note that the results are robust to changes in the sample size n, the value of K, and the number of replicates B.

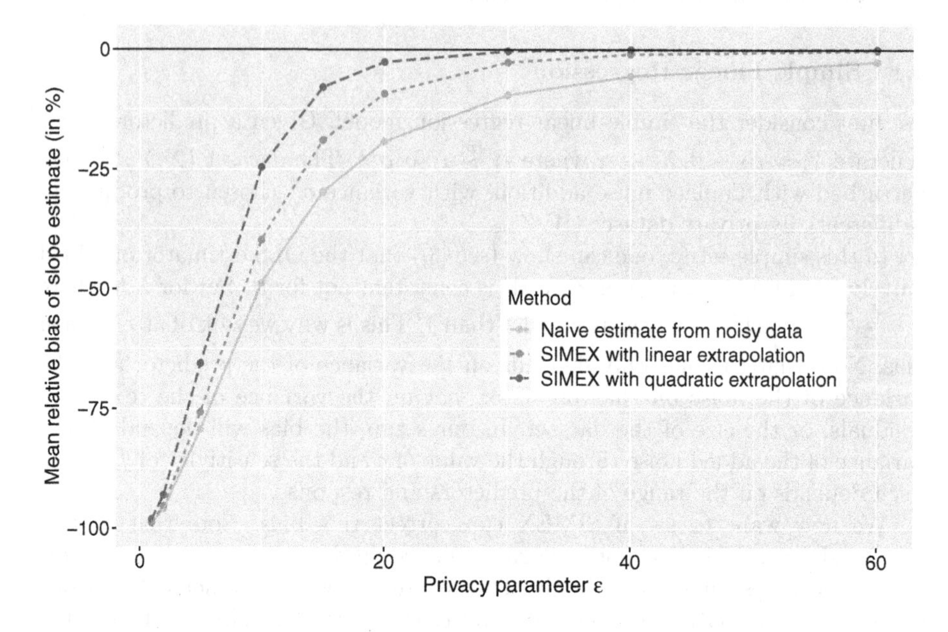

Fig. 1. Mean relative bias of slope estimates for different ϵ and estimation methods

We find that SIMEX methodology can reduce the attenuation bias observed in the naive estimates, which is caused by the addition of Laplace noise to obtain differential privacy. It is however also evident from Fig. 1 that ϵ as large as 15 or 20 is required to obtain sufficiently accurate estimates. In this case, although differential privacy still provides a rigorous measure of the privacy promise, this promise is objectively really weak.

Regarding SIMEX, quadratic extrapolation is consistently better than linear extrapolation. Estimates from non-linear extrapolation with SIMEX are not included in the figure because, while accurate for a certain range of ϵ, the models were difficult to fit in other cases. They were extremely unstable for small values of ϵ, and seemed to overfit the data for large values of ϵ, often leading to absurd extrapolations.

4.2 Multiple Linear Regression

Even though our results on simple linear regression suggest that the methodology will not generalize well to multiple linear regression, we still empirically document the impact of increasing the number of variables in the dataset and the model, and of correlation between the variables in the dataset.

The Effect of the Number of Predictors. This simulation is conducted exactly as above, except that we generate p uncorrelated predictors, each with slope $\beta_j = 2$ in the model predicting Y. The main change is thus in the amount of noise added, as increasing the number of variables increases the l_1-sensitivity of the function f used to generate the microdata.

Figure 2 shows how fast the bias increases (in absolute value) with p. Here, ϵ is set to 20. Since variables are uncorrelated, the effect is similar across parameters, and so we report here the average bias over all estimated slopes. Clearly, even for such a large value of ϵ increasing the number of variables in the dataset rapidly diminishes the quality of the estimates.

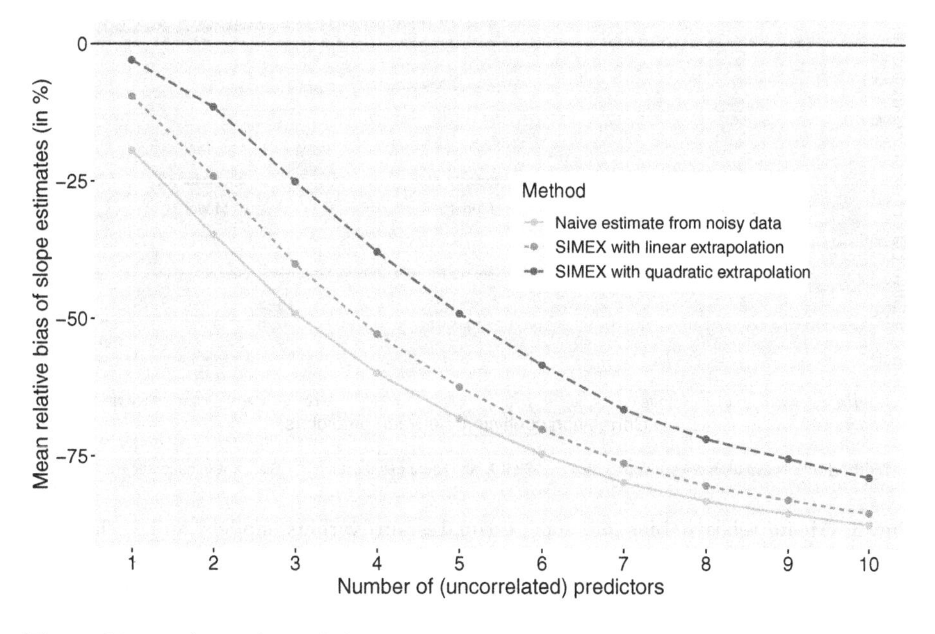

Fig. 2. Mean relative bias of slope estimates for $\epsilon = 20$ and an increasing number of predictors

The Effect of Correlation Between the Predictors. When multiple predictors are perturbed, and they are correlated, the relationship to the bias is quite complicated [2]. We illustrate this in the case of two predictors, generated from correlated uniforms obtained following [10], again for $\epsilon = 20$. We consider three models: in model 1, $\beta_1 = \beta_2 = 2$, in model 2, $\beta_1 = 2$ and $\beta_2 = -2$ and in model 3, $\beta_1 = 2$ and $\beta_2 = 4$.

Figure 3 shows the mean relative bias across simulations separately for each variable as a function of the correlation coefficient between the predictors. For model 1, the SIMEX estimate with quadratic extrapolation is relatively accurate for both parameters, with bias decreasing as correlation increases. For model 2, where variables have effects in opposite directions, the bias for both slopes increases as the correlation between the parameters increases. Finally, model 3 shows that the bias can have different trajectories for different parameters, and that bias can even become positive.

Note that in all three cases, increasing ϵ to 50 allows to recover unbiased estimates for almost all correlations, while clearly greatly diminishing the privacy guarantee.

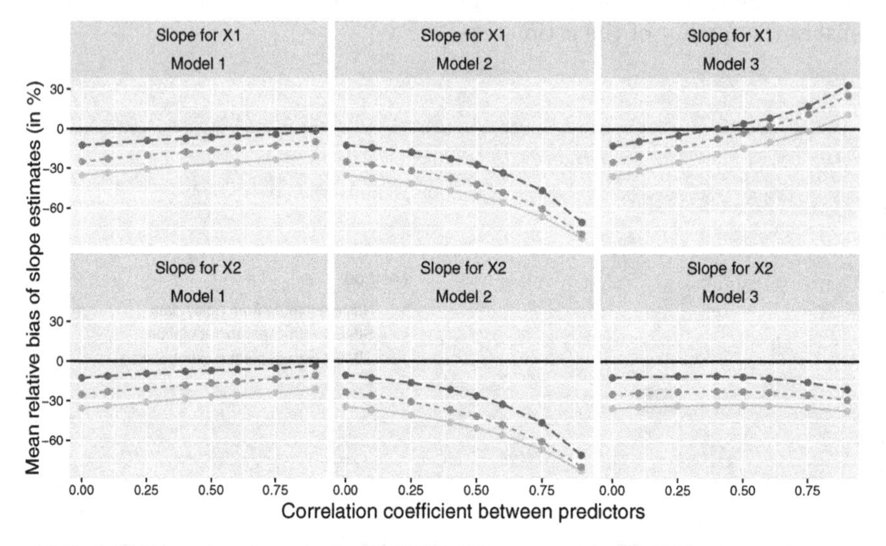

Fig. 3. Mean relative bias of slope estimates for various models and estimation methods, as a function of the correlation between the two predictors. In model 1, $\beta_1 = \beta_2 = 2$, in model 2, $\beta_1 = 2$ and $\beta_2 = -2$ and in model 3, $\beta_1 = 2$ and $\beta_2 = 4$.

5 Discussion

Our simulations results clearly show that even for simple linear regression models SIMEX can not be used to recover estimates with acceptably small bias unless ϵ is taken so large that the privacy guarantee of differential privacy does not mean much anymore. The main difficulty is that when generating microdata with Laplace noise addition the l_1 sensitivity does not decrease with the sample size, as is often the case when publishing statistics under the constraint of differential privacy.

We note that the results are dependent on our choice of distribution for the predictors. Indeed, the variance of the Laplace noise required for a certain privacy level depends on the range of possible values for the predictors, while the attenuation bias depends on the variance of the predictors. The overall effect on the bias of the SIMEX estimate may thus differ for other distributional assumptions on the predictors, although we do not expect that the strategy would work much better in any case.

A question of interest is whether adding normal noise and applying the original SIMEX algorithm would prove more successful. We believe that indeed this approach should be considered but that we would loose the differential privacy guarantee. Indeed, while in theory one may achieve (ϵ, δ)-differential privacy in this way, using the Gaussian mechanism proposed in [9], this requires $\epsilon \in (0, 1)$. Given our results here, we doubt that changing from Laplace noise to Gaussian noise will allow for unbiased estimates of the regression parameters for such small values of ϵ.

Nevertheless, the strategy could be used if one does not require differential privacy, in which case the size of the noise variance would have to be established using other measures of privacy, perhaps based on re-identification estimates from various user scenarios, and may be smaller than that needed for differential privacy. While it requires a bit of computer time, SIMEX is relatively easy to implement and can also be adapted for correlated errors [20], and so it could still provide a way for end-users to improve the estimates from such perturbed microdata. Future work will investigate this possibility, and study more carefully some aspects which were not explored here, such as variance estimation of the SIMEX estimates. Generalizations of SIMEX-style approaches, such as Monte Carlo corrected score methods described in [3] should also be explored.

Acknowledgements. This work was supported by the Natural Sciences and Engineering Research Council of Canada grant No. RGPIN-435472-2013.

References

1. Acs, G., Castelluccia, C., Chen, R.: Differentially private histogram publishing through lossy compression. In: 2012 IEEE 12th International Conference on Data Mining, pp. 1–10. IEEE (2012)
2. Carroll, R.J., Ruppert, D., Stefanski, L.A., Crainiceanu, C.M.: Measurement Error in Nonlinear Models: A Modern Perspective. CRC Press (2006)
3. Carroll, R.J., Ruppert, D., Stefanski, L.A., Crainiceanu, C.M.: Measurement Error in Nonlinear Models: A Modern Perspective, 2nd edn. (2006)
4. Charest, A.-S.: Empirical evaluation of statistical inference from differentially-private contingency tables. In: Domingo-Ferrer, J., Tinnirello, I. (eds.) PSD 2012. LNCS, vol. 7556, pp. 257–272. Springer, Heidelberg (2012). https://doi.org/10.1007/978-3-642-33627-0_20
5. Cook, J.R., Stefanski, L.A.: Simulation-extrapolation estimation in parametric measurement error models. J. Am. Stat. Assoc. **89**, 1314–1328 (1994)

6. Drechsler, J.: Synthetic Datasets for Statistical Disclosure Control: Theory and Implementation, vol. 201. Springer, Heidelberg (2011). https://doi.org/10.1007/978-1-4614-0326-5
7. Drechsler, J.: Some clarifications regarding fully synthetic data. In: Domingo-Ferrer, J., Montes, F. (eds.) PSD 2018. LNCS, vol. 11126, pp. 109–121. Springer, Cham (2018). https://doi.org/10.1007/978-3-319-99771-1_8
8. Dwork, C., McSherry, F., Nissim, K., Smith, A.: Calibrating noise to sensitivity in private data analysis. In: Halevi, S., Rabin, T. (eds.) TCC 2006. LNCS, vol. 3876, pp. 265–284. Springer, Heidelberg (2006). https://doi.org/10.1007/11681878_14
9. Dwork, C., Roth, A., et al.: The algorithmic foundations of differential privacy. Found. Trends Theor. Comput. Sci. **9**(3–4), 211–407 (2014)
10. Falk, M.: A simple approach to the generation of uniformly distributed random variables with prescribed correlations. Commun. Stat.-Simul. Comput. **28**(3), 785–791 (1999)
11. Hardt, M., Ligett, K., McSherry, F.: A simple and practical algorithm for differentially private data release. In: Advances in Neural Information Processing Systems, pp. 2339–2347 (2012)
12. Hundepool, A., et al.: Statistical Disclosure Control. Wiley, Hoboken (2012)
13. Koul, H.L., Song, W., et al.: Simulation extrapolation estimation in parametric models with Laplace measurement error. Electr. J. Stat. **8**(2), 1973–1995 (2014)
14. Li, H., Xiong, L., Jiang, X.: Differentially private synthesization of multi-dimensional data using copula functions. In: Advances in database technology: proceedings. In: International Conference on Extending Database Technology, vol. 2014, p. 475. NIH Public Access (2014)
15. Polzehl, J., Zwanzig, S.: SIMEX and TLS: an equivalence result. AMS 2000 Subject Classification: 62F12, 62J05 (2014)
16. R Core Team: R: a language and environment for statistical computing. R Foundation for Statistical Computing, Vienna, Austria (2020). https://www.R-project.org/
17. Raab, G.M., Nowok, B., Dibben, C.: Practical data synthesis for large samples. J. Priv. Confidentiality **7**(3), 67–97 (2016)
18. Raghunathan, T.E., Reiter, J.P., Rubin, D.B.: Multiple imputation for statistical disclosure limitation. J. Off. Stat. **19**(1), 1 (2003)
19. Reiter, J.P.: Inference for partially synthetic, public use microdata sets. Surv. Methodol. **29**(2), 181–188 (2003)
20. Ronning, G., Rosemann, M.: SIMEX estimation in case of correlated measurement errors. AStA Adv. Stat. Anal. **92**(4), 391–404 (2008)
21. Stefanski, L.A., Cook, J.R.: Simulation-extrapolation: the measurement error Jackknife. J. Am. Stat. Assoc. **90**, 1247–1256 (1995)
22. Tantipongpipat, U., Waites, C., Boob, D., Siva, A.A., Cummings, R.: Differentially private mixed-type data generation for unsupervised learning. arXiv preprint arXiv:1912.03250 (2019)
23. Vega-Márquez, B., Rubio-Escudero, C., Riquelme, J.C., Nepomuceno-Chamorro, I.: Creation of synthetic data with conditional generative adversarial networks. In: Martínez Álvarez, F., Troncoso Lora, A., Sáez Muñoz, J.A., Quintián, H., Corchado, E. (eds.) SOCO 2019. AISC, vol. 950, pp. 231–240. Springer, Cham (2020). https://doi.org/10.1007/978-3-030-20055-8_22
24. Willenborg, L., De Waal, T.: Elements of Statistical Disclosure Control, vol. 155. Springer, Heidelberg (2012). https://doi.org/10.1007/978-1-4613-0121-9

An Analysis of Different Notions of Effectiveness in k-Anonymity

Tanja Šarčević[1], David Molnar[2], and Rudolf Mayer[1,2(✉)]

[1] SBA Research, Vienna, Austria
{TSarcevic,RMayer}@sba-research.org
[2] Vienna University of Technology, Vienna, Austria

Abstract. k-anonymity is an approach for enabling privacy-preserving data publishing of personal, sensitive data. As a result of this anonymisation process, the utility of the sanitised data is generally lower than on the original data. Quantifying this utility loss is therefore important to estimate the usefulness of the resulting datasets. In this paper, we analyse several of these utility aspects.

Data utility can be measured as a direct property of the resulting, anonymised dataset, or via the effectiveness that a statistical analysis, such as a machine learning model, achieves upon this dataset, as compared to the original data. While the latter is more tailored to the specific dataset, it is also generally less efficient. We therefore analyse whether there is a correlation between these two types of measures, and whether the measurement on the effectiveness can be substituted by a measurement of the data properties. Further, we evaluate to what extent different solutions for the same level of k-anonymity differ in regards to effectiveness.

Keywords: k-anonymity · Utility evaluation · Utility metrics · Machine learning

1 Introduction

Day after day we generate more and more data in every sector of our daily life. There are different types of data regarding the domain, but one of the most valuable is personal data, since they contain information about people, which is relevant for commercial as well as other purposes, such as healthcare.

For any statistical analysis such data is the key ingredient. However, individuals' privacy can be compromised even if direct personal identifiable information is removed. The Netflix Prize from 2007 is a famous example of how customer privacy can be threatened without any identifiers, by matching two related datasets.

Distributing personal data is highly regulated by law, especially so in the European Union with the General Data Protection Regulation (GDPR), which

J. Domingo-Ferrer and K. Muralidhar (Eds.): PSD 2020, LNCS 12276, pp. 121–135, 2020.
https://doi.org/10.1007/978-3-030-57521-2_9

came into effect in May 2018. For many purposes, datasets have to be therefore anonymised before distributing them, in order to avoid identification of the people contained in these datasets.

k-Anonymity is a privacy model that can be applied to sensitive datasets by obfuscating information that can be utilised to re-identify individual records in a dataset from which direct identifiers have been removed. Besides weaknesses in the privacy guarantee of k-anonymity, and other proposed models such as Differential Privacy [1], or also approaches such as synthetic data generation [2], these also do have practical downsides. k-anonymity as a model that facilitates easy data sharing is thus still considered in several settings.

Another aspect to consider for datasets that have been such treated is the utility of the resulting data. While anonymisation techniques such as k-anonymity, as well as its extensions such as l-diversity and t-closeness, provide individuals' anonymity in datasets, it at the same time disturbs the effectiveness of machine learning algorithms. This is due to that, when sanitising a dataset, via anonymisation or other approaches, some sensitive information at the level of individual records is invariably removed [3].

Given a candidate for anonymised data, a *utility metric* quantifies the utility (or sometimes called the quality) of this release candidate (resp. the information loss due to the anonymisation process). Data utility can in principal be evaluated via two approaches. One is to utilise one or more quantitative measures of information loss (see [3] and Sect. 2 for an overview). Another approach is to measure the effectiveness of the final statistical analysis to be carried out on the data, such as a predictive machine learning model, compared to an analysis that would have been using the original, unabridged data. The latter is a very task-specific approach, and further less efficient, as it is generally more resource consuming (time, computing power, etc.) than the quantitative measures on the data itself.

We are therefore specifically interested in to what extent these two approaches correlate, and whether one can be used as a proxy measure for the other. We are estimating this in an experimental evaluation, utilising different machine learning models on different classification tasks. We thus utilise correlation analysis to find relationships between classifier behaviour and utility metrics of the anonymised datasets. As part of this evaluation, we generally compare the utility of the anonymised datasets to the original, source data.

Another aspect of our investigation is centred along the fact that there is generally not only one solution for achieving a certain sanitised version of an original dataset that fulfills the desired level of k-anonymity. In contrast, often a large number of candidate solutions exists, and finding the optimal solution is generally solved via heuristic approaches. Therefore, most algorithms utilise implicitly some data utility metric when deciding which solution to find. We want to investigate to what extent this influences the utility of the final, resulting dataset. To this end, we carry out experiments not only on the "best" found candidate, but also on different candidates covering the entire range of the solution space.

The remainder of this paper is organised as follows. Section 2, before Sect. 3 will detail our evaluation methodology. Section 4 discusses and analyses our results, before Sect. 5 provides conclusions and future work.

2 Related Work

The concept of k-anonymity was first introduced in the paper of Samarati and Sweeney[4]. This privacy model can be used to obfuscate sensitive datasets in order to be able to share them with other parties, thus also fulfilling regulations such as the EU's GDPR by anonymising data.

In a dataset, we can generally distinguish different types of attributes (sometimes called variables, or features). On the one hand, *(directly) identifying attributes* directly reveal the identity of a data record. Examples are the full name (to some extent), an e-mail address, or a social security number. As a general pre-processing steps, these are in practices removed from the dataset before publishing, or at least replaced with a pseudonym as identifier.

Quasi-identifiers (QIs) do not directly identify a person, but may become uniquely identifying when used in combination with other quasi-identifiers. An example can be a date of birth in combination with information on the residence of a person, even if in the relatively coarse form of a ZIP code. It has to be noted that this will not apply for all records in the dataset, but in some settings, a large number of them can become re-identifiable. For instance, [5] mentions that 87% of U.S. citizens in 2002 could be re-identified by using attributes zip code, sex and date of birth.

Besides potentially helping in identification, quasi-identifiers often hold significant, demographic information, which is required in analysis processes for differentiating between different groups. In medical analysis, for example, it is often important to differentiate between age groups, the type of job, or information on the location of the residence of patients. Thus, this information cannot simply be omitted as well.

Sensitive data is contained in attributes that for example hold information about a certain type of illness, or the salary of an individual. These are generally the main target in statistical analysis, and can therefore not be omitted or obfuscated.

k-anonymity is a property of the dataset, which ensures that for the identified quasi-identifiers, there are at least k records in the dataset that are indistinguishable in regards to the quasi identifiers. These records that share the same quasi-identifier values are called equivalence groups (or classes) or Q-blocks.

k-anonymity can be achieved by suppression and generalisation, where by suppression we mean simple deletion of values, whereas generalisation refers to a decrease in a value's granularity.

Generalisation utilises so-called *generalisation hierarchies*, which run from leaf nodes denoting particular values via internal nodes to their most general root. In the generalisation process for k-anonymity, one traverses the tree from a leaf node of the original input value upwards until we can construct an equivalence group with all quasi-identifiers being duplicates of one another.

One further needs to distinguish between a global or local generalisation. Global generalisation means that an attribute is put to the same generalisation level for each data record. Local generalisation on the other hand optimises the generalisation by choosing a minimal required loss of precision for each equivalence group. As each level of generalisation invokes an increasing loss of specificity, we want to minimise a dataset's overall information loss. This makes k-anonymisation an NP-hard problem due to an exponential number of possible data-row combinations one can examine. For local generalisation, the search space becomes even larger.

Based on k-anonymity, several related concepts have been proposed, each addressing potential attack vectors for disclosure that the original model did not consider. l-diversity [6] and t-closeness [7] are among the most prominent of those, ensuring diversity among the sensitive attributes. We do not evaluate these at this stage of our work, however.

Several different, mostly heuristic, approaches have been proposed for finding an optimal level of suppression and generalisation for achieving a specific level of k-anonymity. Samarati [8] introduces a concept of *minimal generalisation* that captures the property of the release process not to distort the data more than needed to achieve k-anonymity. One globally-optimal anonymisation algorithm is Flash [9], which we utilise in the implementation provided with the anonymisation software ARX[1]. We further utilise an algorithm providing local generalisation, using a version of a greedy clustering algorithm called SaNGreeA (Social network greedy clustering), [10], as implemented for relational data by [11].

Measuring the quality of the output datasets is a complex aspect. It can be addressed by supporting multiple quality models which can be used as objective function in the optimisation process of the output data. These can include cell-oriented, attribute-oriented and record-oriented general-purpose models.

In the Flash algorithm we utilise, the default objective function of the anonymisation process is *Loss*, which "summarises the degree to which transformed attribute values cover the original domain of an attribute."[2] Since the anonymisation is based on this metric, the prime interest is the correlation of this measurement with the classification results. However, we further compute additional utility metrics that describe the output dataset, namely:

- *Record-level squared error:* This utility metric is the sum of squared errors in groups of indistinguishable records in the transformed dataset. The error is the attribute distance between records in the original dataset and anonymised dataset according to the normalised Euclidean Distance. The higher the error, the greater is the information loss. This metric can take values in the interval of [0, 1] [12].

[1] https://arx.deidentifier.org/.

[2] https://arx.deidentifier.org/overview/metrics-for-information-loss/.

- *Non-uniform entropy:* This metric tries to evaluate and quantify the differences within attribute value distributions. To calculate the non-uniform entropy for a transformed dataset, the non-uniform entropy of each column has to calculated and summed up. Non-uniform entropy compares the frequency of each feature value in the original dataset and the transformed dataset. This basic idea does not work well for local recording. Therefore, this utility metric will be calculated as follows: First, the generalisation level for each record will be calculated, which is followed by identifying the records that are affected by that generalisation level. Finally, the information loss according to non-uniform entropy will be calculated for each generalisation level. Additionally, the calculated value will be scaled into the interval [0, 1] [13].
- *Granularity:* This utility metric captures the granularity of the data. For numerical attributes, the granularity of the generalisation intervals will be determined by the possible interval end points created during the discretisation. This metric can take values between 0 and 1 [14].

Measuring the effectiveness of anonymised data via statistical analysis tasks, such as a predictive machine learning model, is investigated e.g. in [15]. The authors compare applying six different algorithms, with very diverse results. The authors only evaluated the setting of 2-anonymous datasets, which would generally be regarded as too low.

A scheme for controlling over-generalisation of less identity-vulnerable QIs in diverse classes by determining the importance of QIs is presented in [16]. Comparing this scheme to others (such as Mondrian [17]), the authors measure accuracy on Decision Trees, Random Forests and SVMs. Their performance on large factors of k not only remains stable, but in some cases increases with k.

Effects of suppressing records costly to anonymise, instead of generalising several other records as well, has been studied in [18], on a number of binary classification problems. Multi-class problems are addressed in [11], with a focus of selectively deleting outliers to reduce the information loss during the anonymisation process. The authors consider Logistic Regression, SVMs with linear kernel, Random Forest, as well as Gradient Boosting.

3 Methodology

Data In our experiments, we use the Adult Data Set[3] from the UCI Machine Learning Repository. The dataset is prepared with the same steps as described in [11]. The dataset contains some missing values which will be eliminated due to its small number and therefore the dataset has 30162 data entries. The dataset has 15 attributes, only 14 of them will be used for the experiments since the attribute "education" represents the same information as "education-num", only differently encoded. To ensure a proper distribution of each attribute, we modify the column "native-country" to only contain *US-States* and *Non-US* since the

[3] https://archive.ics.uci.edu/ml/datasets/Adult.

value *US-States* dominated the attribute distribution over 90% over all other countries.

Contrary to the default task in this dataset, which is a binary prediction of the income level, we evaluate a more challenging multi-class task. To this end, we define two different target variables, "education" and "marital-status". For "education" we group the 16 continuous "education" levels into four groups, while for "marital-status" we leave the dataset unmodified.

k-Anonymity. We utilise the Flash and SaNGreeA algorithms described above, which use global and local generalisation, respectively. For Flash, to evaluate the effectiveness of different candidate datasets, we created a multitude of these datasets, namely the ten best, one from the middle of the solution space and the worst found solution after the anonymisation process for a given k. We produce perturbations with $k = 3, 7, 11, 15, 19, 23, 27, 31, 35, 100$, and compare with the original, unmodified dataset.

Classification. In order to measure the quality of the anonymised datasets for practical use, we train multiple classification algorithms with the dataset. We use Gradient Boosting, Random Forest, Logistic Regression and Linear SVC as classification algorithms of the python scikit-learn framework[4]. To avoid any optimisation bias towards a specific dataset, only a limited hyper-parameter optimisation has been conducted.

We primarily use the F1 score as the evaluation metric in our experiments. F1 measures the test's accuracy by taking both recall and precision into account. All exported anonymised dataset will be executed with the defined machine learning pipeline.

Correlation Analysis. Beside comparing classification results directly, this paper aims to find relationships between the classification results and the utility metrics which characterise the anonymised datasets. To this end, we calculate the correlation between F1 score and the utility measurements. Our method calculates correlation based on the Pearson correlation coefficient as implement in Python library *scipy*[5].

We compare the earlier mentioned objective function *Loss*, *Record-level squared error*, *Non-uniform entropy*, and *Granularity*.

4 Evaluation and Analysis

In this section, we describe and discuss our experiments. We start with a general comparison of the effectiveness of the k-anonymous data, as seen in Fig. 1.

We can see in all plots that there is a decline in classification effectiveness when anonymising the data, compared to the baseline of no anonymisation. However, there are very large differences in how the single classifiers are affected. For

[4] https://scikit-learn.org/stable/index.html.
[5] https://docs.scipy.org/doc/scipy-0.14.0/reference/generated/scipy.stats.pearsonr.html.

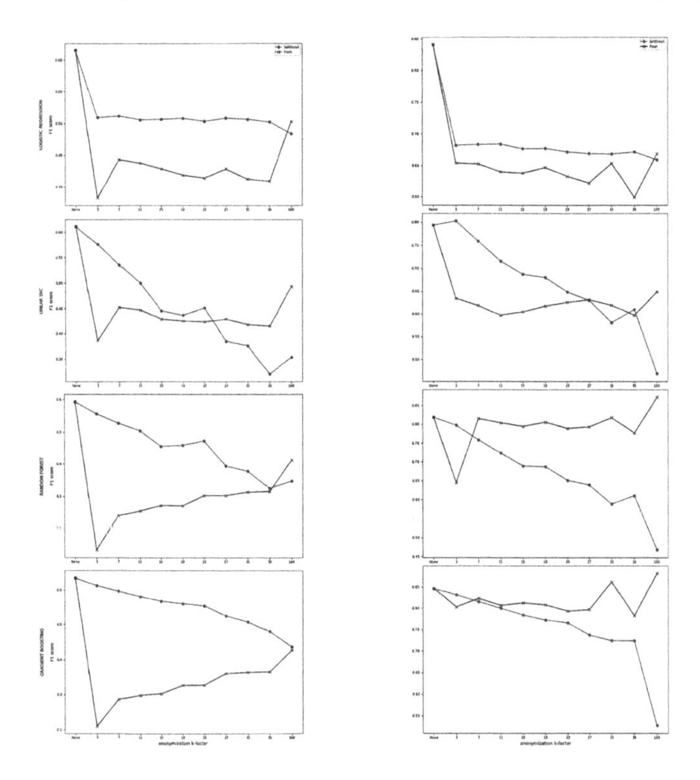

Fig. 1. Classification results for Flash and SaNGreeA, for several k values and two classification targets. (left: *education*, right: *marital status*)

example, Logistic Regression immediately drops in F1 score when performing an anonymisation of $k = 3$. However, for higher values of k, the further deterioration is rather low. It can be observed that the local generalisation provided by SaNGreeA performs better in this case. A similar observation can be made by the Linear Support Vector classifier (SVC), which is not surprising, as these two classification models have a rather similar objective function they minimise. However, for SVC increasing k, and at some point, the global generalisation of Flash becomes superior.

For the ensemble methods of Random Forests and Gradient Boosting, the results are somewhat different. In general, there is a large deterioration in effectiveness between the baseline and $k = 3$, however the effectiveness appears to increase for larger values of k. For the target "marital-status", the global anonymisation of Flash is performing better for these classifiers. Especially for the bagging method of Random Forests for the same target, the drop in effectiveness is relatively low.

As a conclusion, the overall performance of the ensembles on the anonymised data is at a comparable level to the unabridged data, even for relatively large values of k.

We now specifically analyse the difference in utility when not using just the best, but also multiple solutions found by the Flash algorithm.

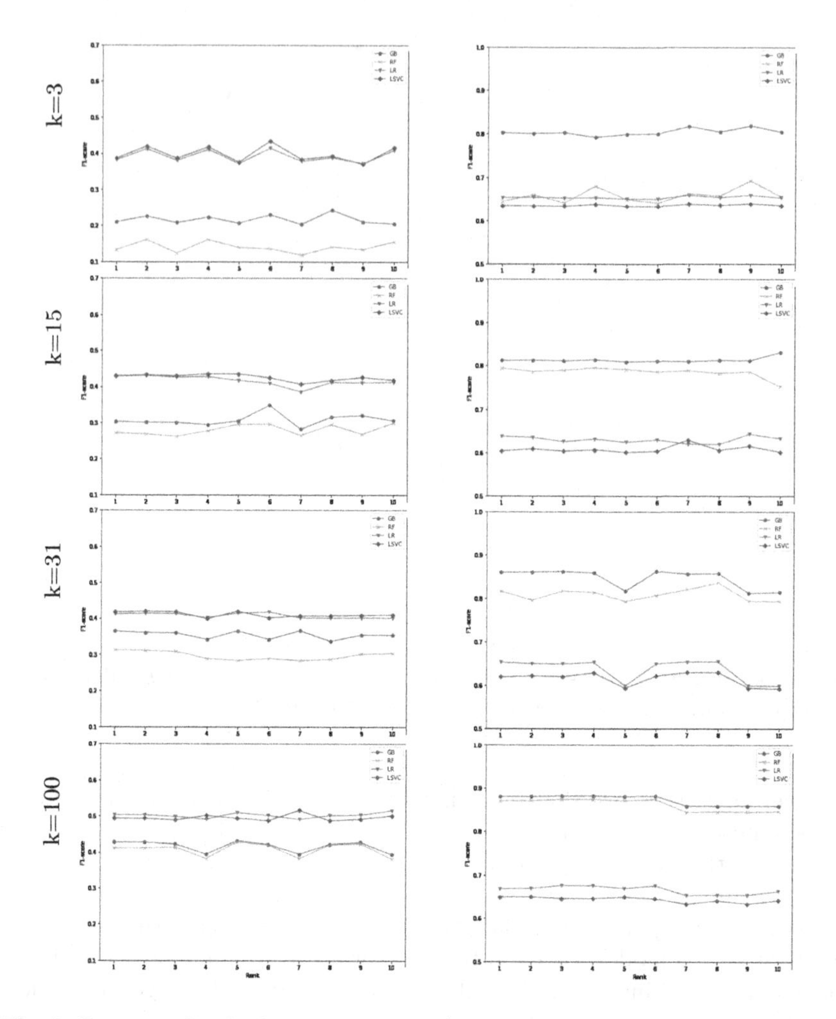

Fig. 2. F1 scores for the best ten datasets (left: education; right: marital status)

Figure 2 shows the classification results measured by the F1-score of the best ten output dataset for each k value and classification method, for "marital-status" and "education", respectively. In general, the fluctuations on F1-score are rather minute. It is visible from the plots that there are no significant differences between the classification results along the best ten datasets. While for some values of k, there is a slight decrease in the classification effectiveness (e.g. $k = 100$ for marital status), for other values, such as $k = 31$ on the marital status target, the tenth solution is actually the best.

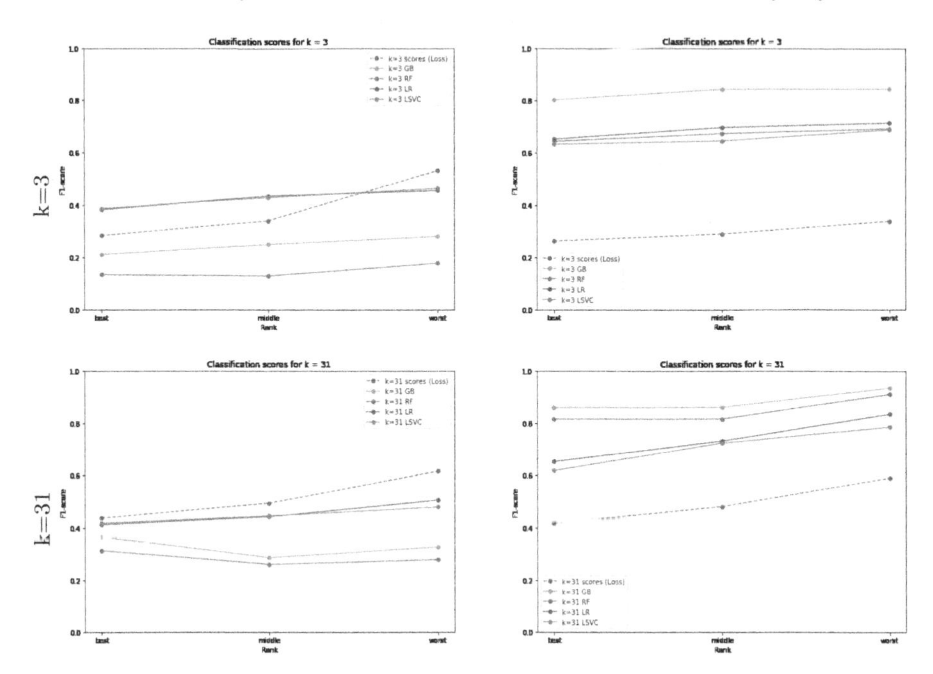

Fig. 3. Classification results of the best, middle and worst found dataset of the solution space

In order to investigate a wide spectrum of the solution space we also analysed the worst found dataset for each k, and one from the middle. Figures 9 and 10 show the overall rankings for these datasets (i.e. they indicate how many different solutions were found). The left column of Figures 3 and 8 shows the classification results for the best, middle and worst found dataset in the solution space for "marital-status", whereas the right column for "education". As we can see, there are no significant degradation in the classification performances along these datasets. Moreover, in some cases we found a better classification performance for the "worst" dataset than for the best. The dashed blue line on each diagram shows the objective function score results for each dataset.

The left columns of Figs. 4, 5, 11 and 12 show the investigated utility metrics for each k value. On the right side of the figures, we see the correlation results between F1 score and Loss for each k value and each classifier; since Loss was the objective function, we examined the correlation of this value. The scores are computed via the Pearson correlation coefficient, where 1 means strong relationship, while -1 means negative correlation. In order to find reasonable correlations, we multiplied the correlation value by -1, since if Loss is higher, we expect worse classification results. As we can see, there is no clear relation between the classification results (F1 score) and the Loss score for "education". For $k = 100$, we observe a clear trend and moderate correlation, and further for the linear models (Logistic Regression and linear SVC) for $k = 31$. For "marital-status",

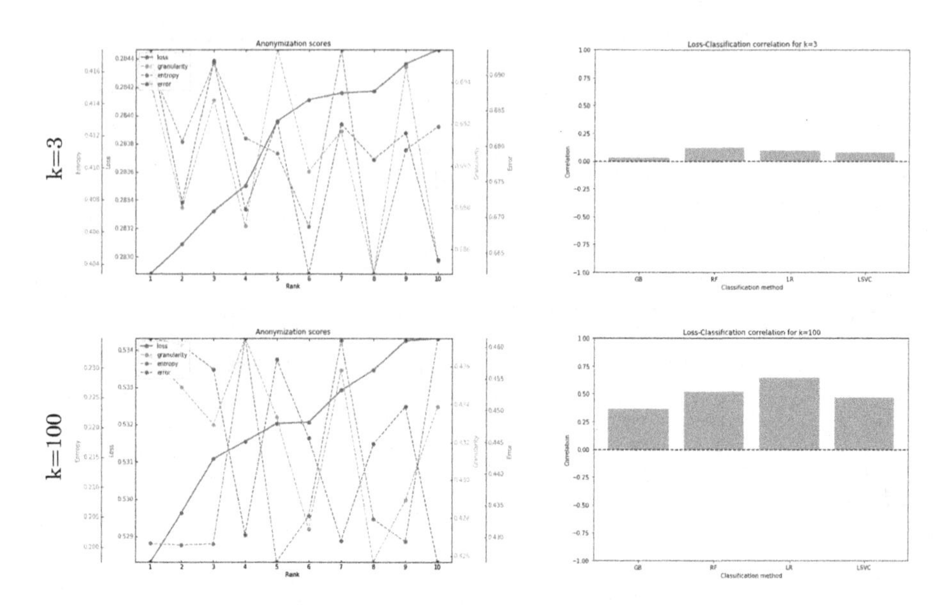

Fig. 4. Utility metrics and loss correlation for *education*

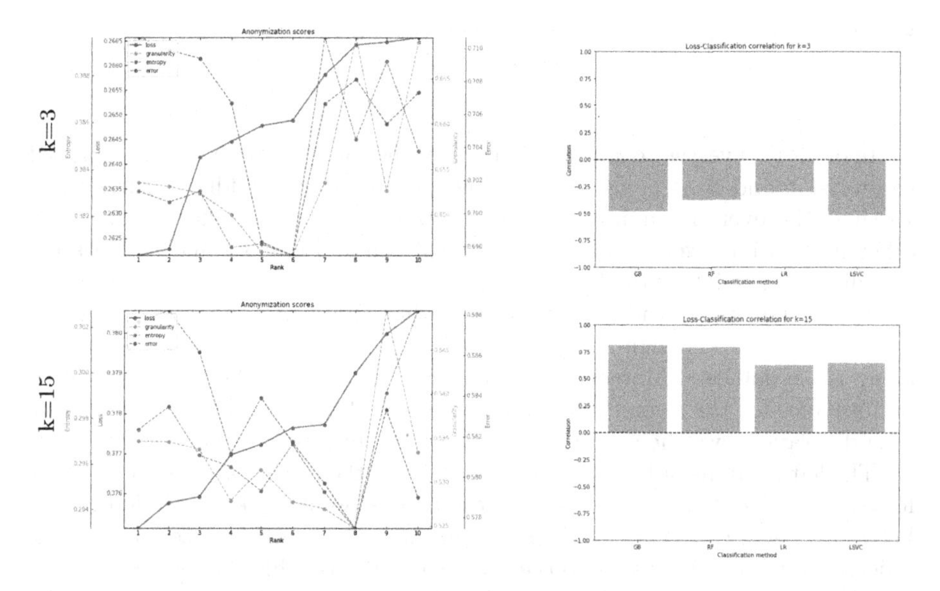

Fig. 5. Utility metrics and loss correlation for *marital-status*

we can observe an overall correlation between all metrics and the F1 score for "marital-status" with $k = 3$, and to some extent also for $k = 100$. There's an overall indirect correlation, though not that strong, for "marital-status" and $k = 1$. The other settings show either none, or no clear trend of correlation.

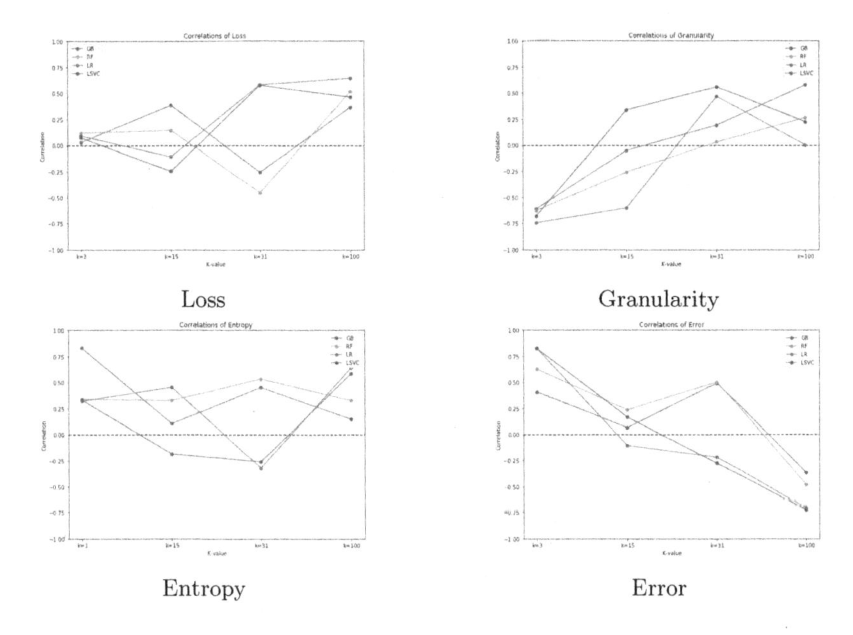

Fig. 6. Correlation results for all investigated utility metrics for *education*

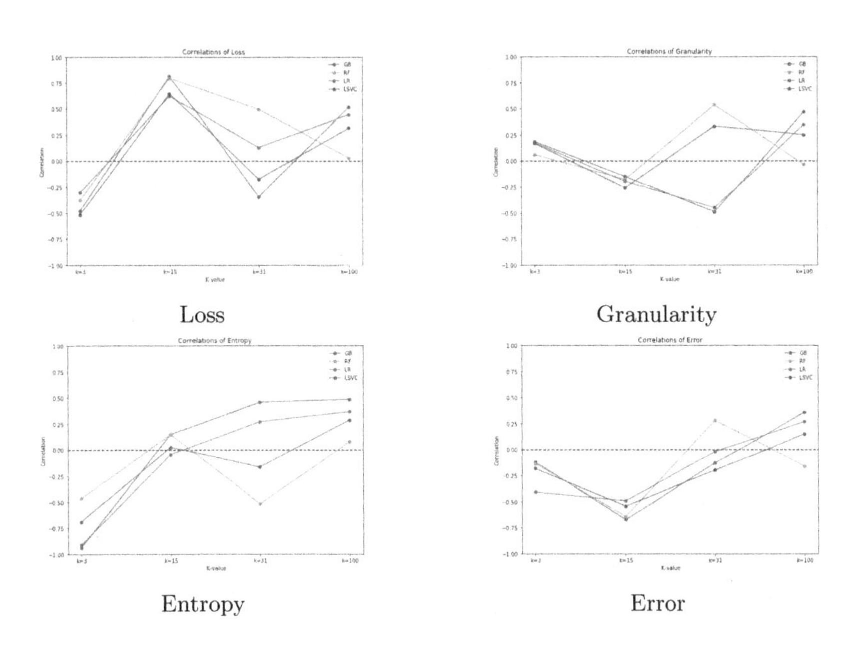

Fig. 7. Correlation results for all investigated utility metrics for *marital-status*

In order to investigate not only the Loss but also other utility metrics, we correlated the classification results (F1 score) with all scores. Figure 7 shows the correlation results for "marital-status", and Fig. 6 for "education". As the plots show, we cannot derive global rules for the correlation. However, we can observe some case specific strong correlations. The record-level squared error and the non-uniform entropy correlates strongly for $k = 3$ on the "education" target attribute with the classification results, while granularity shows also a strong relationship for $k = 31$ and $k = 100$ on the same target variable. Loss also correlates for $k = 15$ on "marital-status" and some on "education". We further observe correlation for Logistic Regression and Linear SVC for $k = 31$.

5 Conclusion and Future Work

In this paper, we performed an analysis on the utility of k-anonymous datasets for specific classification tasks. We investigated (i) the differences between the multiple (syntactically valid) solutions found by heuristic anonymisation techniques, considering the ten best, as well as one from the middle of the solution space and the worst generated dataset. We can conclude that there is very little difference in these solutions, which entails that the effectiveness of the resulting dataset is rather stable, and not influenced by potentially minute aspects in the heuristic. In some cases, even the supposedly worse solutions marginally outperform the best solution. We further investigated (ii) whether there is a correlation between measure that estimate the data utility directly on the dataset, versus the utility for the specific classification task. We specifically analysed Loss, Granularity, Non-uniform Entropy and Record-level squared error. Although, we could not derive any global rule of these correlations that can be applied independently of the task or the k value, we could see some specific correlations between classification and utility metrics. We can conclude that there is no overall, reliable correlation between these two measures, and it is thus not generally possible to estimate the classification performance based on the measures from the dataset alone.

Future work will focus on extending this analysis to further machine learning tasks such as regression, and will include further datasets. We will also extend the analysis to multiple solvers of the k-anonymity problem.

Acknowledgements. This work was partially funded by the BRIDGE 1 programme (No 871267, "WellFort") of the Austrian Research Promotion Agency (FFG), the EU Horizon 2020 research and innovation programme under grant agreement No. 826078 (Project "FeatureCloud"). SBA Research (SBA-K1) is funded within the framework of COMET—Competence Centers for Excellent Technologies by BMVIT, BMDW, and the federal state of Vienna, managed by the FFG.

Appendix

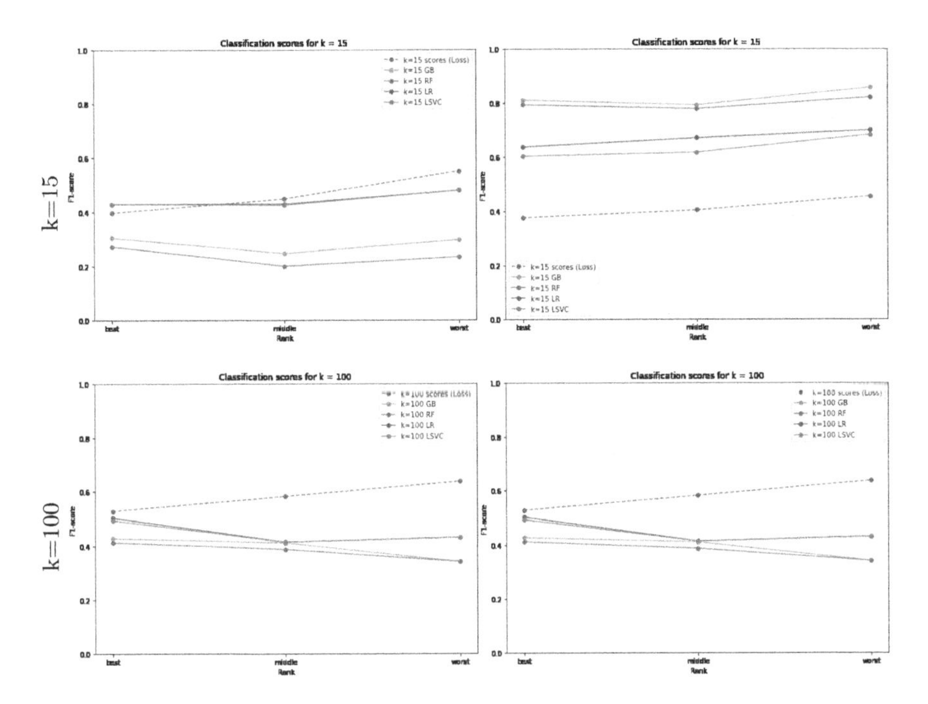

Fig. 8. Classification results of the best, middle and worst found dataset of the solution space

k value	middle rank	worst rank
3	46	92
15	48	96
31	40	80
100	16	33

Fig. 9. Rankings for *education*

k value	middle rank	worst rank
3	36	72
15	36	72
31	40	81
100	33	66

Fig. 10. Rankings for *marital-status*

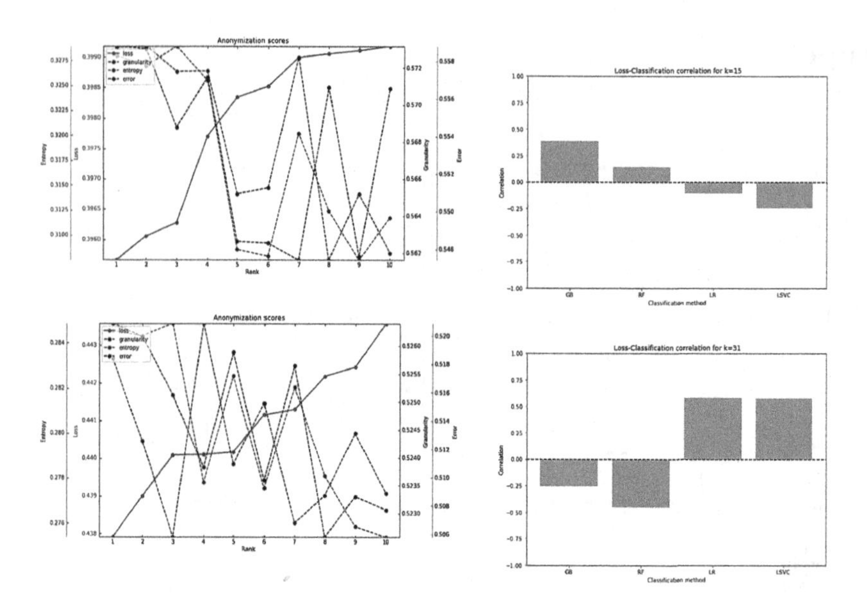

Fig. 11. Utility metrics and loss correlation for *education*

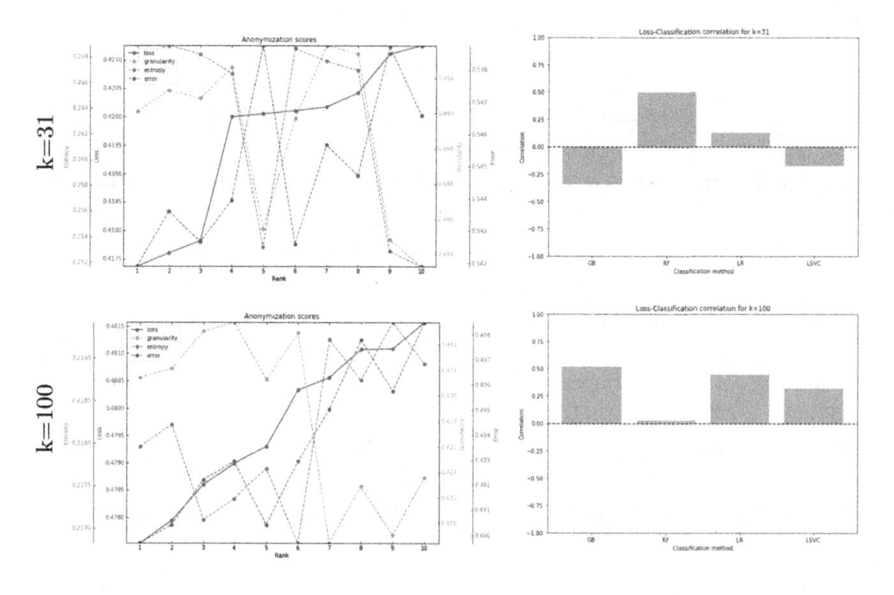

Fig. 12. Utility metrics and loss correlation for *marital-status*

References

1. Dwork, C.: Differential privacy. In: Bugliesi, M., Preneel, B., Sassone, V., Wegener, I. (eds.) ICALP 2006. LNCS, vol. 4052, pp. 1–12. Springer, Heidelberg (2006). https://doi.org/10.1007/11787006_1

2. Hittmeir, M., Ekelhart, A., Mayer, R.: On the utility of synthetic data: an empirical evaluation on machine learning tasks. In: International Conference on Availability, Reliability and Security (ARES), Canterbury, UK. ACM (2019)
3. Chen, B.-C., Kifer, D., LeFevre, K., Machanavajjhala, A.: Privacy-preserving data publishing. Found. Trends Databases **2**(1–2), 1–167 (2009)
4. Samarati, P., Sweeney, L.: Protecting privacy when disclosing information: k-anonymity and its enforcement through generalization and suppression (1998)
5. Sweeney, L.: K-anonymity: a model for protecting privacy. Int. J. Uncertainty Fuzziness Knowl.-Based Syst. **10**(5), 557–570 (2002)
6. Machanavajjhala, A., Gehrke, J., Kifer, D., Venkitasubramaniam, M.: L-diversity: privacy beyond k-anonymity. In: 22nd International Conference on Data Engineering (ICDE 2006) (2006)
7. Li, N., Li, T., Venkatasubramanian, S.: t-closeness: privacy beyond k-anonymity and l-diversity. In: 2007 IEEE 23rd International Conference on Data Engineering, Istanbul. IEEE (2007)
8. Samarati, P.: Protecting respondents identities in microdata release. IEEE Trans. Knowl. Data Eng. **13**(6), 1010–1027 (2001)
9. Kohlmayer, F., Prasser, F., Eckert, C., Kemper, A., Kuhn, K.A.: Flash: efficient, stable and optimal k-anonymity. In: International Conference on Privacy, Security, Risk and Trust and International Confernce on Social Computing, Amsterdam, Netherlands. IEEE (2012)
10. Campan, A., Truta, T.M.: Data and structural *k*-anonymity in social networks. In: Bonchi, F., Ferrari, E., Jiang, W., Malin, B. (eds.) PInKDD 2008. LNCS, vol. 5456, pp. 33–54. Springer, Heidelberg (2009). https://doi.org/10.1007/978-3-642-01718-6_4
11. Malle, B., Kieseberg, P., Holzinger, A.: DO NOT DISTURB? Classifier behavior on perturbed datasets. In: Holzinger, A., Kieseberg, P., Tjoa, A.M., Weippl, E. (eds.) CD-MAKE 2017. LNCS, vol. 10410, pp. 155–173. Springer, Cham (2017). https://doi.org/10.1007/978-3-319-66808-6_11
12. Jordi, S.-C., Josep, D.-F., David, S., Sergio, M.: t-closeness through microaggregation: strict privacy with enhanced utility preservation. IEEE Trans. Knowl. Data Eng. **27**(11), 3098–3110 (2015)
13. Fabian, P., Raffael, B., Kuhn, K.A.: A generic method for assessing the quality of de-identified health data. Stud. Health Technol. Inform. **228**, 312–316 (2016)
14. Iyengar, V.S.: Transforming data to satisfy privacy constraints. In: ACM SIGKDD International Conference on Knowledge Discovery and Data Mining (KDD), Edmonton, Alberta, Canada. ACM Press (2002)
15. Wimmer, H., Powell, L.: A comparison of the effects of k-anonymity on machine learning algorithms. In: Proceedings of the Conference for Information Systems Applied Research (2014)
16. Abdul, M., Farman, U., Lee, S.: Vulnerability- and diversity-aware anonymization of personally identifiable information for improving user privacy and utility of publishing data. Sensors **17**(5), 1059 (2017)
17. LeFevre, K., DeWitt, D.J., Ramakrishnan, R.: Mondrian multidimensional k-anonymity. In: 22nd International Conference on Data Engineering (ICDE 2006), Atlanta, GA, USA. IEEE (2006)
18. Malle, B., Kieseberg, P., Weippl, E., Holzinger, A.: The right to be forgotten: towards machine learning on perturbed knowledge bases. In: Buccafurri, F., Holzinger, A., Kieseberg, P., Tjoa, A.M., Weippl, E. (eds.) CD-ARES 2016. LNCS, vol. 9817, pp. 251–266. Springer, Cham (2016). https://doi.org/10.1007/978-3-319-45507-5_17

Multivariate Top-Coding for Statistical Disclosure Limitation

Anna Oganian[1(✉)], Ionut Iacob[2], and Goran Lesaja[2,3]

[1] National Center for Health Statistics, 3311 Toledo Road, Hyattsville, MD 20782, USA
aoganyan@cdc.gov

[2] Department of Mathematical Sciences, Georgia Southern University, P.O. Box 8093, Statesboro, GA 30460, USA
{ieiacob,goran}@georgiasouthern.edu

[3] Mathematics Department, United States Naval Academy, 121 Blake Road, Annapolis, MD 21402, USA

Abstract. One of the most challenging problems for national statistical agencies is how to release to the public microdata sets with a large number of attributes while keeping the disclosure risk of sensitive information of data subjects under control. When statistical agencies alter microdata in order to limit the disclosure risk, they need to take into account relationships between the variables to produce a good quality public data set. Hence, Statistical Disclosure Limitation (SDL) methods should not be univariate (treating each variable independently of others), but preferably multivariate, that is, handling several variables at the same time. Statistical agencies are often concerned about disclosure risk associated with the extreme values of numerical variables. Thus, such observations are often top or bottom-coded in the public use files. Top-coding consists of the substitution of extreme observations of the numerical variable by a threshold, for example, by the 99th percentile of the corresponding variable. Bottom coding is defined similarly but applies to the values in the lower tail of the distribution. We argue that a univariate form of top/bottom-coding may not offer adequate protection for some subpopulations which are different in terms of a top-coded variable from other subpopulations or the whole population. In this paper, we propose a multivariate form of top-coding based on clustering the variables into groups according to some metric of closeness between the variables and then forming the rules for the multivariate top-codes using techniques of Association Rule Mining within the clusters of variables obtained on the previous step. Bottom-coding procedures can be defined in a similar way. We illustrate our method on a genuine multivariate data set of realistic size.

Keywords: Statistical disclosure limitation (SDL) · Top-coding · Hierarchical clustering · Association Rule Mining · Dimensionality reduction · Genetic algorithm

J. Domingo-Ferrer and K. Muralidhar (Eds.): PSD 2020, LNCS 12276, pp. 136–148, 2020.
https://doi.org/10.1007/978-3-030-57521-2_10

1 Introduction

Many national surveys conducted by government agencies have a large number of attributes of different types. Some examples of such surveys in the USA are the National Health Interview Survey [15], the Behavioral Risk Factor Surveillance System [4], the Current Population Survey [7], and the American Community Survey [1]. Government statistical agencies have an obligation by law to protect the privacy and confidentiality of their respondents who can be individuals or enterprises. This is usually done by altering—we use the term *masking*—the original data before release, for example, by aggregating categorical values, swapping data values for selected records, adding noise to numerical values, or synthesizing some or all of the responses. See [12,13] for more details.

Records that have extreme or very large values of numerical attributes are often a subject of concern about disclosure risk associated with these values. One way of addressing such a risk is to top code numerical attributes which are considered as "visible" or possibly known from other publicly available data sources and which are not a subject to very frequent variation. For example, a person's height can be top-coded to 75 in., so all the individuals who are taller than 75 in. are recorded in the category "75 in. and above". Such top-coding thresholds are chosen by the data protectors. Typically these thresholds are the estimates of the upper percentiles of the corresponding variable, for example, 95th, 97th, or 99th percentiles.

However, when top-coding thresholds are determined independently of other variables, protection may be inadequate for some groups of individuals. For example, assume the attribute weight is top-coded to 300 pounds for all the respondents. However, a female respondent with such a top-coded weight whose race/ethnicity is Asian could be more extreme as opposed to a respondent with the same weight who is a white male [14]. Being more extreme and rare, these individuals are more likely to be subject to re-identification. Thus, from the disclosure risk perspective it would be desirable to determine appropriate top-codes for the individuals in this group, different from those for the rest of the population. First, such subgroups should be identified. In some cases, as in the example above, it may be intuitive and easy. However, in general, in data sets with a large number of attributes, such a task is not always trivial. In this paper we propose a procedure that we call *multivariate top-coding*. It consists of identifying sub-populations/groups of records that require adjusted top-codes and then computing such top-codes for these groups.

1.1 Contribution and Plan of the Paper

The main contribution of the paper is a new multivariate top-coding procedure which is based on clustering variables and using techniques of Association Rule Mining (ARM) [2] to determine the sub-populations that should be top-coded differently from others. In Sect. 2 our multivariate top coding procedure is described. In Sect. 3 we illustrate the application of this procedure to a genuine data set of realistic size. Concluding remarks are given in Sect. 4.

2 The Description of the Multivariate Top-Coding Method

Assume there is a microdata set D with p variables and the data protector decides to top-code numerical variables $T = \{T_1, \cdots, T_k\} \in D$, $k < p$. If there are many variables in D, then the number of possible combinations of categories of variables can be extremely high, and each such combination defines a potential sub-population or group of individuals. Thus, identification of the groups of individuals which require adjusted top-codes can be computationally very demanding. To make it feasible, we propose first to cluster the variables in D into groups, where each group is formed around each numerical variable T_i that is selected for top-coding. Next, we perform the search of the sub-populations that should have special top-codes for T_i within the vertical partition corresponding to the cluster of variables around T_i.

2.1 Clustering Approach

In [16], several methods of clustering of variables were described and compared within the framework of disclosure limitation. These are hierarchical clustering methods that operate on the dissimilarity matrix which represent pairwise dissimilarities or "distances" between the variables. The metric of distance is based on the squared canonical correlation which can be computed for variables of different types (see [6,16]). The dissimilarity matrix is created as a lower triangular $p \times p$ matrix DM with elements $DM[i,j] = 1 - r[i,j]$ for $i > j$, and 0 otherwise, where $r[i,j]$ is a squared canonical correlation between variables V_i and V_j.

In [16], *K-Link* and *Single-Link* methods were ranked high among the best performing clustering methods within the framework of disclosure limitation. These methods, however, may produce big clusters where some of the variables within the cluster may have low correlation, which is not optimal for our case. For example, if the variable income is being top-coded, then variables that are not correlated with income most likely will not be included by the subsequent ARM in the description of those sub-populations which need special top-codes for income.

A better way to group the variables for multivariate top-coding is to include in each cluster only the closest variables to T_i, which are no further than $1 - h$ from T_i. The cut-off value h depends of the preferences of the data protector, intuitively representing a trade-off between accuracy/utility and computational complexity of the procedure. In this approach multiple cluster membership is allowed so the same variables may be used to describe different sub-populations. For example, sex and race could define different subpopulations such as "Asian females" and "white males" that should have different top-codes for a person's weight. This simple variable grouping is much faster than other clustering algorithms as it does not even require computation of the whole dissimilarity matrix DM, but only those rows of DM which correspond to the variables in T. Once variables are clustered in k groups, each one centered at some $T_i \in \{T_1, \cdots, T_k\}$, the search of sub-populations that require special top-codes for each of T_i will be done within the corresponding cluster. To accomplish this search we propose to use Association Rule Mining (ARM), a popular machine learning rule-based methodology for discovering interesting relationships between the variables. There are several reasons why we decided to use ARM. First, the problem of multivariate top-coding, as we outlined it above, can be expressed as a search of association rules for variables T_i. An association rule [2] is an expression of the form $X \rightarrow Y$, where X and Y express conditions on the attributes of the following form:

$$V_i = cat_{i_l} \wedge \cdots V_f \in [l_f, u_f] \cdots \wedge V_j = cat_{j_m} \tag{2.1}$$

where $V_i, V_j, \cdots V_f$ are the variables from the data set D, $cat_{i_l}, \cdots cat_{j_m}$ are the categories of the categorical variables, and $[l_f, u_f]$ are specific intervals within the domains of the corresponding continuous variables. In the paper we call the antecedent of the rule, X, a "LHS of the rule", and the consequent of the rule, Y, a "RHS of the rule".

The association rules that we are proposing for multivariate top-coding are of the form:

$$(V_i = cat_{i_l}) \wedge (\cdots V_j = cat_{j_m}) \rightarrow T_i < threshold \tag{2.2}$$

For example, $(Sex = Female) \wedge (Height < 65\,\text{in.}) \rightarrow (Weight < 200)$.

Another reason for using ARM is that these techniques are designed to work well for large data bases. ARM algorithms are implemented in many software packages, including R.

2.2 Background on ARM

Association rule $X \rightarrow Y$ is characterized by its support and confidence. According to the original definition and notation used in [2], a support of X, the antecedent of the rule, is defined as the proportion of records in the database D that satisfy the expression X:

$$Supp(X) = |\{r \in D | X \subseteq r\}| / |D|$$

where r denotes a record in D and $|\cdot|$ means cardinality.

A confidence of the rule is defined as the proportion of the records in D that contains X which also contains Y:

$$Conf(X \rightarrow Y) = Supp(X \cup Y) / Supp(X)$$

The standard Apriori [3] algorithm (or other well known algorithms, for example, ECLAT [20], FP_GROWTH [11] or ASSOC [10]) can be used to mine association rules where all the attributes are categorical. The procedure usually consists of two steps. The first step is to mine the so called set of frequent itemsets, that is, to find expressions X with support higher than a predefined minimal support of the rule, $MinSupp$. The second step is to discover all the rules with the confidence higher than a predefined minimal confidence of the rule, $MinConf$.

Mining association rules on both categorical and numerical attributes, often called mining quantitative association rules, have been covered significantly less in the literature. There is no method that is considered a "gold standard" for quantitative association rules. The difficulty of mining these rules stems from the fact that numerical attributes are usually defined on a wide range of different values. It's not practical to work on all possible numeric values, as is done for categorical values, because in most cases, there are many such values and each numeric value does not appear frequently.

In [18], a genetic-based algorithm called QuantMiner for mining quantitative association rules was proposed. QuantMiner works directly on a set of rule templates - preset formats - specifying which attributes occur in the LHS and the RHS of the rule. Templates can be chosen by the user or computed by the system.

For categorical variables QuantMiner computes frequent itemsets similar to Apriori; that is, finds frequently occurring instantiations $V_i = cat_{i_l} \wedge \cdots V_j = cat_{j_m}$. Then it generates a rule template for each such instantiation. For each rule template, the algorithm looks for the best intervals of the numerical attributes occurring in that template, which is achieved using the Genetic Algorithm.

The algorithm starts with an initial population of rules for each rule template. Different rules in the initial population have different intervals for continuous variables, randomly chosen within their domains. In the following generations, the intervals are subject to change by genetic operators of mutation and crossover [18]. The mutation operator changes the lower or the upper bound of the interval. The crossover operator consists of taking two intervals, called parents, at random and generating new intervals in such a way so that the new interval is either inherited from one of the parents or formed by mixing the bounds of the two parents. These operators are applied to the rules of each generation. After each application, the fitness of each rule is evaluated and the best rules, according to the chosen fitness function, are selected for the next generation. This process repeats over *GenN* generations (*GenN* is a parameter of the algorithm). After the last generation is created the best rule for each rule template is selected from the corresponding population of rules and included in the output.

The fitness function used in QuantMiner is proportional to the Gain of the rule [9] and the length of the intervals in the rule:

$$Fitness(Rule) = Gain(Rule) * \prod_j (1 - Prop_j)^2 \tag{2.3}$$

where *Gain* is defined as follows:

$$\begin{aligned} Gain(Rule) &= Gain(LHS \rightarrow RHS) \\ &= (Conf(LHS \rightarrow RHS) - MinConf) * Supp(LHS) \end{aligned} \tag{2.4}$$

and where $Prop_j$ is the ratio of the interval length to the length of the domain of V_f.

2.3 Multivariate Top-Coding Using ARM

Let P be a percentile rank chosen by the data protector to compute top-code thresholds for variable T_i. For example, if $P = 99$ then 99th percentile of T_i serves as a top-code threshold for this variable. For each variable T_i, let $Clust_i$ be the cluster of variables that contains T_i. We propose the following procedure to determine which sub-populations may need special top-codes (that is, lower than the rest of the population) for each variable T_i in T.

1. Compute the P-th percentile for the variable T_i using all the records in the data set. Denote this marginal percentile as Z_i.
2. Mine the following type of association rules on the vertical partition of the data that corresponds to the cluster of variables $Clust_i$:

$$X \rightarrow T_i < (Z_i - \Delta) \tag{2.5}$$

   The LHS of the rule X represents any combination of the variables/categories from $Clust_i$, in the form given by expression (2.1). The RHS of the rule is the expression that makes the implication (that is, the rule) true. In the RHS of the rule we introduce parameter Δ which is the minimal difference between Z_i and the percentile for a particular sub-population that should "get" its own top-coding threshold, different from Z_i. Δ can be chosen by the data protector for practical reasons in order not to have too many top-codes which are not very different from Z_i.
3. Choose the rules with the confidence equal to $P/100$ or higher. Denote this set of rules as S. Note that, the confidence of a rule is the probability

$$P(T_i < (Z_i - \Delta)|X) \tag{2.6}$$

   Hence, the LHS of such rules defines sub-populations for which the P-th percentile of the variable T_i is at most $Z_i - \Delta$. Thus, extreme observations in these subpopulations might need to be protected by adjusting, that is lowering, their top-code thresholds.
4. For each subpopulation defined by the LHS of the rules mined on the previous step, compute the P-th percentile for T_i using the records that belong to these subpopulations. The computed percentiles may serve as the top-codes for these subpopulations.

To find quantitative association rules (step 2 of the procedure above) we used a modified QuantMiner procedure: we changed the way how interval boundaries of numerical variables that appear on the LHS of the rules are calculated. We also changed the form of the fitness function. Regarding the calculation of interval boundaries, in the original version of QuantMiner both ends of the intervals are subject to change by the operators of crossover and mutation and the shortest intervals are being sought. However, we fixed the lower end of the intervals at the minimal value of the domain for those numerical variables that appear on the LHS of the rule and are positively correlated with the top-coded variable T_i. If the numerical variable on the LHS of the rule is negatively correlated with T_i, then the lower end of the interval is subject to change and the upper end is fixed. This is done in order not to exclude the individuals with values of numerical variables close to the boundaries of the domain from protection by top-coding who should otherwise be protected. For example, assume the variable $Income$ is being top-coded and the numerical variable $Hours$, hours worked per week, is positively correlated with $Income$. So, the association rule has the form: $Hours \in [l, u] \rightarrow Income \leq threshold$. If $l \neq min(Hours)$, then those individuals with $Hours \in [min(hours), l]$ will not be top-coded, but those with $hours \in [l, u]$ will. This, however, leaves the former groups of individuals unprotected. It also does not make sense given the nature of relationship between $Hours$ and $Income$.

As mentioned above, we also modified the form of the fitness function. Contrary to [18] our fitness function favors larger intervals of numerical variables V_f on the LHS of the rule, subject to the resulting rule satisfying minimal confidence and minimal support.

$$Fitness(Rule) = \begin{cases} \prod_j (Prop_j)^2, & \text{if } Supp(Rule) \geq MinSupp \\ & \text{ and } Conf(Rule) \geq MinConf \\ 0, & \text{otherwise} \end{cases} \tag{2.7}$$

The reason of this modification is again not to exclude any individuals from protection that otherwise should be protected. Indeed, larger intervals typically correspond to larger groups of individuals having values of numerical variables within these intervals. Thus, the largest intervals on the LHS of the rule in our algorithm define the largest sub-population for which expression (2.6) is true. Hence, these individuals need lower top-codes for variable T_i than the top-codes for rest of the population.

Finally, it is important to note that the procedure of bottom-coding is a straightforward conversion of the top-coding procedure described above.

3 Numerical Experiments

We applied our approach of multivariate top-coding to a genuine multivariate data set that was downloaded from the UCI Machine Learning Repository [8]. This is a sample drawn from the Public Use Microdata Samples (PUMS) person-level 1990 US Census file. We will refer to this file as Census in the paper. In our experiments we used 66 numerical and categorical variables from this data set. Full description of the variables can be found in [5]. Some variables were excluded from the experiments, such as allocation flags, serial number and some others because they would not be used in practice. There are 1.8 million records in our data set.

To illustrate our approach we choose the variables $Income1$ - wages or salary earned by the individuals in 1989 and Age for top-coding. These types of variables are usually top-coded. As outlined in Sect. 2, we first found clusters of variables around these two variables. For Age, the cluster consisted of the following variables: *Relat*1 - relationship of the respondent to the householder (householder is defined later in the text) with 13 categories, $Marital$ - marital status with 5 categories, *Disable*2 - work prevented status with two categories, *Income*5 - social security income in 1989 (a numerical variable), *Rlabor* - employment status with 7 categories, $Work89$ - worked or not in 1989 with two categories, and $Yearsch$ - educational attainment with 18 categories. In our experiments $Yearsch$ was treated as a numerical variable, hence, the output rules were given in the form if $Yearsch < i \rightarrow Income1 < Y$, which is more meaningful than potentially a large number of rules, each one differing by a particular category in $Yearsch$.

The cluster of variables around *Income*1 includes the following variables: $Class$ - class of worker with 10 categories, $IndustryClass$ - industry class with 13 categories, *Ocupclass* - occupation class with 8 categories, *Relat*1 - relationship within the household with 13 categories, *Disable*1 - work limitation with three categories, *Rlabor* - employment status with 7 categories, $Hour89$ - numerical variable denoting usual hours worked per week the year before the interview, $Week89$ - weeks worked the year before the interview, and $Yearsch$ - educational attainment with 18 categories.

The default minimal support of the rules in QuantMiner is set up to be 10%, but in our experiments, we lowered the minimal support to 1% in order to be able to identify small sub-populations (of the size of 1% of the data set or larger) which may require their own top-codes. For the data set of this size, it means that the size of these sub-populations should be at least $18,000$. The main constraint on lowering support of the rules is the computational burden, because many more subpopulations need to be checked, and, as a consequence, many more potential rules should be tested by the algorithm.

It should be noted that the main purpose of the proposed procedure is to assist the data protector in the otherwise daunting task of going through the large number of possible combinations of the relevant attributes in a big data set in order to find rarely observed extreme observations of top-coded variables for certain groups of records or sub-populations. These sub-populations are usually associated with lower values of the numerical variables subject to top-coding. Our rules are meant to bring such special cases to the data protector's attention. However, the decision about whether to use these rules to apply top-codes or not depends on many factors, such as a particular scenario of data release, SDL practice at a particular institution, and preferences of data protectors. In any case, such decisions are usually made together with the subject area specialists. Furthermore, some of the rules may be obvious, or they may be always observed in the data; for example, the rules that have confidence equal to 100%. Thus, not every automatically mined rule should imply top-coding. In some instances, the rules that have confidence equal to 100% may be used with the goal to check and find incorrectly recorded observations or the values that are not plausible.

Due to space limitation, below we present a selection of rules for *Age* and *Income*1 that are representative for this data set. They have attribute categories that appear most frequently. It should be noted that, the rules presented below are not our recommendations for top-coding for this particular data set nor any similar data set. The rules and results presented in this section are for the illustration of our method of multivariate top-coding only. In our experiments we used a 99-th percentile as a parameter for top-coding thresholds. So, for the groups of individuals that fit the description that appears in the LHS of the rules, at least 99% of individuals in the data set have $Income1$ (or Age) below the threshold that appears on the RHS of the rules. It is worth noting, that univariate top-code thresholds for this data set using the same parameter P, that is, the 99-th marginal percentile rank, would be $88,000 for $Income1$ and 87 years old for Age. Hence, univariate top-coding would imply that these thresholds would apply to all the individuals, regardless of their other characteristics. Below we list several age- and income- related rules as an illustration.

Age-related rules:

$Relat1$ = Son/daughter of the householder $\rightarrow Age < 60$
$Relat1$ = Other persons in group quaters $\wedge$ $Work89$ = Yes $\rightarrow$
$Age < 60$
$Relat1$ = Housemate $\wedge$ $Work89$ = Yes $\rightarrow Age < 65$
$Marital$ = Never married $\wedge$ $Relat1$ = Housemate $\wedge$ $Work89$ = Yes $\rightarrow$
$Age < 60$
$Marital$ = Never married $\wedge$ $Income5 = 0$ $\wedge$ $Work89$ = Yes $\rightarrow Age < 60$
$Marital$ = Never married $\wedge$ $Rlabor$ = Civilian employee, at work $\rightarrow$
$Age < 65$
$Marital$ = Never married $\wedge$ $Rlabor$ = Civilian employee, at work $\wedge$
$Income5 \in [0.0; 2500.0] \rightarrow Age < 60$
$Marital$ = Never married $\wedge$ $Disable2$ = No, not prevented from working $\wedge$ $Work89$ = Yes $\rightarrow Age < 65$

Income1-related rules:

$Hour89 < 35 \rightarrow Income1 < 28,000$
$Week89 < 40.0 \rightarrow Income1 < 40,000$
$Relat1$ = Son/daughter of the householder $\rightarrow Income1 < 40,000$
$Relat1$ = Grandson/granddaughter of the householder $\rightarrow Income1 < 33,000$
$Relat1$ = Persons in group quarters $\rightarrow Income1 < 35,000$
$Relat1$ = Other nonrelative of the householder $\rightarrow Income1 < 47,000$
$Relat1$ = Other relative of the householder $\rightarrow Income1 < 45,000$
$Relat1$ = Householder $\wedge$ $Hour89 < 35 \rightarrow Income1 < 35,000$
$Disabl2$ = Yes, limited in kind or amount of work $\rightarrow Income1 < 50,000$
$Rlabor$ = Institutionalized persons $\rightarrow Income1 < 30,000$
$Occupclass$ = Service $\rightarrow Income1 < 40,000$
$Occupclass$ = Farming $\rightarrow Income1 < 55,000$
$Class$ = Employee of private for profit company $\wedge$
$Yearsch$ = High school diploma or less $\rightarrow Income1 < 55,000$
$Relat1$ = Husband/wife $\wedge$ $Yearsch$ = High school diploma or less $\rightarrow$
$Income1 < 40,000$

Some of the rules presented above seem intuitive or common sense. One example of such rules are those that have income on the RHS and *Hour*89 (usual hours worked per week in 1989) and *Week*89 (weeks worked in 1989) on the LHS. These two variables are positively correlated with income. These rules, in essence, describe part-time workers in the previous year. Therefore, the rules suggests lower top-codes for income for these individuals compared to the rest of the population.

Another example of rules that are intuitive are the rules that involve *Relat*1 (relationship of the respondent to the householder) on the LHS of the rules. For instance, when *Relat*1 = *son/daughter*, then the threshold for the *Income*1 and *Age* may be lower comparative to other groups of individuals. According to the documentation on 1990 Census data files [19], in most cases, a householder is the person, or one of the persons, in whose name the home is owned, being bought, or rented. Higher income respondents may be expected to be householders themselves, rather than living with a parent-householder, which may be one of the reasons for lower income and possibly younger age for these types of respondents. Similar reasoning may be applied to the rules that involve other relatives of the householder and their respective top-codes. Note, that in some (possibly rare) instances when several members of the family can be linked together, the advanced age of the son may allow the intruder to get a good estimate of the age of a parent, despite the fact that the age of the parent was top-coded. For example, if the age of a son of the householder is 75 years old (which is above the threshold limit in the corresponding rule above), and the age of a parent-householder is 95 years old, then univariate top-coding, at 87 years old will only apply to the householder, but not to the son. However, based on the age of the son, the intruder would know that the parent must be older than the threshold value of 87 years old, and most likely around 95 years old. Such an extreme age and such a rare combination (if present in the data) could lead to the re-identification of these individuals if univariate top-coding is used. Presence of other variables could improve the assessment of the intruder even further.

Rules that include a combination of the following three characteristics: marital status = "Never married" combined with zero or small values of social security income in the previous year (variable *Income*5), no disability, and worked during the previous year (*Work*89 = *yes*) for the most part describe a younger group of respondents in this data set; thus, the 99-th percentile of age for this group of individuals found by the rules is generally smaller than for the rest of the population.

Another characteristic that is related to income is the occupation of the respondent (*Occupclass* variable). The rules identified some occupation classes with lower values of *Income*1 in this data set. For example, *Occupclass* = *Service* which includes cooks, waiters and waitresses, housekeepers, cleaners, maids and housemen, hairdressers, welfare service aides and some others, has a lower 99-th percentile of income than the others, which agrees with the literature on the subject [17]. Also, according to the rules *Occupclass* = *Farmers* has a lower 99-th percentile of *Income*1 as well.

As expected, rules that included the variable $Yearsch$, educational attainment, indicated that if educational attainment is less than high school, then $Income1$ is limited, especially for certain categories of individuals in the data set.

We conclude this section by emphasizing that the focus of the paper is not the discussion and analysis of particular rules, but the development and description of the methodology to obtain such rules. Deeper analysis of the rules obtained by our procedure should be done by the data protector and subject area specialist for each particular data set and the scenario of data release.

4 Concluding Remarks and Future Work

In this paper we propose a new approach for multivariate top-coding for disclosure limitation in large databases with many attributes of different types. We outlined an automated procedure that can help the data protector to find sub-populations that may need their own top-codes, lower than the rest of the population. Such a procedure may be used as an aid for the data collecting organizations in the disclosure review process as an alternative, or in addition, to their regular procedures. Such procedures often involve identification of risky combinations of the variables, which is often based on intuition as well as knowledge of a particular data set. In big data sets these procedures may be complicated and computationally involved as they require computation of many tabulations to identify potentially rare/risky combinations of the categories of these attributes. Thus, an automated procedure to identify such cases can be helpful especially when the data protector intends to release data sets with many attributes of different types, such as big government surveys.

To reduce the complexity of the problem we outlined a two-step approach which consists first of clustering the variables around the top-coded variables, using squared canonical correlations, then running our association rule mining algorithm on a vertical partition of the data that consist of the variables that are in the same cluster with the top-coded variables. This two-step approach makes association rule mining and the subsequent work with the rules by subject area specialists computationally feasible.

We would like to note that the association rules found by the proposed approach are meant to bring to the data protector's attention particular combinations of the attributes that are rarely associated with the extreme values of the numerical variable that is subject to protection. Data protectors can choose top-coding or some other technique for protection of these groups of individuals. For example, synthesis can be used to impute safer values of numerical attributes.

Our future work consists of finding efficient ways for further reduction of the number of association rules. Another direction of future research is to investigate possible ways of incorporation of the data protector's preferences and knowledge in the algorithm. For example, certain individual characteristics are more visible or noticeable than the others; for instance, amputations/missing limbs, walking aids and some others. So, we will investigate the best way of weighting the variables/characteristics on the clustering step and the association rule mining algorithm as well.

Acknowledgments. The findings and conclusions in this paper are those of the authors and do not necessarily represent the official position of the Centers for Disease Control and Prevention. The first author would like to thank Ellen Galantucci from the Bureau of Labor Statistics for the helpful discussion on the content of Sect. 3. Also we would like to thank John Pleis from the National Center for Health Statistics for the careful review of the paper.

A Appendix. Variables in the Census Data Set Mentioned in the Paper

Class - Class of worker. Categories: 0 N/a, Unemployed who never worked. 1 Employee of a private for profit company. 2 Employee of a private not for profit company. 3 Local government employee. City, county, etc. 4 State government employee. 5 Federal government employee. 6 Self employed in own not incorporated business. 7 Self employed in own incorporated business. 8 Working without pay in family business or farm. 9 Unemployed, last worked in 1984 or earlier.

IndustryClass - Industry class. Categories: 1 Agriculture. 2 Mining. 3 Manufacturing. 4 Transportation. 5 Wholesale trade. 6 Retail trade. 7 Finance. 8 Business. 9 Personal services. 10 Entertainment. 11 Professional. 12 Public administration.

Occupclass - occupation class. Categories: 1 Managerial. 2 Professional. 3 Technical. 4 Service. 5 Farming. 6 Precision. 7 Operators. 8 Military.

*Relat*1 - Relationship to the householder. Categories: 0 Householder. 1 Husband/wife 2 Son/daughter 3 Stepson/stepdaughter 4 Brother/sister 5 Father/mother 6 Grandchild 7 Other relative 8 Roomer/boarder/foster child 9 Housemate/roommate 10 Unmarried partner 11 Other non related. 12 Institutionalized person. 13 Other person in group quarters.

*Disable*1 - Work limitation. Categories: 0 N/a. 1 Yes, Limited in kind or amount of work. 2 No, not Limited.

Rlabor - Employment status. Categories: 0 N/a 1 Civilian employee, at work. 2 Civilian employee, with a job but not at work. 3 Unemployed. 4 Armed forces, at work. 5 Armed forces, with a job but not at work. 6 Not in labor force.

*Hour*89 - Usual hours worked per week the year before the interview. This is a numerical variable with range from 0 to 99.

*Week*89 - Weeks worked the year before the interview. This is a numerical variable with range from 0 to 52.

Yearsch - educational attainment. Categories: 0 N/a. 1 No school completed. 2 Nursery school. 3 Kindergarten. 4 1st, 2nd, 3rd, or 4th grade. 5 5th, 6th, 7th, or 8th grade. 6 9th grade. 7 10th grade. 8 11th grade. 9 12th grade, No diploma. 10 High school graduate, diploma or GED. 11 Some College, But no degree. 12 Associate degree in College, Occupational. 13 Associate degree in College, Academic Program. 14 Bachelors degree. 15 Masters degree. 16 Professional degree. 17 Doctorate degree.

References

1. ACS: American community survey. United States Census Bureau. https://www.census.gov/programs-surveys/acs

2. Agrawal, R., Imieliński, T., Swami, A.: Mining association rules between sets of items in large databases. In: Proceedings of the 1993 ACM SIGMOD International Conference on Management of Data - SIGMOD 1993, pp. 207–216, June 1993
3. Agrawal, R., Srikant, R.: Fast algorithms for mining association rules. In: Proceedings of the 20th International Conference on Very Large Data Bases, VLDB, Santiago, Chile, pp. 487–499, September 1994
4. BRFSS: Behavioral risk factor surveillance system. Centers for Disease Control and Prevention (CDC). https://www.cdc.gov/brfss/index.html
5. Census: US census (1990) data set. UCI Machine Learning Repository (2017). https://archive.ics.uci.edu/ml/datasets/US+Census+Data+%281990%29
6. Chavent, M., Kuentz-Simonet, V., Liquet, B., Saracco, J.: ClustOfVar: an R package for the clustering of variables. J. Stat. Softw. **50**(i13), 1–16 (2012)
7. CPS: Current population survey. United States Census Bureau. https://www.census.gov/programs-surveys/cps.html
8. Dheeru, D., Karra Taniskidou, E.: UCI machine learning repository. University of California, Irvine, School of Information and Computer Sciences (2017). http://archive.ics.uci.edu/ml
9. Fukuda, T., Morimoto, Y., Morishita, S., Tokuyama, T.: Mining optimized association rules for numeric attributes. In: Proceedings of the 15th ACM SIGACTSIGMOD - SIGART PODS96, pp. 182–191. ACM Press (1996)
10. Hájek, P., Havránek, T.: Mechanizing Hypothesis Formation: Mathematical Foundations for a General Theory. Springer, Heidelberg (1978). https://doi.org/10.1007/978-3-642-66943-9
11. Han, J.: Mining frequent patterns without candidate generation. In: Proceedings of the 2000 ACM SIGMOD International Conference on Management of Data. SIGMOD 2000, pp. 1–12 (2000)
12. Hundepool, A., et al.: Handbook on Statistical Disclosure Control (version 1.2). ESSNET, SDC project (2010). http://neon.vb.cbs.nl/casc
13. Hundepool, A., et al.: Statistical Disclosure Control. Wiley, Hoboken (2012)
14. NHANES: National Health and Nutrition Examination Survey. Centers for Disease Control and Prevention (CDC). National Center for Health Statistics (NCHS). https://www.cdc.gov/nchs/data/factsheets/factsheet_nhanes.htm
15. NHIS: National Health Interview Survey. Centers for Disease Control and Prevention (CDC). National Center for Health Statistics (NCHS). https://www.cdc.gov/nchs/nhis/index.htm
16. Oganian, A., Iacob, I., Lesaja, G.: Grouping of variables to facilitate SDL methods in multivariate data sets. In: Domingo-Ferrer, J., Montes, F. (eds.) PSD 2018. LNCS, vol. 11126, pp. 187–199. Springer, Cham (2018). https://doi.org/10.1007/978-3-319-99771-1_13
17. Ross, M., Bateman, N.: Meet the low-wage workforce. Technical report, Brookings (2019)
18. Salleb-Aouissi, A., Vrain, C., Nortet, C., Xiangrong Kong, X., Vivek Rathod, V., Cassard, D.: QuantMiner for mining quantitative association rules. J. Mach. Learn. Res. **14**(61), 3153–3157 (2013). http://jmlr.org/papers/v14/salleb-aouissi13a.html
19. U.S. Department of Commerce Economics and Statistics Administration. BUREAU OF THE CENSUS: 1990 Census of Population and Housing. Public Use Microdata Samples. United States (1990)
20. Zaki, M.J.: Scalable algorithms for association mining. IEEE Trans. Knowl. Data Eng. **12**(3), 372390 (2000)

Protection of Statistical Tables

Calculation of Risk Probabilities for the Cell Key Method

Tobias Enderle, Sarah Giessing(✉), and Reinhard Tent

Federal Statistical Office of Germany, 65180 Wiesbaden, Germany
{Tobias.Enderle,Sarah.Giessing,Reinhard.Tent}@destatis.de

Abstract. When deciding on the parameters of the Cell Key method (c.f. [2, 3, 5, 8, 10]), agencies should take into account disclosure risk issues connected to candidate parametrizations. The present paper offers suggestions for analytical calculation of certain risk probabilities using candidate noise parameters for both, the case of tables of counts, as well as for the case of magnitude tables.

1 Introduction

The cell key method (CKM) for statistical disclosure limitation by random noise is a well-known, and recently widely discussed though not yet very widely used post-tabular perturbative disclosure control method. In this paper we therefore assume some familiarity with the basic CKM concept. It is implemented for example in the package τ–Argus and as separate R package cellKey, c.f. [8]. Both packages rely on the R-package *p*table [2] to compute random distributions by maximizing entropy [4, 7].

When deciding on the parameters of the Cell Key method (CKM), agencies cannot focus on information loss alone – they also have to take into account disclosure risk issues connected to candidate parametrizations. One such issue is the conditional probability for a particular perturbed cell value to be indeed identical or very close to the original cell value which – even though intruders may not be aware of the issue – should not be too high, for sake of sufficient protection. After recalling some basic principles of the CKM implementation of [8] in Sect. 3 we explain how to compute such conditional probabilities for a candidate set of parameters using Bayes' theorem for both cases: counts and magnitudes of continuous data. Section 4 considers a very special form of a differencing attack in the context of published means of a continuous variable when the underlying frequencies are protected by noise. Because after CKM protection interior cells of a table often do not add up exactly to the perturbed margins there is a general risk of disclosure by differencing connected to the method. While [1] suggests rigorous randomized computation of feasibility intervals to compare risk avoidance potentials of different parametrizations, the present paper takes a different approach: we study how to analytically compute probabilities of typical differencing risks using the candidate noise parameters for the case of tables of counts (Sect. 5) and of magnitudes (Sect. 6).

J. Domingo-Ferrer and K. Muralidhar (Eds.): PSD 2020, LNCS 12276, pp. 151–165, 2020.
https://doi.org/10.1007/978-3-030-57521-2_11

2 Recalling Principles of Noise Design and Lookup

The R-package *p*table supports the noise design of the implementation of CKM for frequency and magnitude tables in τ–Argus and cellKey. It computes perturbation tables (the "p-tables") used by τ–Argus and cellKey to obtain a perturbation value v in a so called "lookup" using the cell key of a table cell.

When generated to protect *tables of counts*, perturbation tables define conditional probabilities $p_{ij} = P$(perturbed cell value is j | original cell value is i). As suggested in [3] and [4], the noise distributions resulting from the algorithm implemented in the *p*table package ensure that the perturbations will have zero mean, take integer values, not produce negative cell values and that the following criteria hold:

(1) the perturbations have a constant variance σ^2;
(2) the absolute value of any perturbation is less than a specified integer value D.

For *tabulations of continuous (magnitude)* variables, the principle of a constant noise variance will usually not be suitable: it will lead to either not enough protection for table cells with large contributions, or to too much protection (e.g. high information loss) for table cells with small contributions. Building on ideas of [10], τ–Argus and cellKey computes the noise by adding the sum of *topK* random variables $\hat{X}_j (j = 1, .., topK)$, where each component $\hat{X}_j$ has a fixed conditional distribution, i.e. conditional on the (weighted) top-*k*th contribution $w_j y_j (= x_j)$ to a table cell with ordered contributions ($|x_1| \geq |x_2| \geq \ldots$) as proposed in [5]. In the present paper we focus on the simplest case, $topK = 1$, only and do not consider an extra provision of [5] to ensure a minimum noise variance also for x_1 tending to zero.

To summarize, in this paper we assume a perturbed value for original value $x = \sum_{i=1}^{n} w_i y_i$ to be computed according to ((2.2) in [5]) as

$$\hat{x} := x + x_\delta \cdot V, \text{ with } x_\delta = x_1 m(x_1) \text{ for } x_1 > z_f, \text{ and } x_\delta = \sigma_1 x_1, \text{ for } x \leq z_f, \text{ and } V \text{ a random variable defining the noise.} \quad (1)$$

Moreover, in this paper we generally assume $x_\delta \leq x$. This allows us to ignore the special technique implemented to avoid changes of sign due to the perturbation, and to assume realizations of V to be $v \in \left\{-D, \frac{1}{l} - D, \frac{2}{l} - D, \ldots, D - \frac{1}{l}, D\right\}$, obtained by a "lookup" in a suitable perturbation table (the "p-table" supplied by the *p*table package for the case of continuous data), using the cell key of x. The reciprocal of the "step-width", l, is a user defined number, like 2, 8, 10, 100. The p-table defines, e.g., probabilities $(p_{D,k})_{k=0,1,\ldots,2 \cdot l \cdot D}$ for an original value $x \geq D$ to be perturbed by $\frac{k}{l} - D$.

The so called flex-function $m(x_1)$ computes the coefficient $m(x_1)$ of $x_1 \cdot V$ (c.f. (1)) as a decreasing function of x_1 (c.f. [5], (2.2)), with $m(x_1) \approx \sigma_0$ for larger arguments, and $m(z_f) = \sigma_1$ at the "flex-point" $z_f \geq 0$. Parameters σ_0 and σ_1 are to be defined such that $0 < \sigma_0 \leq \sigma_1$.

3 Probability for Original Values Too Close to the Perturbed Ones

For any given set of observed and perturbed frequencies x and $\hat{x}$, the disseminator can easily compute the probability for a particular perturbed frequency to be indeed identical or very close to the original frequency. Denoting for example $B := \{\hat{x} = 1\}$ the event of a perturbed frequency to be 1 and $A := \{x = 1\}$ the event that the original frequency in a given population is indeed 1, according to Bayes' theorem we have $P(A|B) = \frac{P(B|A) \cdot P(A)}{P(B)}$, hence in our case: $P(x = 1|\hat{x} = 1) = \frac{1}{P(\hat{x}=1)} P(\hat{x} = 1|x = 1) \cdot P(x = 1)$.

Using the conditional probabilities p_{i1} = P(perturbed cell value is 1 | original cell value is i) defined in a perturbation table, we compute $P(\hat{x} = 1)$ as $\sum_{i=1,\dots,D} p_{i1} P(x = i)$, to derive $P(x = 1|\hat{x} = 1) = \frac{p_{11} \cdot P(x=1)}{\sum_{i=1,\dots,D} p_{i1} P(x=i)}$. Obviously, the probability of the perturbed count of 1 to actually relate to a true count of 1 can be quite high, if, compared to p_{11}, the probabilities p_{i1} are small for $i \neq 1$, or if, compared to the frequency of 1's there are very few observations $x = i$; $i = 2, \dots, D$ in the set, i.e. if $P(x = 1)$ is much larger than $P(x = i)$; for $i = 2, \dots, D$.

This kind of risk assessment is also possible in the case of the generalization of the cell key method to the case of a continuous variable, like for example the size of a dwelling. For the assessment, we consider only the most critical scenario of observations relating to only a single contribution, i.e. the case $x = x_1$. The aim is to compute for a given set of original and perturbed observations x and $\hat{x}$ the probability for a particular perturbed observation to be indeed identical or very close to the original observation. Denote $B := \{\hat{x} = y\}$ the event of a perturbed observation to be y and $D_\varepsilon := \{y - \varepsilon \leq x \leq y + \varepsilon\}$ the event that the original observation x is within a small range of $\pm\varepsilon$ of this perturbed observation.

According to Sect. 2, (c.f. (1)), the perturbed observation $\hat{x}$ is computed as $\hat{x} = x + x_\delta \cdot V$, with $x_\delta = x_1 m_1(x_1)$, and assuming $x_\delta \leq x$, and V a random variable defining the noise. We focus here on the case $x_\delta = x_1 m_1(x_1)$. For ease of illustration we simplify further, assuming for m_1 a linear function, i.e. $m_1(x_1) = m \cdot x_1$. Because in our critical scenario $x = x_1$ we can write $\hat{x} = x + m \cdot x \cdot v = x \cdot (1 + m \cdot v)$.

Allowing for some rounding in the publication of $\hat{x}$, like e.g. to integers, we consider $\hat{x} = y$ to be "matched" by $x \cdot (1 + m \cdot v)$, if $y - 0.5 \leq x \cdot (1 + m \cdot v) < y + 0.5$, for a realization $v \in \{-D, \frac{1}{l} - D, \frac{2}{l} - D, \dots, D - \frac{1}{l}, D\}$.

Appendix A.1 provides an illustrative example for an instance with $y = 50$, $l = 2$, $D = 3$, noise probability distribution $(p_{D,k})_{k=0,1,\dots,2Dl}$ as displayed in Fig. 1, and two different factors m defining the strength of the perturbation. Using Bayes' theorem, Appendix A.1 shows how to use these intervals to construct a partition of B which can be used to compute $P(D_\varepsilon|B)$ as ratio of two weighted sums of the noise distribution probabilities $p_{D,k}$, where the weights are probabilities of partition events corresponding to $v = \frac{k}{l} - D$. Using formula (A.4) from the appendix, and assuming for x a uniform distribution, with a strong perturbation parameter $m = 0.25$ we obtain a low probability $P(\{48.5 \leq x \leq 51.5\}|\hat{x} = 50) = 18\%$ for the risk of an original value x to be within ± 1.5 of the perturbed value $\hat{x} = 50$, whereas with a weak parameter $m = 0.015$ we end up with rather high probability 94%.

4 An Attacker Strategy to Disclose Original Frequencies Using Published Means

In the following example we assume the agency to publish noisy counts along with original means of a magnitude variable, like e.g. rent per m^2. The illustrative example presented in Table 1 displays original as well as noisy counts, along with the original magnitudes and means.

Table 1. Illustrative Example Mean rent per m^2

	Area A	Area B	Total
No. of dwellings (orig)	8	12	20
No. of dwellings (pertubed)	9	14	19
Sum of rent per m^2	90	95	185
Mean of rent per m^2	11.25	7.9167	9.25

▒ not to be published

Again, we assume the worst case: the intruder knows the maximum deviation D. In this example, let $D = 2$. Knowing that the original counts for areas A and B sum up to the Total, the intruder can easily generate all possible combinations of counts (six in this instance), listed in the first three columns of Table 2. In the next step, the intruder estimates the respective sums of rent per m^2 (c.f. Table 1 – not meant to be published by the agency), by multiplying the frequencies from the six combinations with the respective published mean rent per m^2 figures. Actually, for the total of the two areas A and B, there will be two estimates, one directly obtained using the candidate counts for the total of the two areas, and a second one which results from summing the estimates for area A and area B. Obviously, for the correct candidate combination of possible counts (in the example: $8 + 12 = 20$), the difference of the two estimates will be zero (apart from eventual minor effect due to rounded publication of the mean figures). For the other candidate combinations those differences tend to be non-zero, as can be seen in the last column of Table 2.

Therefore, when an agency releases perturbed counts along with unperturbed means on a general basis, for an attacker interested in disclosing the original counts, it might be a promising strategy to systematically implement the approach explained here, considering the candidate count combination with the smallest difference in the two different magnitude estimates as disclosed correct combination of original counts.

Even when the summed magnitudes themselves in the enumerator of mean figures to be released along with the respective perturbed frequencies are not meant to be published, it is therefore maybe advisable for the disseminator, to anyway perturb those summed magnitudes and compute the released means using the perturbed magnitudes.

Notably, when following this strategy, rounding to multiples of a fixed rounding base b would be a rather ill-advised technique for perturbing the magnitudes. As pointed out in [9], of course now an attacker following the approach above will find the difference

Table 2. Candidate count combinations and estimates for summed rent per m^2

Candidate counts			Area estimates for summed rent per m^2		Total area estimates for summed rent per m^2		Difference: Total-(A+B)
A	B	Tot.	A	B	A+B	Total	
7	12	19	78.75	95	173.75	175.75	2
7	13	20	78.75	102.917	181.667	185	3.333
8	12	20	90	95	185	185	0
7	14	21	78.75	110.833	189.583	194.25	4.667
8	13	21	90	102.917	192.917	194.25	1.333
9	12	21	101.25	95	196.25	194.25	-2

between the two estimates to be usually non-zero then also for the correct original count combination. However, she may realize that unlike in case of the other combinations, for the correct combination the difference tends to be a multiple of *b*.

Perturbation by random noise, on the other hand, can solve the problem. Appendix A.2 presents design and results of a little simulation study to establish how much perturbation is needed for this purpose. A first observation of the study is that in contrast to rounding of the magnitudes prior to calculation of the means (as briefly mentioned above) rounding of the original mean is indeed an efficient strategy to reduce the risk: rounding to 3 decimals for example reduces the attacker success rate from 97.5% to 88% in the most risky constellation considered in the study. Noise, even with very low $\sigma = 0.001$, applied to the enumerator of the mean, can – in combination with rounding of the mean to 3-decimals – reduce the success rate to below 70, which should be enough to discourage potential intruders.

5 Differencing Risk Probabilities for Tables of Counts

As mentioned in the introduction, there is always a certain risk of disclosure by differencing connected to any non-additive SDC method like cell key based noise. When deciding on the noise parameters, one should therefore take into account the potential of a candidate set of parameters to avoid such risks.

A typical instance of a disclosable constellation is the following:

Assume a maximum deviation of $D = 2$ to be known to the intruder who is also informed that zero counts are never changed into non-zero counts. Assume a table row consisting of two original counts of 1 (and original margin of 2). Assume both 1's remain unperturbed, and the margin 2 is perturbed to 0. The intruder can then conclude that the margin must be at least 2 because none of the 1's could have been a zero. On the other hand, the margin can be at most 2, because the perturbed value (0) plus D is 2. Therefore, the original count of the margin must be 2. If any of the interior cells would be 2 or more, this would lead to a margin larger than 2, hence the original interior counts must both be 1.

So, if we publish a table row which consists of only two entries which are both original counts of 1, it can be disclosed, if both 1 s remain unchanged, and the margin is perturbed into 0. The probability of this event is given by $p_{11} \cdot p_{11} \cdot p_{20}$, where p_{ij} are the probabilities defined in the p-table for an original count of i to change into a perturbed count of j. So, the risk of this case to occur in a published set of tables is $P(\{1, 1\}) \cdot p_{11} \cdot p_{11} \cdot p_{20}$; $P(\{1, 1\})$ denoting here the – table, or publication dependent – probability of a row with only two interior cells, both with an original entry of 1, to occur in the tables.

This instance leads us to the following strategy for a (partly) analytical disclosure risk benchmark: First, identify the most typical constellations C_1, $C_{2,\ldots}$ of interior cell combinations that can be disclosed through computation of feasibility intervals, if interior cells and margins are perturbed in a particular way[1]. In the second step, compute for each constellation its probability $\pi(\{C_k\})$ using the respective probabilities of the p-table under consideration. In the third step, analyze every table relation (row, column,...) of the test table set used for the benchmarking and establish, if it is an instance of one of the typical constellations C_1, $C_{2,\ldots}$. Notably, such categorization can usually be based on the following information on a table relation: number of interior cells, number of zeros, and number of 1's, 2's,..., up to a certain count (depending on the p-table under consideration). This categorization of table relations leads to empirical estimates of the probabilities $P(C_k)$. Finally, compute the risk indicator for a proposed p-table as $\sum_k P(C_k)\pi(C_k)$.

5.1 Test Results

In order to demonstrate to some degree the efficiency of the strategy explained above in comparison to the alternative, discussed for example in [1], of randomized feasibility interval (c.f. [6], 4.3.1) computation, we systematically generated pseudo test instances of table relations (i.e. single relation tables), all belonging to either of two typical disclosable constellations C_1 and C_2, varying the number of (up to 15) non-zero internal cells (and their values) in the relations. This way we ended up with thirty constellations C_{1k} and C_{2k}; $1 \leq k \leq 15$, nineteen test instances relating to every C_{2k} and one instance for every C_{1k}. While the probabilities for the C_{2k} only depend on probabilities in the tails of the noise distribution, for the C_{1k} they strongly depend on the probability of original 1 s not being perturbed to 0.

In the next step, we generated n perturbed versions of each table, drawing cell keys for the respective instance n times. Finally we computed the feasibility intervals for the n perturbed versions of each instance, using four different candidate parameter sets, considering an instance as disclosed, if upper and lower bound of the feasibility interval coincide[2]. With $n = 100$ only those instances were disclosed (sometimes even in more than one perturbed version) which relate to $C_{.k}$ with $\pi(C_{.k})$ at least ca. 0.3%. Actually all those instances relate to constellations of the C_{1k} type.

[1] Out of sensitivity concerns we refrain from any further description of such constellations here.

[2] Due to the construction of the instances, if the bounds coincide for one cell of an instance, they always coincide for all cells of the instance.

Table 3 reports for those instances for the two parameter sets with higher risk the frequency of disclosed perturbed versions along with the probability $\pi(C_{1k})$ of the event.

Table 3. Disclosed perturbed versions of test tables for constellations C_{1k}

k	Parameter set 1		Parameter set 2	
	Disclosed table versions	$\pi(C_{1k})$	Disclosed table versions	$\pi(C_{1k})$
7	1	0.008	-	0.002
6	1	0.010	1	0.003
5	2	0.013	1	0.005
4	2	0.017	2	0.008
3	4	0.023	3	0.013
2	2	0.015	4	0.021

For the third parameter set, we observed only two disclosed instances, both relating to $C_{12}, \pi(C_{12}) = 0.024$ for this parameter set. For the fourth parameter set, none of the test table versions was disclosed. For this set, the largest of the probabilities was $\pi(C_{12}) = 0.0009$. The respective probabilities for the (never disclosed) C_{2k} constellations range from $C_{22} = 0.006$ (for the third parameter set) to zero, quickly decreasing for all parameter sets with increasing k.

Obviously, the theoretical probabilities $\pi(C_{.k})$ offer a more direct comparison of the risk avoidance potential of different parametrizations as compared to randomized feasibility interval computation.

6 Differencing Risk Probabilities for Tables of Continuous Data

Similar to the frequency counts, also in the continuous data case there are typical constellations where a row (or column) of a table can be disclosed exactly, when an intruder who knows the noise coefficient x_δ (c.f. (1) in Sect. 2) for all cells, and is aware of the maximum deviation parameter[3], computes the feasibility intervals for original interior and margin cell values using the noisy cell values, as illustrated by the following example:

Assume two internal cells x_1, x_2 with margin cell $x_1 + x_2 = x$. Table 4 shows for all three cells, in line 1: the random noise v assumed to result from the p-table lookup in a p-table with maximum deviation D; in line 2: the upper bound ub computed as $\hat{x} + x_\delta D = x + x_\delta(v + D)$ for the margin cell; in line 3: the lower bound lb computed as $\widehat{x_j} - x_{\delta_j} D = x_j + x_{\delta_j}(v - D)$ for the interior cells. The shaded cell in line 3 shows

[3] This may seem to be a silly scenario, because our super knowledgeable intruder has all the information (e.g. on the maximum contributions) he is supposed trying to gain in his attempt of breaking protection in advance of the attack. Still, for the sake of benchmarking risk avoidance potentials of different parameter settings it is a useful scenario. Otherwise results of the benchmark would depend largely on the parameters used to model the attacker knowledge, hampering their interpretability.

the lower bound for the margin computed as $lb(x_1) + lb(x_2)$. The shaded cells in line 2 present *ub* for the interior cells x_j computed as $ub(x) - lb(x_i)(i \neq j)$. Line 4 presents the resulting width of the interval $ub - lb$.

Table 4. Example: upper and lower attacker bounds in a highly risky scenario

	x	x_1	x_2
v	$-D$	D	$(D-\varepsilon)$ *)
ub	x	$x - (x_2 - x_{\delta_2}\varepsilon) = x_1 + x_{\delta_2}\varepsilon$	$x - x_1 = x_2$
lb	$x_1 + x_2 - x_{\delta_2}\varepsilon$ $= x - x_{\delta_2}\varepsilon$	x_1	$x_2 - x_{\delta_2}\varepsilon$
$ub - lb$	$x - (x - x_{\delta_2}\varepsilon)$ $= x_{\delta_2}\varepsilon$	$x_1 + x_{\delta_2}\varepsilon - x_1 = x_{\delta_2}\varepsilon$	$x_2 - (x_2 - x_{\delta_2}\varepsilon)$ $= x_{\delta_2}\varepsilon$

*) $\varepsilon = k/l, 0 \leq k < 2 \cdot l \cdot D$

The example is constructed in such a way that the noise of internal cells and margin is the maximum possible amount (D), but in opposite direction for internal cells and margin, and with a small deviation of ε from D for the second internal cell. The resulting differences $ub - lb$, i.e. the widths of the feasibility intervals, show that the true observations are exactly disclosed in this example, if $\varepsilon = 0$. Otherwise, the width of the interval for x_1 relative to its largest observation x_{1_1} is $x_{\delta_2}\varepsilon / x_{1_1} = \varepsilon \cdot m(x_{2_1}) \cdot x_{2_1} / x_{1_1}$. This means it depends on the ratio of the largest observation x_{2_1} in the second interior cell to that of the first interior cell. Hence x_{1_1} would be disclosed almost exactly, if x_{2_1} is very small compared to x_{1_1}. Assuming that $m(x)$ is defined in such a way that sufficient protection is provided, if

$$ub - lb\big(\text{here} : \varepsilon \cdot m(x_{2_1}) \cdot x_{2_1}\big) \geq \big(m(x_{1_1}) \cdot x_{1_1}\big)/l, \tag{2}$$

e.g. if $x_{\delta_2}\varepsilon \geq x_{\delta_1}/l$, there is a risk of disclosure for x_{1_1}, if $\varepsilon \leq x_{\delta_1}/(l \cdot x_{\delta_2})$, or (substituting ε by k/l) if $1 \leq k \leq \text{floor}(x_{\delta_1}/x_{\delta_2})$. Obviously, if $x_{\delta_1}/x_{\delta_2} > 2 \cdot l \cdot D$ then the probability of this event is 1. If, on the other hand, $x_{\delta_1}/x_{\delta_2} < 1$, then it is given by the probability π_{2Dl} for $v = D$ (which is equivalent to $\varepsilon = 0$). In the other cases it can be computed by cumulating those probabilities $p_{D,k}$ in the p-table for perturbation values v between D and $D - \text{floor}(x_{\delta_1}/x_{\delta_2})/l$, i.e. by the cumulated probability $\phi_{\text{floor}(x_{\delta_1}/x_{\delta_2})} := \sum_{j=0,\ldots,\text{floor}(x_{\delta_1}/x_{\delta_2})} p_{D,2Dl-j}$.

We could now compute for every element of a test set of pairs of two interior cells used for the benchmarking of different candidate parametrizations $f := \text{floor}(x_{\delta_1}/x_{\delta_2})$, obtain this way empirical probabilities p_f from the test set which can be used to obtain a risk indicator $R := \sum_{f=0,\ldots,2Dl} p_f \cdot \phi_f$. Notably, before using the indicator for comparing different parameter sets, the indicator has to be multiplied by the probability $p_{D,2Dl} \cdot p_{D,0}$ of the constellation, to take into account the general probability of the setting, i.e. that the larger of the two cells is perturbed by D (i.e. $k = 0$) and the sum of the two cells by $D - 2 \cdot D \cdot l/l = -D$. Because $p_{D,2Dl} \cdot p_{D,0}$ may differ between the different candidate parametrizations.

6.1 Illustrative Example

We demonstrate the above approach by comparing a weaker and a stronger set of parameters, each one in combination with three different, standardized (i.e. variance of 1) p-tables with maximum deviations of $D = 1, 3$ and 5, and step-width $1/l = 1/2$. In the weaker setting, the flex function used to compute the noise CV parameters was designed with parameter σ_0 for large values of 0.01, and 0.1 in the stronger setting.

The line on top of Table 5 presents the data used in the example, i.e. the largest observations for the two interior cells. The example keeps the largest observation x_{1_1} for the first interior cell x_1 at 200, but varies the largest observation x_{2_1} for the second interior cell. Below this line, the upper half of the table refers to the stronger parameter setting, the lower half to the weaker setting. Col. 2 of the table displays the noise CV parameters σ_1 for smaller values (below $z_f = 120$ in the weaker setting, and below $z_f = 12$ in the strong setting). For both settings, there are lines in the table showing the noise coefficients x_{δ_2} for the four variants of x_{2_1} along with the noise coefficients x_{δ_1} of x_{1_1} computed according to (1) in Sect. 2 with parameters of the flex-function just mentioned, and floor($x_{\delta_1}/x_{\delta_2}$). The last lines in the upper and in the lower half of the table present the disclosure risk probabilities $\cdot\phi_f$, along with the factor $p_{D,2Dl} \cdot p_{D,0}$, and finally the weighted risk indicator R computed as mean of the four $\cdot\phi_f$, weighted by the constellation probability $p_{D,2Dl} \cdot p_{D,0}$.

Table 5. Example: Disclosure risk probabilities for a typical, disclosable constellation

		x_{2_1}				x_{1_1}	
		100	60	40	20	200	
Setting:	x_{δ_j}	11.8	8.7	7.4	6.5	21	
	f	1	2	2	3	-	
σ_0=0.1 σ_1=0.5	**D**	$\phi_{\text{floor}(x_{\delta_1}/x_{\delta_2})}$				$p_{D,2Dl}\, p_{D,0}$	**R**
	1	0.4	0.6	0.6	0.8	0.04	0.024
	3	0.01141	0.03897	0.03897	0.10433	5.46E-06	2.64E-07
	5	0.00001	0.00008	0.00008	0.00051	5.48E-13	9.20E-17
Setting:	x_{δ_j}	5	3	2	1	6.5	
	f	1	2	3	6	-	
σ_0=0.01 σ_1=0.05	**D**	$\phi_{\text{floor}(x_{\delta_1}/x_{\delta_2})}$				$p_{D,2Dl}\, p_{D,0}$	**R**
	1	0.4	0.6	0.8	1	0.04	0.028
	3	0.01141	0.03897	0.10432	0.59920	5.46E-06	1.03E-06
	5	0.00001	0.00008	0.00051	0.03849	5.48E-13	5.35E-15

The results clearly show that the risk for x_{1_1} of being not sufficiently protected, in the high risk constellation considered here, depends in the first place on the maximum deviation parameter D. This is not surprising – after all our definition (2) of sufficient protection is based on the flex-function parameter, assuming they are defined properly. Still, comparing the risk indicator relating to same D, but different noise CV parameter setting, the results show a slight tendency for higher risk connected to the weaker parameter set.

Considering the order of magnitude of the risk probabilities especially at $D = 3$, and $D = 5$, it becomes clear that observing such cases of too small feasibility intervals through randomized feasibility interval computation is like finding a needle in a hay-stack.

7 Summary and Conclusions

This paper has investigated how to use the parameters defining the noise applied by CKM to protect tables of counts and magnitudes to compute indicators for several types of disclosure risk: Sect. 3 suggests an approach to calculate the probability for a true observation to be (too) close to the published perturbed observation – the disseminator should of course select the parameters in such a way that this probability is not excessively high! In Sect. 4 we observe that perturbation of the magnitudes in the enumerator of means published along with CKM protected counts is well suited to reduce attacker success rates of a scenario targeted to this special case even with rather weak parameter choice.

The main contribution of the paper consists of concepts proposed in Sects. 5 and 6 for analytical calculation of typical differencing risk probabilities to compare the risk avoidance potential of candidate noise parameters. The proposed concepts include suggestions for weighted indicators: risk probabilities relating to certain constellations shall be weighted by the empirical probability of the respective constellation to occur in a given test data set. As demonstrated by examples presented in those sections, for rather risk averse parametrizations the order of magnitude of the risk probabilities is extremely low. Thus, observing such cases through randomized feasibility interval computation as proposed in [1] is like finding a needle in a hay-stack, hampering interpretability of results. The analytical risk probabilities allow clear and direct comparison even of risk averse candidate parametrizations.

However, in the present paper, we have only considered differencing risks of univariate tables, taking into the scenario only one relation between interior and margin cell. In multivariate tables those relations are not independent. Moreover, the approach of Sect. 6 concerning tables of continuous data is restricted to relations with only two non-zero cells. If and how it could be extended to the case of more non-zero cells in a relation is not yet clear and has to be studied in future work. This holds as well for the effects of relation interdependency in multiway tables on risk probabilities.

Appendix

A.1 How to Use $p_{D,k}$ to Compute the Probability of $(y - \varepsilon \leq x \leq y + \varepsilon | \hat{x} = y)$ Applying Bayes' Theorem?

As pointed out in Sect. 3, we assume $\hat{x} = y$ to be "matched" by $x \cdot (1 + m \cdot v)$, if $y - 0.5 \leq x \cdot (1 + m \cdot v) < y + 0.5$ for $v \in \left\{-D, \frac{1}{l} - D, \frac{2}{l} - D, \ldots, D - \frac{1}{l}, D\right\}$.

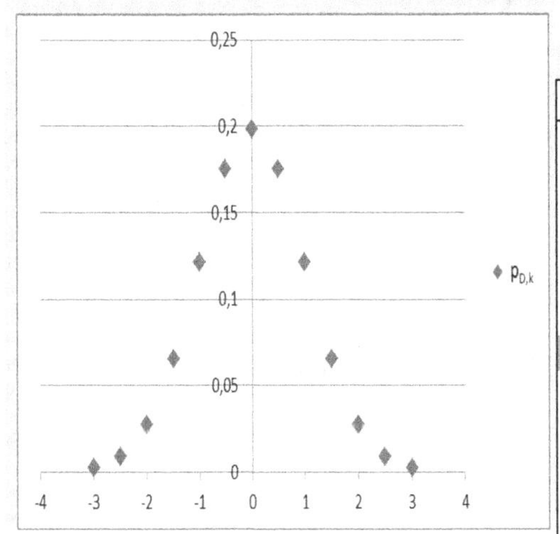

Fig. 1. Noise distribution probabilities $(p_{D,k})_{k=0,1,\ldots,12}$ of the example

Table 6. Interval bounds of $\left\{z_k^L \le x < z_k^U\right\}$ corresponding to $v = \frac{k}{l} - D$

m = 0.25		m = 0.015	
z_k^L	z_k^U	z_k^L	z_k^U
198,0	202,0	51,8	52,9
132,0	134,7	51,4	52,5
99,0	101,0	51,0	52,1
79,2	80,8	50,6	51,7
66,0	67,3	50,3	51,3
56,6	57,7	49,9	50,9
49,5	50,5	49,5	50,5
44,0	44,9	49,1	50,1
39,6	40,4	48,8	49,8
36,0	36,7	48,4	49,4
33,0	33,7	48,1	49,0
30,5	31,1	47,7	48,7
28,3	28,9	47,4	48,3

Now let $z_k^L := \frac{y-0.5}{\left(1+m\cdot\left(\frac{k}{l}-D\right)\right)}$, $z_k^U := \frac{y+0.5}{\left(1+m\cdot\left(\frac{k}{l}-D\right)\right)}$, and $E_k := \left\{z_k^L \le x < z_k^U\right\}$. Then $(E_k)_{k=1,\ldots,2\cdot l\cdot D}$ is a sequence of intervals where for $x \in E_k$ perturbation $v = \frac{k}{l} - D$ results in $\hat{x} = y$ being matched. Figure 1 and Table 6 provide an illustrative example for an instance with $y = 50$, $l = 2$, $D = 3$, two different factors m defining the strength of the perturbation, and Table 6 presenting the interval boundaries z_k^L, z_k^U for those two settings.

For the stronger perturbation with $m = 0.25$ the intervals E_k do not overlap. But for the weaker perturbation with $m = 0.015$, they do. Here, for Bayes' theorem to apply, we need a suitable partition $\left\{C_j\right\}_{j\in J}$. It can be constructed from the list of interval bounds $\{z_m^L, z_n^U\}_{m,n\in\{0,\ldots,2Dl\}}$: after sorting this list, and removal of possibly tied entries we get a sorted list $\{c_s\}_{s=1,\ldots,I}$, and construct $\left\{C_j\right\}_{j\in J}$ as $\{[c_1; c_2), [c_2; c_3), \ldots, [c_{I-1}; c_I)\}$. $\left\{C_j\right\}_{j\in J}$ is a partition of the union $\{E_k\}_{k=0,\ldots,2Dl}$ such that either $C_j \subseteq E_k$ or $C_j \cap E_k = \emptyset$, $\bigcup_{j\in J} C_j = \{E_k\}_{k=0,\ldots,2Dl}$, $C_{j_1} \cap C_{j_2} = \emptyset$ for $j_1 \ne j_2$.

For illustration, consider the entries of Table 6 corresponding to $m = 0.015$ and $k = 1$: As there are no ties in the list of interval bounds, $I = 2 \cdot (2Dl + 1) = 26$, and $J = \{1, \ldots, 25\}$. The first 6 elements in $\{c_s\}_{s=1,\ldots,26}$ for example are 47.4, 47.7, 48.1, 48.3, 48.4, and 48.7, providing the interval bounds of C_1 to C_5. The union of C_1 to C_3 is E_{12}, and E_{11} is the union of C_2 to C_5.

Recalling the event B of an original observation x being perturbed into $\hat{x}$ where $y - 0.5 \le \hat{x} < y + 0.5$, conditional on $x \in C_j$, the probability of B is defined by the probabilities $p_{D,k}$ specified by the perturbation table: For C_j, as well as for any subset $S(C_j)$ of C_j, we have

$$P\big(B|x \in S(C_j)\big) = \sum\nolimits_{\substack{\{k=0,\ldots,2Dl;\\ S(C_j)\cap E_k \neq \emptyset\}}} p_{D,k}. \tag{A.1}$$

Recalling now the event D_ε defined in Sect. 3 of an original observation x to be within a (too) small range of $\pm\varepsilon$ of the perturbed observation, of course also $\left\{(C_j \backslash D_\varepsilon) \cup (C_j \cap D_\varepsilon)\right\}_{j\in J}$ defines a non-overlapping partition of $\{E_k\}_{k=0,\ldots,2Dl}$. and of course $C_j \backslash D_\varepsilon$ and $C_j \cap D_\varepsilon$ are subsets of C_j and hence subsets of (one or more) of the intervals E_k.

According to the law of total probability, we can write $P(B) = \sum\limits_{j\in J}\big(P(B|C_j\backslash D_\varepsilon)P(C_j\backslash D_\varepsilon) + P(B|C_j \cap D_\varepsilon)P(C_j \cap D_\varepsilon)\big)$. Using (A.1) with $C_j \backslash D_\varepsilon$ and $C_j \cap D_\varepsilon$ as subsets $S(C_j)$, and after some rearrangement of summation terms we get

$$P(B) = T_1 + T_2 = T_3, \tag{A.2}$$

$$\text{where } T_1 := \sum\nolimits_{\{k=0,\ldots,2Dl\}} p_{D,k}\left(\sum\nolimits_{\{j\in J;\, C_j\backslash D_\varepsilon \subseteq E_k\}} P\big(C_j \backslash D_\varepsilon\big)\right),$$

$$T_2 := \sum\nolimits_{\{k=0,\ldots,2Dl\}} p_{D,k}\left(\sum\nolimits_{\{j\in J;\, C_j\cap D_\varepsilon \subseteq E_k\}} P\big(C_j \cap D_\varepsilon\big)\right),$$

$$\text{and } T_3 := \sum\nolimits_{\{k=0,\ldots,2Dl\}} p_{D,k}\left(\sum\nolimits_{\{j\in J;\, C_j \subseteq E_k\}} P\big(C_j\big)\right).$$

Of course $P(D_\varepsilon \cap C_j|B) = \frac{1}{P(B)} P(B|D_\varepsilon \cap C_j)P\big(D_\varepsilon \cap C_j\big)$. Now, using (A.1), we compute

$$P\big(D_\varepsilon \cap C_j|B)\big) = \frac{1}{P(B)} \sum\nolimits_{\substack{\{k=0,\ldots,2Dl;\\ C_j\cap D_\varepsilon \subseteq E_k\}}} p_{D,k}\big(P\big(C_j \cap D_\varepsilon\big)\big) \text{ for } j \in J \tag{A.3}$$

By construction, only observations in $\left\{C_j\right\}_{j\in J}$ can be perturbed into $\hat{x}$, hence $P(D_\varepsilon \backslash \left\{C_j\right\}_{j\in J}|B) = 0$. So we have $P(D_\varepsilon|B) = P(D_\varepsilon \cap \left\{C_j\right\}_{j\in J}\}|B) = P(\left\{D_\varepsilon \cap C_j\right\}_{j\in J}|B) = \sum\limits_{j\in J} P(D_\varepsilon \cap C_j|B)$. Because of (A.3) it follows $P(D_\varepsilon|B) = \frac{1}{P(B)} \sum_{j\in J} \sum_{\substack{\{k=0,\ldots,2Dl;\\ C_j\cap D_\varepsilon \subseteq E_k\}}} p_{D,k} \cdot P\big(C_j \cap D_\varepsilon\big)$, and thus $P(D_\varepsilon|B) = \frac{1}{P(B)} \sum_{\{k=0,\ldots,2Dl\}} p_{D,k}\Big(\sum_{\{j\in J;\, C_j\cap D_\varepsilon\subseteq E_k\}} P\big(C_j \cap D_\varepsilon\big)\Big)$. Because of (A.2) it follows

$$P(D_\varepsilon|B) = \frac{T_2}{P(B)} = \frac{T_2}{T_1 + T_2} = \frac{T_2}{T_3} \tag{A.4}$$

The probabilities $P\big(C_j\big)$ in the term T_3 and $P\big(C_j \cap D_\varepsilon\big)$ in the term T_2 depend on the distribution of the original magnitudes x. For the remainder of the example, we assume

x to be uniform distributed on some interval $[\alpha, \beta]$ covering all the C_j. Then $P(C_j)$ and $P(C_j \cap D_\varepsilon)$ can be computed as $\frac{r-s}{\beta-\alpha}$; r, s denoting interval bounds of a C_j or $C_j \cap D_\varepsilon$. The term $\beta - \alpha$ cancels out in the computation of $(D_\varepsilon|B)$ as T_2 divided by T_3.

In our instance, we find T_3 to be $1.1/(\beta - \alpha)$ in the case of $m = 0.25$ and $1.0/(\beta - \alpha)$ in the case $m = 0.015$. The shaded cells in Fig. 1 indicate which of the boundaries z_k^U, z_k^L are contained in the critical interval defined by $D_{1.5} = \{48.5 \le x \le 51.5\}$. For the stronger perturbation with $m = 0.25$, this is the case only for $k = 6$ which is the center of the distribution where the perturbation $\frac{k}{l} - D$ is zero, meaning that here only zero perturbation leads to a perturbed value within $D_{1.5}$. Using (A.4), with this strong perturbation parameter we obtain a low probability $P(\{48.5 \le x \le 51.5\}|\hat{x} = 50) = \frac{p_{D,6}(50.5-49.5)}{1.1} = 18\%$ for the risk of an original value x to be within ± 1.5 of the perturbed value $\hat{x} = 50$, whereas with the weak parameter $m = 0.015$ we end up with rather high probability 94%.

A.2. Simulation Study to Estimate Attacker Success Rates for the Scenario of Sect. 4, Before and After Noise

In the study, we consider random noise generated by truncated normal distributions V with $\mu = 0$, and several combinations of standard deviation σ and truncation (e.g., maximum deviation) parameter D.

While the illustrative example of Tables 1 and 2 of Sect. 4 is based on the instance of counts and means evaluated for a variable with two categories (the two areas A and B), the study also considers instances with categories $j = 2, \ldots, n_c$; $n_c \in \{2, 3, 4\}$. For each combination of n_c and parameters σ and D, we randomly generated 1000 times a test with counts between 2 and 100. Using a two stage mechanism, in the first stage we drew for each category j a count f_j from uniform distribution $U([2, 100])$. In the second stage, for each category j, f_j observations were drawn independently from $U([1, 75])$ (*viz.* realizations of a variable "age" truncated at 75). Denoting x_j the sum of the f_j observations, and $x_{j,Top1}$ the largest of them, $x := x_1 + \ldots + x_{n_c}, f := f_1 + \ldots + f_{n_c}$ and $x_{Top1} = \max\{x_{j,Top1}; j = 1, \ldots n_c\}$, perturbed means were computed as $\hat{\bar{x}}_j := \bar{x}_j + \frac{x_{j,Top1}}{f_j} \cdot V$ and $\hat{\bar{x}} := \bar{x} + \frac{x_{Top1}}{f} \cdot V$. The frequencies f_i and f were perturbed using a p-table with maximum deviation of 2.

After generating the instances, the study ran the attack illustrated by Tables 1 and 2 for every instance for all of the 1000 cases, i.e. establishing for every case candidate original counts and then candidate original means by multiplication of perturbed means with candidate counts, obtaining this way two different estimates for the original observation total x, as described in Sect. 4.

A first observation of the study is that in contrast to rounding of the magnitudes prior to calculation of the means (as briefly mentioned above), rounding of the original mean is indeed an efficient strategy to reduce the risk: rounding to 3 decimals for example reduces the attacker success rate from 97.5% to 89% in the most risky constellation, i.e. the case with $n_c = 2$ categories only. A second observation is that – after rounding to 3 decimals – the success rate decreases with the number of categories n_c of the variable: while for $n_c = 2$ the success rate is 89%, for $n_c = 3$ it is 62%, and for $n_c = 4$ it is only about 20%.

Table 7 below presents success rates of the first strategy after rounding to 3 decimals and perturbation by random noise with different noise variance and maximum deviation. Obviously, success rates shrink with rising noise variation, and maximum perturbation. Even in the $n_c = 2$ categories case, where the intruder is still quite successful even when true rounded means are published, noise with $D = 3$ and very low $\sigma = 0.001$ reduces the success rate to below 70.

Table 7. Attacker success rate (in %) after 3-decimals rounding of means and noise perturbation (of the mean enumerator)

n_c	no noise	$D = 1$			$D = 3$		
		Noise std. dev. σ			Noise std. dev. σ		
		0.001	0.01	0.05	0.001	0.01	0.05
2	89	78	14	2	68	6	1
3	62	48	32	17	43	19	8
4	22	12	0	0	5	0	0

References

1. Enderle, T., Giessing, S., Tent, R.: Taking into account disclosure risk in the selection of parameters for the cell key method. In: Domingo-Ferrer, J., Montes, F. (eds.) Privacy in Statistical Databases, LNCS, pp. 28–42, vol. 11126. Springer (2018a)
2. Enderle, T., Giessing, S.: Implementation of a 'p-table generator' as separate R-package. In: Deliverable D3.2 of Work Package 3 "Prototypical implementation of the cell key/seed method" within the Specific Grant Agreement "Open Source tools for perturbative confidentiality methods" (2018b)
3. Fraser, B., Wooton, J.: A proposed method for confidentialising tabular output to protect against differencing. In: Monographs of Official Statistics. Work session on Statistical Data Confidentiality, Eurostat-Office for Official Publications of the European Communities, Luxembourg, pp. 299–302 (2006)
4. Giessing, S.: Computational issues in the design of transition probabilities and disclosure risk estimation for additive noise. In: Domingo-Ferrer, J., Pejić-Bach, M. (eds.) PSD 2016. LNCS, vol. 9867, pp. 237–251. Springer, Cham (2016). https://doi.org/10.1007/978-3-319-45381-1_18
5. Giessing, S., Tent, R.: Concepts for generalising tools implementing the cell key method to the case of continuous variables. In: Paper Presented at the Joint UNECE/Eurostat Work Session on Statistical Data Confidentiality (The Hague, 29-31 October 2019) (2019). http://www.unece.org/fileadmin/DAM/stats/documents/ece/ces/ge.46/2019/mtg1/SDC2019_S2_Germany_Giessing_Tent_AD.pdf
6. Hundepool, A., et al.: Statistical Disclosure Control. Wiley, Chichester (2012)
7. Marley, J.K., Leaver, V.L.: A method for confidentialising user-defined tables: Statistical properties and a risk-utility analysis. In: Proceedings of 58th World Statistical Congress, pp. 1072–1081 (2011)

8. Meindl, B., Kowaric, A., De Wolf, P.P.: Prototype implementation of the cell key method, including test results and a description on the use for census data. In: Deliverable D3.1 of Work Package 3 "Prototypical implementation of the cell key/seed method" within the Specific Grant Agreement "Open Source tools for perturbative confidentiality methods" (2018)
9. Ramic, A.A.: Internal Communication (2019)
10. Thompson, G., Broadfoot, S., Elazar, D.: Methodology for the automatic confidentialisation of statistical outputs from remote servers at the Australian bureau of statistics. In: Paper Presented at the Joint UNECE/Eurostat Work Session on Statistical Data Confidentiality (Ottawa, 28–30 Oktober 2013) (2013). http://www.unece.org/fileadmin/DAM/stats/documents/ece/ces/ge.46/2013/Topic_1_ABS.pdf

On Different Formulations of a Continuous CTA Model

Goran Lesaja[1,2](✉), Ionut Iacob[1], and Anna Oganian[3]

[1] Department of Mathematical Sciences, Georgia Southern University, 65 Georgia Avenue, Statesboro, GA 30460-8093, USA
{goran,ieiacob}@georgiasouthern.edu
[2] Department of Mathematics, US Naval Academy, 121 Blake Road, Annapolis, MD 21402-1300, USA
[3] National Center for Health Statistics, 3311 Toledo Road, Hyattsville, MD 20782, USA
aoganyan@cdc.gov

Abstract. In this paper, we consider a Controlled Tabular Adjustment (CTA) model for statistical disclosure limitation of tabular data. The goal of the CTA model is to find the closest safe (masked) table to the original table that contains sensitive information. The measure of closeness is usually measured using ℓ_1 or ℓ_2 norm. However, in the norm-based CTA model, there is no control of how well the statistical properties of the data in the original table are preserved in the masked table. Hence, we propose a different criterion of "closeness" between the masked and original table which attempts to minimally change certain statistics used in the analysis of the table. The Chi-square statistic is among the most utilized measures for the analysis of data in two-dimensional tables. Hence, we propose a *Chi-square* CTA model which minimizes the objective function that depends on the difference of the Chi-square statistics of the original and masked table. The model is non-linear and non-convex and therefore harder to solve which prompted us to also consider a modification of this model which can be transformed into a linear programming model that can be solved more efficiently. We present numerical results for the two-dimensional table illustrating our novel approach and providing a comparison with norm-based CTA models.

Keywords: Statistical disclosure limitation · Controlled tabular adjustment models · Linear and non-linear optimization · Interior-point methods · Chi-square statistic

1 Introduction

Minimum-distance controlled tabular adjustment (CTA) methodology for tabular data was first introduced in [7,15]. It is one of the effective statistical disclosure limitation (SDL) methods for the protection of sensitive information in tabular data. An overview of SDL theory and methods can be found in the monograph [17], and for tabular data only, in the survey [8].

J. Domingo-Ferrer and K. Muralidhar (Eds.): PSD 2020, LNCS 12276, pp. 166–179, 2020.
https://doi.org/10.1007/978-3-030-57521-2_12

CTA problem can be formulated as follows: given a table with sensitive cells, compute the "closest" additive safe (masked) table to the original table ensuring that adjusted (masked) values of all sensitive cells are safely away from their original value and that adjusted values are within a certain range of the real values. The additivity of the masked table means in most cases the requirement that the sum of cell values in each row and column of the table remains the same in the original and masked table, [17,19].

In the standard formulation of the CTA model, the closeness of the original and masked table is measured by the weighted distance between the tables with respect to a certain norm. Most commonly used norms are ℓ_1 and ℓ_2 norms. Thus, the problem can be formulated as a minimization problem with the objective function being a particular weighted distance function and constraints being derived from the requirements stated above.

In general, the CTA problem is a Mixed Integer Optimization Problem (MICOP) which is a difficult problem to solve especially for large dimension tables. The MICOP CTA problem involves binary variables that characterize sensitive cells and their values indicate whether the value of the sensitive cell is adjusted upward or downward. Apriori fixing the values of binary variables reduces the problem to the continuous optimization problem which is easier to solve, however, the quality of the solution may be reduced since we are no longer searching for the global optimal solution but its approximation. In addition, the values of the binary variables have to be assigned carefully otherwise the problem may become infeasible. Some strategies to fix the binary variables while preserving the feasibility of the problem were discussed in [11,12]. In the paper we assume that the binary variables are fixed upfront according to one of the strategies presented in these papers, hence we consider continuous CTA models.

As indicated above, the objective function in the continuous CTA model is based on either the ℓ_1-norm or ℓ_2-norm. The formulation of ℓ_2-CTA model leads to the Quadratic Programming (QP) problem, while ℓ_1-CTA model can be formulated as the Linear Programming (LP) problem and, as a Second-Order Cone (SOC) problem, which has recently been proposed in [19].

However, in the standard norm-based CTA model, there is no control of how well the statistical properties of the data in the table are preserved. The numerical experiments summarized in Table 2 in Sect. 2.2 suggest that there is no pattern which would indicate that one CTA model consistently produces the values of the Chi-square statistic, or other statistics, of the masked and original table that are closer to each other than for any other model.

This observation motivated us to consider different criteria of "closeness" between masked and original table which attempts to minimally change certain statistical properties of the table. For example, the Chi-square statistic is an important statistical measure often used to analyze tabular data [10]. Hence, we propose, what we call *Chi-square* CTA model, which minimizes the objective function that depends on the difference of Chi-square statistics of the original and masked table. The Chi-square CTA model is smooth, non-linear, and non-convex which makes it harder to solve the problem. This motivated us to also consider a modification of this model, called *Chi-linear* CTA which can be transformed into a LP problem that can be solved more efficiently.

The Chi-square and Chi-linear CTA models are applied to the two-dimensional table used previously in the literature [5,19] as a two-dimensional test table to compare solutions of different CTA models. Chi-square and Chi-linear CTA models for this table are solved using interior-point methods (IPMs) and compared with results obtained in [19] when norm-based CTAs models were applied to the same table. The Chi-square statistic, Cramer V, and Chi-linear measures were calculated for the original table and for masked tables and compared to illustrate the validity of our approach.

The paper is organized as follows. In Sect. 2 the norm-based CTA models are outlined. In Sect. 3 a novel continuous Chi-square CTA model is presented, as well as its modification, Chi-linear CTA model, and the transformation to LP problem is derived. Section 4 contains numerical results of applying Chi-square, Chi-linear, and norm-based CTA models to the two-dimensional table. The concluding remarks and possible directions for future research are given in Sect. 5.

2 Preliminaries

2.1 Norm-Based CTA Models

In this section, we review the standard norm-based CTA model as it is presented in [19]. Given the following set of parameters:

(i) A set of cells $a_i, i \in \mathcal{N} = \{1, \ldots, n\}$. The vector $a = (a_1, \ldots, a_n)^T$ satisfies certain linear system $Aa = b$ where $A \in \mathbb{R}^{m \times n}$ is an $m \times n$ matrix and and $b \in \mathbb{R}^m$ is m-vector.
(ii) A lower, and upper bound for each cell, $l_{a_i} \leq a_i \leq u_{a_i}$ for $i \in \mathcal{N}$, which are considered known by any attacker.
(iii) A set of indices of sensitive cells, $\mathcal{S} = \{i_1, i_2, \ldots, i_s\} \subseteq \mathcal{N}$.
(iv) A lower and upper protection level for each sensitive cell $i \in \mathcal{S}$ respectively, lpl_i and upl_i, such that the released values must be outside of the interval $(a_i - lpl_i,\ a_i + upl_i)$.
(v) A set of weights, $w_i, i \in \mathcal{N}$ used in measuring the deviation of the released data values from the original data values.

A standard CTA problem is a problem of finding values $z_i,\ i \in \mathcal{N}$, to be released, such that $z_i,\ i \in \mathcal{S}$ are safe values and the weighted distance between released values z_i and original values a_i, denoted as $\|z - a\|_{l(w)}$, is minimized, which leads to solving the following optimization problem

$$
\begin{aligned}
\min_{z}\ & \|z - a\|_{l(w)} \\
s.t.\ & Az = b, \\
& l_{a_i} \leq z_i \leq u_{a_i},\ i \in \mathcal{N}, \\
& z_i,\ i \in \mathcal{S} \text{ are safe values.}
\end{aligned}
\tag{1}
$$

As indicated in the assumption (iv) above, safe values are the values that satisfy

$$z_i \leq a_i - lpl_i \text{ or } z_i \geq a_i + upl_i,\ i \in \mathcal{S}. \tag{2}$$

By introducing a vector of binary variables $y \in \{0,1\}^s$ the constraint (2) can be written as

$$\begin{aligned} z_i &\geq -M\left(1-y_i\right) + \left(a_i + upl_i\right) y_i, & i \in \mathcal{S}, \\ z_i &\leq M y_i + \left(a_i - lpl_i\right)\left(1-y_i\right), & i \in \mathcal{S}, \end{aligned} \tag{3}$$

where $M \gg 0$ is a large positive number. Constraints (3) enforce the upper safe value if $y_i = 1$ or the lower safe value if $y_i = 0$.

Replacing the last constraint in the CTA model (1) with (3) leads to a mixed-integer convex optimization problem (MICOP) which is, in general, a difficult problem to solve; however, it provides a globally optimal solution [6]. The alternative approach is to fix binary variables upfront which leads to a CTA model which is a continuous convex optimization problem that is easier to solve. It is worth noting that the obtained solution is optimal for the fixed combination of binary variables which is different from the global optimum obtained when solving MICOP CTA problem, however, in most cases, it is quite a good approximation that serves the purpose of protecting sensitive values in the table quite well. It is also important to mention that a wrong assignment of binary variables may result in the problem being infeasible. Strategies on how to avoid this difficulty are discussed in [11,12].

In this paper, we consider a continuous CTA model where binary variables are fixed according to one of the strategies suggested in these papers. Furthermore, vector z is replaced by the vector of *cell deviations*, $x = z - a$.

The CTA (1) model with constraints (3) reduces to the following convex optimization problem:

$$\begin{aligned} \min_x \ & \|x\|_{l(w)} \\ s.t.\ & Ax = 0, \\ & l \leq x \leq u, \end{aligned} \tag{4}$$

where upper and lover bounds for x_i, $i \in \mathcal{N}$ are defined as follows:

$$l_i = \begin{cases} upl_i & \text{if } i \in \mathcal{S} \text{ and } y_i = 1 \\ l_{a_i} - a_i & \text{if } (i \in \mathcal{N} \setminus \mathcal{S}) \text{ or } (i \in \mathcal{S} \text{ and } y_i = 0) \end{cases} \tag{5}$$

$$u_i = \begin{cases} -lpl_i & \text{if } i \in \mathcal{S} \text{ and } y_i = 0 \\ u_{a_i} - a_i & \text{if } (i \in \mathcal{N} \setminus \mathcal{S}) \text{ or } (i \in \mathcal{S} \text{ and } y_i = 1). \end{cases} \tag{6}$$

The two most commonly used norms in problem (4) are the ℓ_1 and ℓ_2 norms. For the ℓ_2-norm the problem, (4) reduces to the following ℓ_2-CTA model which is a QP problem:

$$\begin{aligned} \min_x \ & \sum_{i=1}^{n} w_i x_i^2 \\ s.t.\ & Ax = 0, \\ & l \leq x \leq u. \end{aligned} \tag{7}$$

For the ℓ_1-norm the problem, (4) reduces to the following ℓ_1-CTA model:

$$\begin{aligned} \min_x \; & \sum_{i=1}^{n} w_i \left| x_i \right| \\ s.t. \; & Ax = 0, \\ & l \leq x \leq u. \end{aligned} \tag{8}$$

The above ℓ_1-CTA model (8) is a convex optimization problem; however, the objective function is not differentiable at $x = 0$. Since most of the algorithms require differentiability of the objective function, problem (8) needs to be reformulated. The standard reformulation is the transformation of the model (8) to the following LP model:

$$\begin{aligned} \min_{x^-, x^+} \; & \sum_{i=1}^{n} w_i \left(x_i^+ + x_i^- \right) \\ s.t. \; & A \left(x_i^+ - x_i^- \right) = 0, \\ & l \leq x^+ - x^- \leq u, \end{aligned} \tag{9}$$

where

$$x^+ = \begin{cases} x & \text{if } x \geq 0 \\ 0 & \text{if } x < 0, \end{cases} \qquad x^- = \begin{cases} 0 & \text{if } x > 0 \\ -x & \text{if } x \leq 0, \end{cases} \tag{10}$$

The inequality constraints can further be split into lower and upper bounds constraints for x^+ and x^- separately (see [19]).

Recently, another reformulation of ℓ_1-CTA has been proposed. In [19] it was observed that the absolute value has an obvious second-order cone (SOC) representation

$$t_i = |x_i| \quad \longrightarrow \quad \mathcal{K}_i = \left\{ (x_i, t_i) \in \mathbb{R}^2 \; : \; t_i \geq \sqrt{x_i^2} \right\}$$

which leads to the following SOC formulation of the ℓ_1-CTA (8)

$$\begin{aligned} \min_x \; & \textstyle\sum_{i=1}^{n} w_i t_i \\ s.t. \; & Ax = 0, \\ & (x_i, t_i) \in \mathcal{K}_i; \quad i = 1, \ldots, n, \\ & l \leq x \leq u. \end{aligned} \tag{11}$$

The three CTA models outlined above can be solved using interior-point methods (IPMs). IPMs have been developed in the past three decades and have proven to be very efficient in solving large linear and non-linear optimization problems that were previously hard to solve. Nowadays almost every relevant optimization software, whether commercial or open-source, contains an IPM solver. For more information on IPMs see [20–24] and references therein. Specifically, for conic optimization problems and methods see [3,4,16].

We conclude this section by listing several references where numerical experiments and comparisons of different methods for norm-based CTA models were presented [5,9,13,19].

2.2 Motivation to Consider Different CTA Models

In traditional, norm-based continuous CTA models we are finding the closest safe table to the original table with respect to a certain norm, usually l_2 or l_1 norm. However, in norm-based CTA models, there is no control of how well the statistical properties of the data in the table are preserved. The analysis of the data in the masked table with respect to the original table is usually done after the masked table is produced using a CTA model. One of the most utilized measures of analysis is the Chi-square statistic. For example, in [10] Chi-square and Cramer V statistical measures were used in assessing information loss of the masked table produced by the LP CTA model (9). See also references therein.

The definitions of Chi-square statistic and Cramer V measure are well known, however, we list them below for the sake of completeness.

Chi-square statistic of a table is

$$\chi^2 = \sum_{i=1}^{n} \frac{(0_i - e_i)^2}{e_i}, \tag{12}$$

where o_i is an observed cell value and e_i is an expected cell value.

Cramer's V statistical measure is derived from Chi-square statistic

$$V = \sqrt{\frac{\chi^2}{n(r-1)(c-1)}} \tag{13}$$

where r is a number of rows and c is a number of columns and n is a number of cells in the table, i.e. $n = rc$.

An absolute value of the differences instead of a square of the differences as in (12) can also be considered. We call this measure a *Chi-linear* measure.

$$\chi_{abs} = \sum_{i=1}^{n} \frac{|0_i - e_i|}{\sqrt{e_i}}, \tag{14}$$

We performed numerical experiments on a set of randomly generated tables of different dimensions and different numbers of sensitive cells and applied the QP, the LP, and SOC (Conic) CTA models listed above to obtain the masked tables. We used different weights, $w_i = 1/a_i$ and $w_i = 1/e_i$ for QP and the square root of these weights for LP -CTA models. We calculated the Chi-square statistic, and Cramer V and Chi-linear measures for each masked table and for the original table. The summary of the results is presented in Table 2 in the Appendix.

The review of the results in Table 2 leads to the following observations:

- There is no pattern that would indicate that the masked table produced by one of the CTA models consistently exhibits values of the Chi-square, or other statistics, closer to the values of these statistics for the original table than for other models.

- This is consistent with the findings in [10]. The authors compared the original and masked tables generated using only LP CTA model. They observed that Chi-square and Cramer's V measures are affected by the size of the table, the way the cell values are generated, the number of sensitive cells, upper and lower safe values for sensitive cells, etc.
- The conclusion is that standard norm-based CTA models do not guarantee that Chi-square value, or values of other statistics, computed on the masked table will be as close as possible to the corresponding values of the original table given the constraints.

Given these observations, we propose to consider a different measure of "closeness" between masked and original table which attempts to minimize an objective function that depends on the difference between values of statistics of the original and masked table. In the sequel, we specifically focus on designing a CTA model for Chi-square statistic.

3 Chi-Square CTA Model and a Modification

3.1 Chi-Square CTA Model

In this section, we propose a CTA model that we call *Chi-Square* CTA where the minimization of the norm-based objective function in (4) is replaced with the minimization of the absolute value of the differences of values of Chi-square statistic of the masked and original table.

The model is as follows:

$$\begin{aligned} \min & \left|\sum_{i=1}^{n} \frac{(z_i-e_i)^2}{e_i} - \sum_{i=1}^{n} \frac{(a_i-e_i)^2}{e_i}\right| \\ s.t.\ & Ax = 0, \\ & l \le x \le u. \end{aligned} \tag{15}$$

In the model above it seems that there is no connection between variables in the objective function and in the constraints. However, variables z and x are connected as follows $x = z - a$, that is, variable x represents cell deviations. Below, the objective function is transformed in terms of cell deviations x, rather than the original masked values z. We have a similar situation in several other models below.

$$\begin{aligned} f(x) &= \left|\sum_{i=1}^{n} \frac{(z_i-e_i)^2}{e_i} - \sum_{i=1}^{n} \frac{(a_i-e_i)^2}{e_i}\right| \\ &= \left|\sum_{i=1}^{n} \frac{(x_i+a_i-e_i)^2}{e_i} - \sum_{i=1}^{n} \frac{(a_i-e_i)^2}{e_i}\right| \rightarrow \quad d_i := a_i - e_i \\ &= \left|\sum_{i=1}^{n} \frac{(x_i+d_i)^2}{e_i} - \sum_{i=1}^{n} \frac{d_i^2}{e_i}\right| \\ &= \left|\sum_{i=1}^{n} \frac{(x_i+2d_i)x_i}{e_i}\right| \\ &= \left|\sum_{i=1}^{n} \frac{x_i^2+2d_ix_i}{e_i}\right| \end{aligned} \tag{16}$$

The difficulty with the Chi-square CTA model (15) is that it is non-linear, non-convex, and non-smooth. The non-convexity is due to the fact that

$d_i == a_i - e_i$ may b negative for some i. The non-smoothness that is caused by absolute value can be removed by replacing the absolute value with a square of the differences.

$$\begin{aligned} \min & \left[\sum_{i=1}^{n} \frac{(z_i-e_i)^2}{e_i} - \sum_{i=1}^{n} \frac{(a_i-e_i)^2}{e_i}\right]^2 \\ s.t.\ & Ax = 0, \\ & l \leq x \leq u. \end{aligned} \tag{17}$$

Using the same substitutions as in (16) we obtain the following non-linear and non-convex but smooth problem with linear constraints that we call *Chi-square* CTA model.

$$\begin{aligned} \min & \left[\sum_{i=1}^{n} \frac{(x_i+2d_i)x_i}{e_i}\right]^2 \\ s.t.\ & Ax = 0, \\ & l \leq x \leq u. \end{aligned} \tag{18}$$

3.2 Chi-Linear CTA Model

The Chi-square CTA model (18) which is a smooth non-linear and non-convex problem can be solved using an appropriate IPM for non-linear problems. However, the non-linearity and non-convexity of the problem make it harder to solve the problem. In other words, the IPM will be able to handle problems of the smaller size and will perform slower than if it is applied to the LP or QP CTA model. At this point, it is still an open question of whether model (18) can be transformed into a more tractable problem that can be efficiently solved by IPMs. One option is to consider the modification of the Chi-square CTA formulation (18) by minimizing the absolute value of the sum of absolute values of differences (errors) rather than squares of differences as in (18).

The model is as follows:

$$\begin{aligned} \min_z & \left|\sum_{i=1}^{n} \frac{|z_i-e_i|}{\sqrt{e_i}} - \sum_{i=1}^{n} \frac{|a_i-e_i|}{\sqrt{e_i}}\right| \\ s.t.\ & Ax = 0, \\ & l \leq x \leq u, \end{aligned} \tag{19}$$

We call this model *Chi-linear* CTA model. In what follows we will show that this model can be transformed into the LP problem which then can be solved efficiently using IPMs or simplex based algorithms.

The objective function in (19) can be transformed in a similar way as in (16) for Chi-square CTA model (18).

$$\begin{aligned} f(x) &= \left|\sum_{i=1}^{n} \frac{|z_i-e_i|}{\sqrt{e_i}} - \sum_{i=1}^{n} \frac{|a_i-e_i|}{\sqrt{e_i}}\right| && \rightarrow \quad x_i := z_i - a_i \\ &= \left|\sum_{i=1}^{n} \frac{|x_i+a_i-e_i|}{\sqrt{e_i}} - \sum_{i=1}^{n} \frac{|a_i-e_i|}{\sqrt{e_i}}\right| && \rightarrow \quad d_i := a_i - e_i \\ &= \left|\sum_{i=1}^{n} \frac{|x_i+d_i|}{\sqrt{e_i}} - \sum_{i=1}^{n} \frac{|d_i|}{\sqrt{e_i}}\right| && \rightarrow \quad G := \sum_{i=1}^{n} \frac{|d_i|}{\sqrt{e_i}} \\ &= \left|\sum_{i=1}^{n} \frac{|x_i+d_i|}{\sqrt{e_i}} - G\right| && \rightarrow \quad g(x) = \sum_{i=1}^{n} \frac{|x_i+d_i|}{\sqrt{e_i}} \\ &= |g(x) - G| \end{aligned} \tag{20}$$

The Chi-linear CTA model (19) can now be written in the form

$$\begin{aligned} &\min_x |g(x) - G| \\ &s.t.\ Ax = 0, \\ &\quad l \le x \le u, \end{aligned} \tag{21}$$

The transformation of (21) to the LP model is derived below.
The objective function transformation:

$$\begin{aligned} f(x) &= \left|\sum_{i=1}^n \frac{|x_i + d_i|}{\sqrt{e_i}} - G\right| \rightarrow\ y_i = x_i + d_i;\ i = 1, \cdots, n \\ &= \left|\sum_{i=1}^n \frac{|y_i|}{\sqrt{e_i}} - G\right| \quad \rightarrow\ |y_i| = y_i^+ + y_i^-;\ \ y_i^+,\ y_i^- \ge 0;\ \ i = 1, \cdots, n \\ &= \left|\sum_{i=1}^n \frac{y_i^+ + y_i^-}{\sqrt{e_i}} - G\right| \rightarrow\ t = \sum_{i=1}^n \frac{y_i^+ + y_i^-}{\sqrt{e_i}} - G \\ &= |t| \qquad\qquad\qquad \rightarrow\ t = t^+ - t^-,\ \ t^+, t^- \ge 0 \\ &= t^+ + t^- \end{aligned} \tag{22}$$

Equality constraints transformations:

$$\begin{aligned} Ax = 0 \rightarrow\ & A(y - d) = 0 \\ \rightarrow\ & A(y^+ - y^-) = Ad \\ \rightarrow\ & Ay^+ - Ay^- = Ad \end{aligned} \tag{23}$$

It is not hard to show that

$$Ad = 0, \tag{24}$$

hence, we have

$$Ax = 0 \ \rightarrow\ Ay^+ - Ay^- = 0. \tag{25}$$

Inequality constraints transformations:

$$\begin{aligned} l \le x \le u \rightarrow\ & l \le y - d \le u \\ \rightarrow\ & l + d \le y \le u + d \\ \rightarrow\ & l + d \le y^+ - y^- \le u + d \end{aligned} \tag{26}$$

The Chi-linear CTA model (21) transforms now to the following LP problem.

$$\begin{aligned} &\min\ (t^+ + t^-) \\ &s.t.\ Ay^+ - Ay^- = 0, \\ &\quad t^+ - t^- = \sum_{i=1}^n \frac{y_i^+ + y_i^-}{\sqrt{e_i}} - G \\ &\quad l + d \le y^+ - y^- \le u + d \\ &\quad t^+, t^- \ge 0 \\ &\quad y^+, y^- \ge 0, \end{aligned} \tag{27}$$

4 Numerical Results

In this section a two-dimensional table stated in Figure 1 in [19] is considered. The table is listed in Fig. 1 below as Table (a).

This table is used to build five different CTA models, ℓ_1-CTA LP formulation (9), ℓ_1-CTA SOC formulation (11), ℓ_2-CTA QP formulation (7), Chi-linear CTA LP formulation (27), and Chi-square CTA formulation (18). These CTA models are solved using appropriate interior-point methods (IPMs). The first four models were solved using MOSEK solver [1] while the last one is solved using IPOPT solver [2]. The results are listed in Fig. 1.

Original

10$_{(3)}$	15	11	9	45
8	10	12	15	45
10	12	11	**13**$_{(5)}$	46
28	37	34	37	136

(a)

ℓ_1-LP

13	15	11	6	45
10	10	12	13	45
5	12	11	18	46
28	37	34	37	136

(b)

ℓ_1-SOC

13.47	15.26	11.22	5.05	45
8.19	10.43	12.43	13.95	45
6.34	11.31	10.35	18	46
28	37	34	37	136

(c)

ℓ_2-QP

13	15.03	11.03	5.94	45
7.66	11.14	13.14	13.06	45
7.34	10.83	9.83	18	46
28	37	34	37	136

(d)

Chi-linear LP

13.	12.77	11.25	7.98	45
9.26	13.47	11.25	11.02	45
5.74	10.76	11.5	18	46
28	37	34	37	136

(e)

Chi-square

13.	12.13	11.15	8.72	45
8.47	13.68	12.57	10.28	45
6.53	11.19	10.28	18.	46
28	37	34	37	136

(f)

Fig. 1. Masked tables produced by different CTA models for table (a)

In the next Table 1 the values of Chi-square, Cramer V, and Chi-linear statistical measures are listed for the original table and related masked tables produced by five CTA models.

Table 1. Values of the three statistical measures

Tables	Statistical measures		
	Chi-square	Chi-linear	Cramer V
Original	2.89	4.70	0.20
ℓ_2-QP	9.49	8.74	0.36
ℓ_1-LP	10.44	8.49	0.32
ℓ_1-SOC	11.30	9.26	0.39
Chi-Square	**6.81**	7.17	0.31
Chi-Linear	7.38	**6.55**	0.32

From Table 1 we observe that the value of the Chi-square statistic of the masked table produced by the Chi-square CTA model indeed differs the least from the value of the Chi-square statistic of the original table. Similarly, the Chi-linear measure of the masked table produced by the Chi-linear CTA model is the closest to the Chi-linear measure of the original table.

The second observation is about p-values of the Chi-square statistic for the tables listed in Table 1. The p-values for the tables are as follows: original: 0.82, ℓ_2-QP: 0.15, ℓ_1-LP: 0.11, ℓ_1-SOC: 0.08, Chi-square: 0.34, Chi-linear: 0.29. As expected, the p-value of the Chi-square table is the closest to the p-value of the original table. However, the discrepancy between the p-values of the original and masked tables is significant and deserves comment.

The Chi-square statistic is very sensitive to the number of sensitive cells in the table and the level of perturbation needed to get the safe values for these cells [6,10]. The larger the number of sensitive cells and the level of perturbation, in general, the larger the discrepancy between Chi-square statistic values and, consequently, p-values. Therefore, the discrepancy is more due to the requirements for the protection of tabular data and less due to the CTA model used to obtain the masked table which satisfies these requirements. Nevertheless, the new Chi-square CTA model proposed in this paper achieves the p-value of the masked table that is the closest to the p-value of the original table among all other CTA models. On the more general note, this is an illustration of the interplay between maximizing the utility of the masked data while keeping disclosure risk under control which is at the heart of the theory and methods of SDL.

5 Concluding Remarks and Future Work

In this paper, a novel approach to building Continuous CTA models for statistical disclosure limitation of tabular data is discussed. The standard norm-based CTA model finds the closest safe (masked) table to the original table while satisfying additivity equations and safe value inequality constraints, as described in Sect. 2.1. The measure of closeness is usually measured using an ℓ_1 or ℓ_2 norm.

The numerical experiments summarized in Table 2 in Sect. 2.2 suggest that there is no pattern which would indicate that one CTA model consistently produces the values of the Chi-square statistic, or other statistics, of the masked and original table that are closer to each other than for any other model. Hence, we propose a CTA model, which we call *Chi-square* CTA model (18), that produces a masked table with Chi-square statistic closest to Chi-square statistic of the original table.

Given the non-linearity and non-convexity of the Chi-square CTA model, we also consider a modification of this model, a *Chi-linear* CTA model (27) that can be transformed to LP problem, hence allowing IPMs to solve high dimensional tables efficiently. The price to pay is that the closeness between Chi-square statistics of an original and masked table may be affected. Further examination of this topic is the subject of future research.

The goal of the paper is mainly theoretical, that is, to present a novel Chi-square CTA model and its modification, Chi-linear CTA model, as an illustration of a possible new approach in building CTA models which produce masked tables that are the closest to the original table in terms of a certain statistic, rather than in terms of a distance between tables.

The rationale behind the new approach is to consider *Analysis Specific* CTA models. On one hand, they may be more narrow in scope, however, they produce the optimal result for the specific analysis. On the other hand, norm-based CTA models may be wider in scope and produce tables that may have "relatively good" results for multiple different statistical measures, but "really good" (optimal) for none. In addition, we have no explicit control of the quality of results in the norm-based CTA approach.

Directions for future research include more extensive numerical experiments on a larger set of randomly generated two-dimensional tables of different sizes and different numbers of sensitive cells. A more theoretical direction for future research is to examine whether the Chi-square CTA model (18) can be transformed into a more tractable problem that can be efficiently solved by IPMs.

Acknowledgments. Any opinions, findings, and conclusions or recommendations expressed in this publication are those of the authors only and do not necessarily represent the official position of the Centers for Disease Control and Prevention.

A Appendix

Values of Statistical Measures for Different CTA Models

Table 2. Values of statistical measures for different CTA models

Percentage of sensitive cells	Tables	Statistical measures		
		Chi-square	Chi-linear	Cramer V
	Original	1790.69	506.40	0.228821
10	LP	2039.68	536.77	0.244212
	$W(a_i) - LP$	2034.97	518.96	0.243930
	$W(e_i) - LP$	2111.99	540.48	0.248504
	QP	1971.56	519.52	0.240100
	$W(a_i) - QP$	1940.21	512.82	0.238183
	$W(e_i) - QP$	1976.04	519.89	0.240372
	Conic	1954.00	520.86	0.239028
15	LP	2008.59	532.59	0.242344
	$W(a_i) - LP$	1960.27	520.67	0.239411
	$W(e_i) - LP$	2046.51	539.46	0.244621
	QP	2012.38	534.54	0.242573
	$W(a_i) - QP$	1952.36	526.04	0.238928
	$W(e_i) - QP$	2019.59	535.99	0.243007
	Conic	1968.98	525.72	0.239942

(*continued*)

Table 1. (*continued*)

Percentage of sensitive cells	Tables	Statistical measures		
		Chi-square	Chi-linear	Cramer V
20	LP	1950.37	513.28	0.238806
	$W(a_i) - LP$	1922.12	511.77	0.237070
	$W(e_i) - LP$	1993.11	522.35	0.241408
	QP	1949.98	516.88	0.238782
	$W(a_i) - QP$	1881.74	505.13	0.234566
	$W(e_i) - QP$	1947.82	516.88	0.238650
	Conic	1957.35	520.91	0.239233

References

1. Andersen, E.D.: MOSEK solver (2020). https://mosek.com/resources/doc
2. Waechter, A., Laird, C.: IPOPT solver. https://coin-or.github.io/Ipopt/
3. Alizadeh, F., Goldfarb, D.: Second-order cone programming. Math. Program. **95**(1), 3–51 (2003)
4. Andersen, E.D., Roos, C., Terlaky, T.: On implementing a primal-dual interior-point method for conic quadratic optimization. Math. Program. **95**(2), 249–277 (2003)
5. Castro, J.: A CTA model based on the Huber function. In: Domingo-Ferrer, J. (ed.) PSD 2014. LNCS, vol. 8744, pp. 79–88. Springer, Cham (2014). https://doi.org/10.1007/978-3-319-11257-2_7
6. Castro, J.: On assessing the disclosure risk of controlled adjustment methods for statistical tabular data. Int. J. Uncertainty Fuzziness Knowl. Based Syst. **20**, 921–941 (2012)
7. Castro, J.: Minimum-distance controlled perturbation methods for large-scale tabular data protection. Eur. J. Oper. Res. **171**, 39–52 (2006)
8. Castro, J.: Recent advances in optimization techniques for statistical tabular data protection. Eur. J. Oper. Res. **216**, 257–269 (2012)
9. Castro, J., Cuesta, J.: Solving ℓ_1-CTA in 3D tables by an interior-point method for primal block-angular problems. TOP **21**, 25–47 (2013)
10. Castro, J., González, J.A.: Assessing the information loss of controlled adjustment methods in two-way tables. In: Domingo-Ferrer, J. (ed.) PSD 2014. LNCS, vol. 8744, pp. 11–23. Springer, Cham (2014). https://doi.org/10.1007/978-3-319-11257-2_2
11. Castro, J., Gonzalez, J.A.: A fast CTA method without complicating binary decisions. Documents of the Joint UNECE/Eurostat Work Session on Statistical Data Confidentiality, Statistics Canada, Ottawa, pp. 1–7 (2013)
12. Castro, J., Gonzalez, J.A.: A linear optimization based method for data privacy in statistical tabular data. Optim. Methods Softw. **34**, 37–61 (2019)
13. Castro, J., Giessing, S.: Testing variants of minimum distance controlled tabular adjustment. In: Monographs of Official Statistics, Eurostat-Office for Official Publications of the European Communities, Luxembourg, pp. 333–343 (2006)

14. Cox, L.H., Kelly, J.P., Patil, R.: Balancing quality and confidentiality for multivariate tabular data. In: Domingo-Ferrer, J., Torra, V. (eds.) PSD 2004. LNCS, vol. 3050, pp. 87–98. Springer, Heidelberg (2004). https://doi.org/10.1007/978-3-540-25955-8_7
15. Dandekar, R.A., Cox, L.H.: Synthetic tabular data: an alternative to complementary cell suppression. Manuscript, Energy Information Administration, U.S. (2002)
16. Gu, G.: Interior-point methods for symmetric optimization. Ph.D. thesis, TU Delft (2009)
17. Hundepool, A., et al.: Statistical Disclosure Control. Wiley, Chichester (2012)
18. Karr, A.F., Kohnen, C.N., Oganian, A., Reiter, J.P., Sanil, A.P.: A framework for evaluating the utility of data altered to protect confidentiality. Am. Stat. **60**(3), 224–232 (2006)
19. Lesaja, G., Castro, J., Oganian, A.: A second order cone formulation of continuous CTA model. In: Domingo-Ferrer, J., Pejić-Bach, M. (eds.) PSD 2016. LNCS, vol. 9867, pp. 41–53. Springer, Cham (2016). https://doi.org/10.1007/978-3-319-45381-1_4
20. Lesaja, G.: Introducing interior-point methods for introductory operations research courses and/or linear programming courses. Open Oper. Res. J. **3**, 1–12 (2009)
21. Lesaja, G., Roos, C.: Kernel-based interior-point methods for monotone linear complementarity problems over symmetric cones. J. Optim. Theory Appl. **150**(3), 444–474 (2011)
22. Nesterov, Y., Nemirovski, A.: Interior-Point Polynomial Algorithms in Convex Programming. SIAM Studies in Applied Mathematics, vol. 13. SIAM, Philadelphia (1994)
23. Roos, C., Terlaky, T., Vial, J.P.: Theory and Algorithms for Linear Optimization. An Interior-Point Approach. Springer, Heidelberg (2005)
24. Wright, S.J.: Primal-Dual Interior-Point Methods. SIAM, Philadelphia (1996)

Protection of Interactive and Mobility Databases

Privacy Analysis of Query-Set-Size Control

Eyal Nussbaum(✉) and Michael Segal

School of Electrical and Computer Engineering,
Communication Systems Engineering Department,
Ben-Gurion University, 84105 Beer-Sheva, Israel
eyalnus@post.bgu.ac.il

Abstract. Vast amounts of information of all types are collected daily about people by governments, corporations and individuals. The information is collected, for example, when users register to or use on-line applications, receive health related services, use their mobile phones, utilize search engines, or perform common daily activities. As a result, there is an enormous quantity of privately-owned records that describe individuals' finances, interests, activities, and demographics. These records often include sensitive data and may violate the privacy of the users if published. The common approach to safeguarding user information is to limit access to the data by using an authentication and authorization protocol. However, in many cases the publication of user data for statistical analysis and research can be extremely beneficial for both academic and commercial uses, such as statistical research and recommendation systems. To maintain user privacy when such a publication occurs many databases employ anonymization techniques, either on the query results or the data itself. In this paper we examine and analyze the privacy offered for aggregate queries over a data structures representing linear topologies. Additionally, we offer a privacy probability measure, indicating the probability of an attacker to obtain information defined as sensitive by utilizing legitimate queries over such a system.

Keywords: Privacy · Datasets · Anonymity · Vehicular network · Linear topology · Privacy measure

1 Introduction

The problem of privacy-preserving data analysis has a long history spanning multiple disciplines. As electronic data about individuals becomes increasingly detailed, and as technology enables ever more powerful collection and curation of these data, the need increases for a robust, meaningful, and mathematically rigorous definition of privacy, together with a computationally rich class of algorithms that satisfy this definition. A comparative analysis and discussion of such algorithms with regards to statistical databases can be found in [1]. One common practice for publishing such data without violating privacy is applying

J. Domingo-Ferrer and K. Muralidhar (Eds.): PSD 2020, LNCS 12276, pp. 183–194, 2020.
https://doi.org/10.1007/978-3-030-57521-2_13

regulations, policies and guiding principles for the use of the data. Such regulations usually entail data distortion for the sake of anonymizing the data. In recent years, there has been a growing use of anonymization algorithms based on *differential privacy* introduced by Dwork et al. [6]. Differential privacy is a mathematical definition of how to measure the privacy risk of an individual when they participate in a database. While ensuring some level of privacy, these methods still have several issues with regards to implementation and data usability, which we discuss in Sect. 2. Due to these issues and restrictions, other privacy preserving algorithms are still in prevalent in many databases and statistical data querying systems.

In this paper, we analyze the effectiveness of queries using the k-query-set-size limitation over aggregate functions in maintaining individual user privacy. This method is similar in its approach to the k-anonymization technique offered by Sweeney and Samarati et al. [12,14], which states that the information of any individual in released data in indistinguishable from that of at least $k-1$ individuals in the same release. This method was later improved upon by Machanavajjhala et al. [11] adding a l-diversity measure to the anonymization of the data. Li et al. [10] further introduce the $t - closeness$ classification aimed at solving issues that arose with the previous two methods. Srivatsava et al. [8] explore several of these privacy preserving techniques and show their vulnerabilities with regards to what they call "composition attacks" and auxiliary information. In the majority of cases, these methods and the research surrounding them are focused on tabular databases that are "sanitized" (anonymized) and distributed for statistical analysis. In our research we examine a different database structure representing entities distributed throughout a linear topology, and instances where the user has access to specific data queries but not to the entire data. We examine aggregate function queries (minimum/maximum/median values) over a linear topology structure in order to determine the privacy level that can be afforded by the query-set-size control method. We define a privacy metric of "Privacy Probability" for the case of minimum and maximum queries on their own. We also analyze the privacy afforded when the attacker has access to median, minimum and maximum queries over the same data set.

The rest of this paper is organized as follows: we discuss related work in Sect. 2, in Sect. 3 we define the problem outline and models, in Sect. 4 we show the privacy afforded when using only the maximum (or minimum) query and offer a privacy probability measure, and Sect. 5 analyzes the level of privacy that can be attained when the attacker has simultaneous access to the minimum, maximum and median queries. We summarize in Sect. 6.

2 Related Work

Much of the recent research into privacy preservation of published data has been surrounding differential privacy algorithms [5]. To construct a data collection or data querying algorithm which constitutes differential privacy, one must add some level of noise to the collected or returned data respectively. Examples of this method can be found in [2,3,7]. There are however still several issues and limitations to this method. Sarwate and Chaudhuri [13] discuss the challenges of differential privacy with regards to continuous data, as well as the trade-off between privacy and utility. In some cases, the data may become unusable after distortion. Lee and Clifton [9] discuss the difficulty of correctly implementing differential privacy with regards to the choice of ϵ as the differential privacy factor. Zhu et al. [16] detail the issue of querying large data sets with multiple query rounds, compounding the noise added from $Lap(1/\epsilon)$ to $10Lap(10/\epsilon)$ over ten rounds for example. Moreover, even many published differentially private algorithms have been shown to violate this privacy claim [4]. For the aggregate queries used in our model, the amount of noise required for differential privacy would distort the data in a manner that would make it unusable. This is due to the sensitivity (i.e. the maximum influence of a single entry in the data set as defined for differential privacy by Dwork et al [6]) being high. In addition, in our model we do not restrict the user (and the attacker) in the number of queries performed. Such a lack of restriction is known to cause differential privacy to fail, since the noise can be averaged out and filtered from the results. Therefore, our work focuses on data sets utilizing the query-set-size control method which is still used for privacy preservation today. For a similar scenario, Venkatadri et al. [15] recently demonstrated a privacy attack on Facebook users by utilizing an exploit in Facebook's targeted advertising API which used the query-set-size control method to assure user privacy.

3 Research Problem and Models

We attempt to show privacy attacks on data gathered from linear topology system. The gathered data is stored in a centralized database which allows a set of queries that are designed to return meaningful information without compromising the privacy of the users. A privacy attack is defined as access to any information gathered from the user that was not made available from standard queries on the database.

A system is comprised of n unique units distributed in the real world and are displayed on a linear topology G_L as a set of nodes V, distributed along discrete coordinates on the X axis between $-\infty, \infty$, such that each node $v_i, i = (1, 2, 3, \ldots, n)$ represents one unit at a single (discrete) point in time t_j. The timestamps are measured as incremental time steps from the system's initial measurement designated $t_0 = 0$.

For each node v_i at each timestamp, some sensitive attribute is measured. We denote this $s_{i,t}$ with t being a discrete value timestamp. As an example these could be vehicles with measured speeds. However, our analysis and results are not limited to a vehicular network and speed values. The same methods we will be describing can be used for any type of system with a similar node distribution, and on any sensitive value held by these nodes on which the aggregate queries described next can be performed.

A query is performed on the system at a given timestamp τ and over a range R in the topology. The queries allowed are F_{max}, F_{min} and F_{med} which return the maximum, minimum and median values in the given range respectively. The range R is defined by a set of boundaries over the linear topology with a starting coordinate x_{start} and end coordinate x_{end}. As stated, a query will only return a result if at least k nodes reside within the given range. The value k is known to the attacker. In the following sections, we will analyze the privacy levels of different aggregate queries over the structure defined above. The analysis is based on an attacker with legitimate access to a subset of queries from $[F_{min}, F_{max}, F_{med}]$ over a linear topology, with the goal of obtaining the sensitive attribute within a group of less than k entities (nodes) as allowed by the query-set-size limitation. The matching of a sensitive attribute to a single entity (node) constitutes a powerful attack.

4 Privacy Analysis of F_{max} or F_{min} over Random Node Distribution

This section explores the level of privacy afforded to a random node in the system when using the query-set-size control method for query limitation. The query available to the attacker is the F_{max} query (this case is of course symmetrical for the F_{min} query as well). The system will consist of n nodes placed randomly (uniformly distributed) along a linear topology. Each node v_i has its attribute value s_i, and we denote m to be the number of nodes in M where $M = \{v_j \mid s_j > s_i\}$ (i.e. the nodes whose attribute value s is larger than that of v_i). For our analysis, we require that $m \leq n - 2k$. A query result will only be returned if at least k nodes are in the queried range, with $2 < k < \frac{n}{2}$. We consider the privacy of v_i to be maintained if the value s_i cannot be determined by an attacker who holds the results of all possible queries over the data set. In the case of a linear topology, this requires that there exist a set of nodes surrounding v_i, whose attribute values are above s_i, and therefore will mask the value of v_i. We name this set the "*privacy umbrella*" of v_i. The relative locations of nodes which make up this *privacy umbrella* depends on the position of v_i. For any given position of v_i, we look at all possible permutations of placement for the remaining nodes in the topology and analyze which set of permutations constitutes a *privacy umbrella*. Denote this set U_i. We define the privacy level p_i of node v_i to be $\frac{|U_i|}{(n-1)!}$, as $(n-1)!$ is the total number of possible permutations of node positioning, given a static node v_i.

Model: A linear topology G_L with n nodes.
Queries: F_{max} (or F_{min} - symmetric case).
Goal: Find the sensitive attribute of a node v_{target} at a given time τ. In this case, we show the probability of the attacker realizing his goal for any target v_i. There are three scenarios for the position of v_i, each of which have a different set U_i:

1. v_i is one of $k-1$ nodes at one edge of the topology.
2. v_i is the kth node from one edge of the topology.
3. v_i is between the kth and the $(n-k)^{th}$ node in the topology.

We note two trivialities for all scenario proofs:

- Any query over a range with less than k nodes will return no values and is therefore not help the attacker learn any new values of nodes.
- If a query over a range containing v_i has returned some value s which does not belong to v_i, any query containing that range within its range will never return the value s_i belonging to v_i.

Due to these trivialities, it is easy to see that the best strategy for an attacker to receive all possible values is to run all possible k-sized queries over the data set. Our analyses will therefore be based on all possible k-sized queries run over the positioning permutations. In addition, we discovered that in most cases it is simpler to calculate the number of permutations in which a "privacy umbrella" does not exist for a given node position. Therefore, we will present the calculation of number of permutations not in U_i. We denote the set of these permutations W_i and note that $|U_i| = (n-1)! - |W_i|$. We define a "discoverability probability" $q = \frac{W_i}{(n-1)!)}$, and finally realize the privacy measure p as $p = 1 - q$. Following is the analysis of the privacy measure for each scenario.

Scenario 1: v_i is one of $k-1$ nodes at one edge of the topology
We look first to the left edge of the topology. Assume that v_i is the first node in the topology. If one of the following $k-1$ nodes is in M, then the value s_i will never be returned in a query. If the kth node after v_i is the first node in the topology which is in M, then there exists one query that will return s_i (the query containing the first k nodes), however any other query containing v_i will return a value of a node in M and the attacker will not be able to distinguish between the values of the first k nodes (they will only know that s_i belongs to one of them). Therefore, if even one of the first k nodes after v_i is a member of M, the value of v_i is considered private. Now we look at the case where none if these nodes is a member of M. In this case, a query over the first k nodes will return the value s_i. Once the attacker "slides" the k-sized query window by one node, and queries over k nodes starting from the second node in the topology, the value returned will be smaller then s_i. In this manner, the attacker can know for certain that s_i belongs to v_i, and the privacy is lost.

From this, we see that the *privacy umbrella* of v_1 (i.e. the first node in the topology) exists in the set of all permutations where at least one node from M is within the first k nodes after v_1. As noted, it is easier to calculate the number of permutations where this does not occur: place m nodes in any order inside $n-k-1$ positions (the first position belongs to v_i, and we don't want any nodes from M in the next k positions), and the remaining $n-m-1$ nodes can be placed in any permutation among the remaining positions. Therefore, the number permutations not in U_i is $\binom{n-k-1}{m} \cdot m! \cdot (n-m-1)!$.

This same *privacy umbrella* holds for any case where v_i is within the first $k-1$ nodes in the topology. This is due to the fact that any query range starting to the left of v_i will always contain at least v_i and v_{i+1}, and therefore even if there are no nodes from M to the left of v_i the attacker will not be able to distinguish between v_i and v_{i+1}, as long as one of the k nodes following v_i is in

M. It can clearly be seen that due to symmetry, the same is true for the right side (i.e. trailing edge) of the topology. Hence the number of permutations not in U_i for any node v_i adhering to scenario 1 is $\binom{n-k-1}{m} \cdot m! \cdot (n-m-1)!$.

Scenario 2: v_i is the kth node from one edge of the topology
Again, we analyze this scenario by looking at the left side of the topology and consider the right side to behave in the same manner due to symmetry. In this scenario, for v_i to retain privacy, there must either be a node from M within the $k-1$ nodes to the right of v_i, or a node from M to the left of v_i and have the kth node to the right of v_i be from M as well. The former is easy to see, since it won't allow v_i to be isolated and cause any query that may return the value of v_i to contain v_{i-1} in its range as well. The latter case is true since having a node from M to the left of v_i will cause the queries to behave the same as Scenario 1. Assume some node v_m from M is at coordinate x_m to the left of v_i. Any query starting from x_m will return a value of at least s_m. Therefore, the attacker must start their queries after x_m, in which case it will behave as though v_i's position in the topology is smaller than k (i.e. Scenario 1). Once more, we will calculate the number of permutations in W_i. These permutations can be categorized as:

- There are no nodes from M between v_1 and $v_{k+(k-1)}$, the $2k-2$ nodes surrounding v_i.
- There exists at least one node from M within the $k-1$ nodes to the left of v_i, but there are none within the k nodes to the left of v_i

These cover all permutations that will allow the attacker to determine the value of v_i. The number of permutations in each category are as follows:

Category 1 $\binom{n-(2k-1)}{m} \cdot m! \cdot (n-m-1)! = \binom{n-2k+1}{m} \cdot m \cdot (m-1)! \cdot (n-m-1)!$

Category 2 $\sum_{x=1}^{k-1} \binom{n-(2k+2-x)}{m-1} \cdot m \cdot (m-1)! \cdot (n-m-1)!$

Total Number of Permutations
Therefore, the number of permutations in W_i for any node v_i adhering to scenario 2 is $\left(\binom{n-2k+1}{m} + \sum_{x=1}^{k-1} \binom{n-(2k+2-x)}{m-1}\right) \cdot m! \cdot (n-m-1)!$

Scenario 3: v_i is between the *k^{th} and the $(n-k)^{th}$ node in the topology*
While this scenario is a little more complicated, the permutations can be somewhat derived from scenarios 1 and 2. If the kth node the left or right of v_i is in M, then it will require the same permutations as if it were a node in Scenario 2 on the left or right side of the topology respectively. Otherwise, it will require at least one node from M within the $k-1$ nodes to its left, and one node from M within the $k-1$ node to its right. Following is a count of all permutations where none of these conditions are met, split into four different node placement categories:

Surrounding Nodes $v_{i-k} \in M, v_{i-(k-1)} \dots v_{i+k} \notin M$
$\binom{n-2k}{m-1} \cdot m \cdot (m-1)! \cdot (n-m-1)! = \binom{n-2k}{m-1} \cdot m! \cdot (n-m-1)!$

Surrounding Nodes $v_{i-(k-x)} \in M, v_{i-(k-x-1)} \cdots v_{i+k} \notin M$ *or* $v_{i+x} \in M, v_{i+(x-1)} \cdots v_{i-k} \notin M$ *for* $x = 1, 2, \ldots, k-1$
$2 \cdot \sum_{x=2}^{k} \binom{n-(2k+2-x)}{m-1} \cdot m \cdot (m-1)! \cdot (n-m-1)! = 2 \cdot \sum_{x=2}^{k} \binom{n-(2k+2-x)}{m-1} \cdot m! \cdot (n-m-1)!$

Surrounding Nodes $v_{i+k} \in M, v_{i-k} \ldots v_{i+(k-1)} \notin M$
$\binom{n-2k-1}{m-1} \cdot m \cdot (m-1)! \cdot (n-m-1)! = \binom{n-2k-1}{m-1} \cdot m! \cdot (n-m-1)!$

Surrounding Nodes $v_{i-k} \ldots v_{i+k} \notin M$
$\binom{n-2k-1}{m} \cdot m! \cdot (n-m-1)!$

Total Number of Permutations
As such, the number of permutations in W_i for any node v_i adhering to scenario 3 is $(\binom{n-2k}{m-1} + 2 \cdot \sum_{x=2}^{k} \binom{n-(2k+2-x)}{m-1} + \binom{n-2k-1}{m-1} + \binom{n-2k-1}{m}) \cdot m! \cdot (n-m-1)!$.

4.1 Privacy Probability Measure

Using the above scenarios, we can now calculate the privacy level of a randomly chosen node v. The probability that the value s can be discovered and matched to v by the attacker can be calculated as the chance that v falls into a given scenario, multiplied by the probability that an attacker can determine s for a node in that scenario. The probability that v falls into any given scenario is:

- Scenario 1: $\frac{2(k-1)}{n}$
- Scenario 2: $\frac{2}{n}$
- Scenario 3: $\frac{n-2k}{n}$

We can now calculate the "privacy probability measure" p, given in the following theorem:

Theorem 1. *Given the number of nodes in the system, n, the number of nodes with a higher value than the target node, m, and the query-set-size limit k, the privacy probability measure of a random node in a linear topology is:* $p = 1 - q =$

$$1 - \left[2(k-1) \cdot \binom{n-k-1}{m} + 2 \cdot \left(\binom{n-2k+1}{m} + \sum_{x=1}^{k-1} \binom{n-(2k-x)}{m-1} \right) + (n-2k) \cdot \left(\binom{n-2k}{m-1} + 2 \cdot \sum_{x=2}^{k} \binom{n-(2k+2-x)}{m-1} + \binom{n-2k-1}{m-1} + \binom{n-2k-1}{m} \right) \right] \cdot \frac{m! \cdot (n-m-1)!}{(n)!}$$

Figure 1 in the Appendix shows p as a function of m and k in a system of size $n = 1000$. We can see that the privacy probability approaches 1 exponentially as m is increased. As expected, the larger the value of k, the fewer nodes we require with higher values than v in order for v to have a higher privacy probability. We note that there is a linear correlation between m and k with regards to their effect on the privacy probability. Multiplying k by a factor of α reduces the number of nodes m required for the same outcome p by approximately a value of α. For example, in order to reach a privacy probability of over 0.9 in a system of 1000 nodes, we require a set M of over 390 nodes if the k is set to 6. Doubling this value of k to 12 reduces the amount of nodes required in M to 218, roughly half of the previously required value. Doubling k again to 24 now reduces the required m to 115.

5 Analysis of F_{min}, F_{max} and F_{med}

In this section we look at possible attacks using the minimum, maximum and median value queries over ranges in the topology as defined previously by F_{min}, F_{max} and F_{med} respectively. Our approach in this section is slightly different, as we attempt to discover a distribution of nodes in the topology in which at least one node's sensitive attribute is hidden from the attacker. As per the query-set-size restriction, we define that the queries will not return a result if the target Range R at time τ contains less than k individual values. In addition, our analysis of potential attacks rests on the following set of assumptions. The data set consists of n unique values. The value k is known to the attacker. If a result is returned, the number of actual values in (R, τ) is not known to the attacker. If (R, τ) contains an even number of values, F_{med} returns the lower of the two median values. The attacker is limited only to the F_{min}, F_{max} and F_{med} queries, but can perform any number of queries over the data set. The data set will be a linear topology G_L representing a snapshot in time τ of recorded sensitive values of nodes in a specified area (such as speeds in a vehicular network). A query of type q (q being min, max or med) at time τ over a range beginning at x_i and ending at x_j (inclusive) will be denoted $F_q([x_i, x_j])$.

Model: A linear topology G_L with n nodes.
Queries: F_{min}, F_{max} and F_{med}
Goal: Hide the sensitive attribute of a single node at a given time τ.
We note that there are two special cases in which a trivial attack can be performed. We will address these cases before moving on to the general case.

Case 1: k – local Min/Max
If a node has the local minimum or maximum value with regards to his k nearest neighbors then their value can be discovered. This is equivalent to not having a "Privacy Umbrella" as defined in Sect. 4.

Case 2: $k = 3$
In this case, since all values are defined to be unique, querying $F_{min}, F_{max}, F_{med}$ on a range containing exactly three nodes returns three values, each belonging to a specific node. An attacker can query over a single coordinate at the left-most side of the topology and increase the range until a result is returned. The first time a result is returned, the minimum $k = 3$ group size has been reached, and the attacker has the value of each of the 3 nodes. Each value cannot be attributed to a specific node, but we will denote these three values s_1, s_2, s_3. The attacker now decreases the range's size from the left until no result is returned, this

indicates the range now only contains two nodes. Increasing the range to the right until a result is returned indicates that a new node has been added to the range. Since all values are unique, one of the values s_1, s_2, s_3 will be missing from the results. This belongs to the left-most node from the previous query results. Continuing this method until the entire range has been scanned will reveal the values of each node in the system.

The General Case: $k \geq 4$
We show that for the general case of $F_{min}, F_{max}, F_{med}$ queries, there exists a linear placement of nodes such that at least one node will have a value that will remain hidden from an attacker. We use an adversarial model and show that for any number $n \geq 2k$ of nodes and any minimal query size $k \geq 4$, a node arrangement can be created in which the attacker, using any combination of the above mentioned queries, lacks the ability to discover the value of at least one node.

Lemma 1. *Let V be a set of n nodes $v_1, v_2, \ldots, v_n$ positioned along a linear topology at coordinates $x_1, x_2, \ldots, x_n$ at time τ. If $k \geq 4$, for any value $n \geq 2k$ there exists a corresponding assignment of values $s_1, s_2, \ldots, s_n$, such that there exists a node v_j with value s_j which cannot be determined by any attacker with access to the F_{min}, F_{max} and F_{med} queries over the data.*

Proof. We prove by induction for $k = 4$ and $n \geq 2k$, then extrapolate for $k \geq 4$ and $n \geq 2k$.

Show Correctness for $k = 4, n = 8$
With nodes $v_1, v_2, v_3, \ldots, v_8$ positioned at $x_1, x_2, x_3, \ldots, x_8$, set the values of $s_1, s_2, s_3, \ldots, s_8$ such that $s_8 > s_4 > s_3 > s_7 > s_6 > s_1 > s_2 > s_5$. The value of s_3 cannot be determined by an attacker even by running all possible query combinations on the data. The results of all such possible queries can be see in Table 1 in the Appendix.

Assume Correctness for $k = 4, n = (N \mid N \geq 2k)$
Given a set of nodes $v_1, v_2, v_3, \ldots, v_N$ positioned at coordinates $x_1, x_2, x_3, \ldots, x_N$, assume there exists an assignment of corresponding values $s_1, s_2, s_3, \ldots, s_N$ such that there exists some value s_j belonging to some node v_j at position x_j, which cannot be determined by an attacker with access to any number of $F_{min}, F_{max}, F_{med}$ queries under a $k = 4$ limitation.

Prove for $k = 4, n = (N + 1 \mid N \geq 2k)$
We assign $s_1, s_2, s_3, \ldots, s_N$ such that for the range $[x_1, x_N]$, for $k = 4$, there exists some value of s_j which is never revealed by any query $F_{min}, F_{max}, F_{med}$. Assume $s_j \in (s_1, s_2, s_3, \ldots, s_{N-3})$. We note two properties regarding s_{N+1} of the node v_{N+1}, placed at $(x_{N+1} \mid x_{N+1} > x_N)$:

1. Regardless of the value of s_{N+1}: $\forall x_i \mid x_i \in [x_1, x_j] \Rightarrow F_{min}([x_i, x_{N+1}]) < s_j$. (i.e. s_j cannot be the result of any F_{min} query in the range $[x_1, x_{N+1}]$)
2. Regardless of the value of $s_{N+1} : \forall x_i \mid x_i \in [x_1, x_j] \Rightarrow F_{max}([x_i, x_{N+1}]) > s_j$. (i.e. s_j cannot be the result of any F_{max} query in the range $[x_1, x_{N+1}]$)

Therefore, we must only assign s_{N+1} such that it does not cause s_j to be the result of any F_{med} query. Define s_{med_i} to be the result of $F_{med}([x_i, x_N])$. Due to the properties of F_{med}, if $med_1 > s_j$ then $\forall i \mid 1 \leq i \leq j \Rightarrow med_i > s_j$. Conversely, if $med_1 < s_j$ then $\forall i \mid 1 \leq i \leq j \Rightarrow med_i < s_j$. Otherwise at least one of those queries would have returned s_j as a result, which contradicts the induction assumption. Define s'_{med} to be the closest median value to s_j from the previously stated queries. $s'_{med} = F_{med}([x_y, x_N]) \mid \forall i, 1 \leq i \leq j \Rightarrow |s_j - s_{med_i}| \geq |s_j - s_y|$ We set s_{N+1} to be some uniformly distributed random value between s'_{med} and s_j. We now look at $F_{med}([x_i, x_{N+1}])$ and note that for any value $i \mid 1 \leq i \leq j$, the results of $F_{med}([x_i, x_N])$ and $F_{med}([x_i, x_{N+1}])$ are either the same value or adjacent values, as the values in the range differ by exactly one value. Since no value s_{med_i} is adjacent to s_j, then s_j cannot be the result of any value $F_{med}([x_i, x_{N+1}], 1 \leq i \leq j)$. There exist no other queries of the type f_{med} which contain both x_j and x_{N+1}, therefore we now have an assignment $s_1, s_2, s_3, \ldots, s_N, s_{N+1}$ such that the value s_j cannot be discovered by an attacker.

The above holds for the assumption $s_j \in (s_1, s_2, s_3, \ldots, s_{N-3})$. It is easy to see that due to symmetry, the case where $s_j \in (s_4, s_5, s_6, \ldots, s_N)$ allows us to shift all values of S one node to the right, and assign the random value between s'_{med} and s_j to s_1. This completes correctness for all positions of v_j.

Extrapolate for $k \geq 4, n \geq 2k$
Increasing k for a given value of n only reduces the amount of information available to the attacker. Therefore, if a value s_j exists for an assignment in a system with n nodes under the $k = 4$ limitation (with $n \geq 2k$), it will exist for any value of k such that $4 \leq k \leq \frac{n}{2}$. Therefore we have the following theorem:

Theorem 2. *There exists a linear placement of nodes such that at least one node will have a value that will remain hidden from an attacker with access to $F_{min}, F_{max}, F_{med}$ queries under the query-set-size limitation for values of $k \geq 4, n \geq 2k$.*

6 Conclusions and Future Work

With more and more user data being stored by companies and organizations, and a growing demand to disseminate and share this data, the risks to security and privacy rise greatly. While some of these issues have been addressed with encryption and authorization protocols, the potential for misuse of access still exists. The need for protecting user privacy while still maintaining utility of databases has given birth to a wide variety of data anonymization techniques. One such technique used today is the query-set-size control method. In this research we have analyzed the behavior and vulnerabilities of instances of the query-set-size control method. For a linear topology based data set we have offered a privacy measure as the probability of an attacker with access only to either the maximum or minimum query determining the sensitive value of a random node. We have also shown the limitations of an attacker to access some nodes even with access to a combination of the minimum, maximum and median queries. We plan on continued exploration of different privacy preserving data querying methods such as the one detailed in this paper. We are also investigating different implementations on multiple data structures as well as other query formats, such as allowing concatenation of several ranges in a single query. We aim to offer privacy measures similar to the privacy probability measure detailed in this paper.

A Appendix

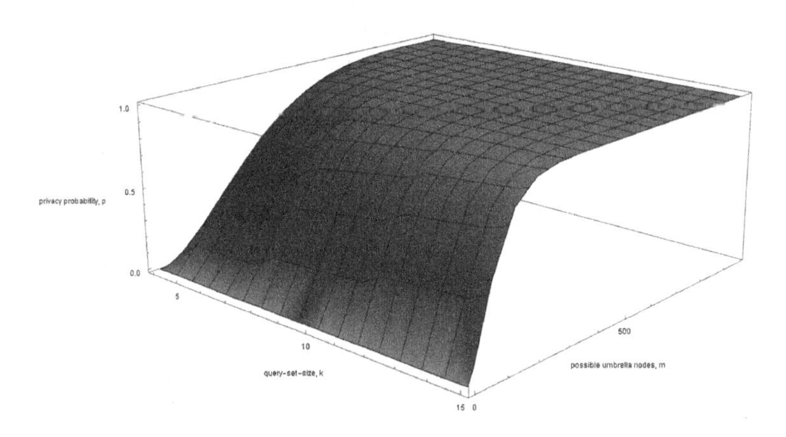

Fig. 1. Privacy Probability of F_{max} under Query-Set-Size Limitation for $n = 1000$.

Table 1. All Possible Results of F_{min}, F_{max} and F_{mid} with $k = 4$ and $n = 8$.

	Containing 4 Nodes				
	$[x_1, x_4]$	$[x_2, x_5]$	$[x_3, x_6]$	$[x_4, x_7]$	$[x_5, x_8]$
F_{min}	s_2	s_5	s_5	s_5	s_5
F_{max}	s_4	s_4	s_4	s_4	s_8
F_{med}	s_1	s_2	s_6	s_6	s_6

	Containing 5 Nodes			
	$[x_1, x_5]$	$[x_2, x_6]$	$[x_3, x_7]$	$[x_4, x_8]$
F_{min}	s_5	s_5	s_5	s_5
F_{max}	s_4	s_4	s_4	s_8
F_{med}	s_1	s_6	s_7	s_7

	Containing 6 Nodes		
	$[x_1, x_6]$	$[x_2, x_7]$	$[x_3, x_8]$
F_{min}	s_5	s_5	s_5
F_{max}	s_4	s_4	s_8
F_{med}	s_1	s_6	s_7

	Containing 7 Nodes	
	$[x_1, x_7]$	$[x_2, x_8]$
F_{min}	s_5	s_5
F_{max}	s_4	s_8
F_{med}	s_6	s_7

	Containing 8 Nodes
	$[x_1, x_8]$
F_{min}	s_5
F_{max}	s_8
F_{med}	s_6

References

1. Adam, N.R., Worthmann, J.C.: Security-control methods for statistical databases: a comparative study. ACM Comput. Surv. **21**(4), 515–556 (1989)
2. Bolot, J., Fawaz, N., Muthukrishnan, S., Nikolov, A., Taft, N.: Private decayed predicate sums on streams. In: ICDT 2013 (2013). http://doi.acm.org/10.1145/2448496.2448530
3. Chan, H.T.H., Shi, E., Song, D.: Private and continual release of statistics. ACM Trans. Inf. Syst. Secur. **14**(3), 1–24 (2011). http://doi.acm.org/10.1145/2043621.2043626
4. Ding, Z., Wang, Y., Wang, G., Zhang, D., Kifer, D.: Detecting violations of differential privacy. In: CCS 2018 (2018). http://doi.acm.org/10.1145/3243734.3243818
5. Dwork, C.: Differential privacy: a survey of results. In: Agrawal, M., Du, D., Duan, Z., Li, A. (eds.) TAMC 2008. LNCS, vol. 4978, pp. 1–19. Springer, Heidelberg (2008). https://doi.org/10.1007/978-3-540-79228-4_1
6. Dwork, C., McSherry, F., Nissim, K., Smith, A.: Calibrating noise to sensitivity in private data analysis. In: Halevi, S., Rabin, T. (eds.) TCC 2006. LNCS, vol. 3876, pp. 265–284. Springer, Heidelberg (2006). https://doi.org/10.1007/11681878_14
7. Dwork, C., Naor, M., Pitassi, T., Rothblum, G.N.: Differential privacy under continual observation. In: STOC 2010 (2010). http://doi.acm.org/10.1145/1806689.1806787
8. Ganta, S.R., Kasiviswanathan, S.P., Smith, A.: Composition attacks and auxiliary information in data privacy. In: KDD 2008 (2008)
9. Lee, J., Clifton, C.: How much is enough? Choosing ε for differential privacy. In: Lai, X., Zhou, J., Li, H. (eds.) ISC 2011. LNCS, vol. 7001, pp. 325–340. Springer, Heidelberg (2011). https://doi.org/10.1007/978-3-642-24861-0_22
10. Li, N., Li, T., Venkatasubramanian, S.: T-closeness: privacy beyond K-anonymity and L-diversity. In: ICDE (2007)
11. Machanavajjhala, A., Kifer, D., Gehrke, J., Venkitasubramaniam, M.: L-diversity: privacy beyond K-anonymity. ACM Trans. Knowl. Discov. Data **1**(1) (2007)
12. Samarati, P., Sweeney, L.: Protecting privacy when disclosing information: K-anonymity and its enforcement through generalization and suppression. Technical report (1998)
13. Sarwate, A.D., Chaudhuri, K.: Signal processing and machine learning with differential privacy: algorithms and challenges for continuous data. IEEE SPM **30**(5), 86–94 (2013)
14. Sweeney, L.: K-anonymity: a model for protecting privacy. Int. J. Uncertain. Fuzziness Knowl.-Based Syst. **10**(5), 557–570 (2002)
15. Venkatadri, G., et al.: Privacy risks with Facebook's PII-based targeting: auditing a data brokers advertising interface. In: Symposium on Security and Privacy (2018)
16. Zhu, T., Li, G., Zhou, W., Yu, P.S.: Differential Privacy and Applications. Springer, Heidelberg (2017). https://doi.org/10.1007/978-3-319-62004-6

Statistical Disclosure Control When Publishing on Thematic Maps

Douwe Hut[1](✉), Jasper Goseling[1,3], Marie-Colette van Lieshout[1,3], Peter-Paul de Wolf[2], and Edwin de Jonge[2]

[1] University of Twente, Enschede, The Netherlands
d.a.hut@student.utwente.nl, j.goseling@utwente.nl
[2] Statistics Netherlands, The Hague, The Netherlands
{pp.dewolf,e.dejonge}@cbs.nl
[3] Centrum Wiskunde & Informatica, Amsterdam, The Netherlands
m.n.m.van.lieshout@cwi.nl

Abstract. The spatial distribution of a variable, such as the energy consumption per company, is usually plotted by colouring regions of the study area according to an underlying table which is already protected from disclosing sensitive information. The result is often heavily influenced by the shape and size of the regions. In this paper, we are interested in producing a continuous plot of the variable directly from microdata and we protect it by adding random noise. We consider a simple attacker scenario and develop an appropriate sensitivity rule that can be used to determine the amount of noise needed to protect the plot from disclosing private information.

1 Introduction

Traditionally, statistical institutes mainly publish tabular data. For the tabular data and underlying microdata, many disclosure control methods exist [10]. A straightforward way to visualise the spatial structure of the tabular data on a map is to colour the different regions of the study area according to their value in a table that was already protected for disclosure control. The connection between disclosure control in tables and on maps is investigated in [16,18], for example.

Drawbacks of giving a single colour to the chosen regions are that the shape of the region influences the plot quite a lot and that the regions might not constitute a natural partition of the study area. This makes it difficult for a user to extract information from the plot. A smooth plot is often easier to work with.

To overcome these disadvantages, more and more publications use other visualisation techniques, such as kernel smoothing, that can be used to visualise data originating from many different sources, including road networks [3], crime numbers [6], seismic damage figures [7] and disease cases [8]. More applications and other smoothing techniques are discussed in [4,5,19].

The views expressed in this paper are those of the authors and do not necessarily reflect the policy of Statistics Netherlands.

J. Domingo-Ferrer and K. Muralidhar (Eds.): PSD 2020, LNCS 12276, pp. 195–205, 2020.
https://doi.org/10.1007/978-3-030-57521-2_14

Research involving the confidentiality of locations when publishing smoothed density maps [14,20] shows that it is possible to retrieve the underlying locations whenever the used parameters are published.

Regarding plots of smoothed averages, [13,22] constructed a cartographic map that showed a spatial density of the relative frequency of a binary variable, such as unemployment per capita. The density was defined at any point, not just at raster points, but the final colouring of the map was discretised, as part of the disclosure control. By the fact that often only one of the values of the variable is considered sensitive information, e.g. being unemployed versus being employed, a practical way to protect locations with too few nearby neighbours is assigning them to the non-sensitive end of the frequency scale. Besides assessing the disclosure risk, some utility measures were constructed.

The starting point for the current research is [23], in which plotting a sensitive continuous variable on a cartographic map using smoothed versions of cell counts and totals is discussed. The authors constructed a $p\%$ rule that used the smoothed cell total and smoothed versions of the largest two contributions per cell.

In this paper, we provide another view on the sensitivity of a map that shows a continuous variable and abandon the idea of explicitly using grid cells, so that the result will be a continuous visualisation on a geographical map. First, in Sect. 2, we will introduce some preliminaries. Then, Sect. 3 will show that the application of disclosure control is needed, after which our method to do so is explained in Sect. 4 and guaranteed to sufficiently protect the sensitive information in Sect. 5. We illustrate our approach by means of a case study in Sect. 6 and make some final remarks in Sect. 7.

2 Preliminaries and Notation

First, we will introduce some notation. Let $\mathcal{D} \subset \mathbb{R}^2$ be an open and bounded set that represents the study region on which we want to make the visualisation. Let the total population be denoted by $\mathcal{U} = \{\boldsymbol{r}_1, \ldots, \boldsymbol{r}_N\} \subset \mathcal{D}$, for $N \in \mathbb{N}$, in which $\boldsymbol{r}_i = (x_i, y_i)$ is the representation of population element i by its Cartesian coordinates (x_i, y_i). We write $\boldsymbol{r} = (x, y)$ for a general point in $\mathcal{D}$ and $||\boldsymbol{r}|| = \sqrt{x^2 + y^2}$ for the distance of that point to the origin. Associated with each population element is a measurement value. By $g_i \geq 0$, we will denote the value corresponding to population element i. As an example, $\mathcal{U}$ could be a set of company locations, where company i has location $\boldsymbol{r}_i$ and measurement value g_i, indicating its energy consumption, as in our case study of Sect. 6.

In order to visualise the population density, one can use kernel smoothing [19]. The approach is similar to kernel density estimation [17], except that no normalisation is applied. Essentially, density bumps around each data point are created and added to make a total density. In our case, the kernel smoothed population density is given by

$$f_h(\boldsymbol{r}) = \frac{1}{h^2} \sum_{i=1}^{N} k\left(\frac{\boldsymbol{r} - \boldsymbol{r}_i}{h}\right),$$

in which $k \colon \mathbb{R}^2 \to \mathbb{R}$ is a so-called kernel function, that is, a non-negative, symmetric function that integrates to 1 over $\mathbb{R}^2$. The bandwidth h controls the range of influence of each data point. The Gaussian kernel $k(\boldsymbol{r}) = (1/2\pi)\exp(-||\boldsymbol{r}||^2/2)$, the Epanechnikov kernel $k(\boldsymbol{r}) = (2/\pi)(1 - ||\boldsymbol{r}||^2)\mathbb{1}(||\boldsymbol{r}|| \leq 1)$ and the uniform kernel $k(\boldsymbol{r}) = (1/\pi)\mathbb{1}(||\boldsymbol{r}|| \leq 1)$ are common choices, but obviously many others kernel functions exist. Some guidelines are given in Sect. 4.5 of [19].

For the measurements values $g_1, \ldots, g_N$, a density can be constructed by multiplying the kernel corresponding to location i with the value g_i:

$$g_h(\boldsymbol{r}) = \frac{1}{h^2} \sum_{i=1}^{N} g_i k\left(\frac{\boldsymbol{r} - \boldsymbol{r}_i}{h}\right).$$

By dividing the two densities f_h and g_h, we get the Nadaraya-Watson kernel weighted average [21]

$$m_h(\boldsymbol{r}) = \frac{g_h(\boldsymbol{r})}{f_h(\boldsymbol{r})} = \frac{\sum_{i=1}^{N} g_i k\left((\boldsymbol{r} - \boldsymbol{r}_i)/h\right)}{\sum_{i=1}^{N} k\left((\boldsymbol{r} - \boldsymbol{r}_i)/h\right)}, \quad \boldsymbol{r} \in \mathcal{D}. \tag{1}$$

Whenever $f_h(\boldsymbol{r}) = 0$, it follows that $g_h(\boldsymbol{r}) = 0$ as well and we define $m_h(\boldsymbol{r}) = 0$. This weighted average is an excellent tool for data visualisation and analysis [5]. The ratio $m_h(\boldsymbol{r})$, $r \in \mathcal{D}$ will be the function of which we will investigate disclosure properties and discuss a possible protection method.

Some remarks are in order. Firstly, the bandwidth h influences the smoothness of m_h. In the limit case of a very large bandwidth, m_h will be constant, while for small h, the plot will contain many local extrema. In the limit case of a very small bandwidth, m_h will be the nearest neighbour interpolation, at least when using a Gaussian kernel. Secondly, note that mass can leak away, since $\mathcal{D}$ is bounded but the kernel is defined on $\mathbb{R}^2$. Consequently, f_h and g_h underestimate the (weighted) population density at $\boldsymbol{r}$ close to the boundary of $\mathcal{D}$. Various techniques to correct such edge effects exist, see [2,9,15].

In this paper, we will frequently use two matrices that are defined in terms of the kernel function, namely

$$\boldsymbol{K}_h = \left(k\left(\frac{\boldsymbol{r}_i - \boldsymbol{r}_j}{h}\right)\right)_{i,j=1}^{N}$$

and

$$\boldsymbol{C}_h = \left(\frac{k\left((\boldsymbol{r}_i - \boldsymbol{r}_j)/h\right)}{\sum_{k=1}^{N} k\left((\boldsymbol{r}_i - \boldsymbol{r}_k)/h\right)}\right)_{i,j=1}^{N}.$$

Lastly, we will write Φ^{-1} for the standard normal inverse cumulative distribution function.

3 Motivation and Attacker Scenario

In this section, we will show that publishing the kernel weighted average reveals exact information on the underlying measurement values. This implies that it is necessary to apply disclosure control before publishing the plot. Our method to do so will be elaborated on in Sect. 4.

Here, we will restrict our attention to the scenario in which an attacker is able to exactly read off the plot of the kernel weighted average (1) at the population element locations $\boldsymbol{r}_i, i = 1, \ldots, N$. Throughout this paper, we will assume that he is completely aware of the method to produce the kernel weighted average and knows what kernel function, bandwidth and population element locations were used.

Using the plot values, the attacker can set up a system of linear equations to obtain estimates of the measurement values, since the kernel weighted average (1) is a linear combination of the measurement values. When the attacker chooses N points to read off the plot of (1) and uses the exact locations $\boldsymbol{r}_i$ for $i = 1, \ldots, N$, he obtains the system

$$\boldsymbol{m}_h = \boldsymbol{C}_h \boldsymbol{g}, \tag{2}$$

with the known plot values $\boldsymbol{m}_h = (m_h(\boldsymbol{r}_i))_{i=1}^N$ and the unknown measurement value vector $\boldsymbol{g} = (\boldsymbol{g}_i)_{i=1}^N$. We know the following about solvability of the system.

Theorem 1. *Whenever $\boldsymbol{K}_h$ is invertible, system (2) can be solved uniquely and the attacker can retrieve all measurement values exactly.*

Proof. Assume that $\boldsymbol{K}_h$ is invertible. Then $\boldsymbol{C}_h$ is invertible as well, as it is created from $\boldsymbol{K}_h$ by scaling each row to sum to 1. Hence, the linear system (2) is uniquely solvable and an attacker can retrieve the vector $\boldsymbol{g}$ of measurement values by left-multiplying $\boldsymbol{m}_h$ with $\boldsymbol{C}_h^{-1}$. □

In particular, Theorem 1 shows that there is at least one configuration of points at which the attacker can read off the plot of (1) to retrieve the measurement values $g_i, i = 1, \ldots, N$ exactly.

For the Gaussian kernel, amongst others, $\boldsymbol{K}_h$ is positive definite and thus invertible, regardless of h, N and $\boldsymbol{r}_i$, $i = 1, \ldots, N$, only provided that all $\boldsymbol{r}_i$ are distinct.

In the remainder of this paper, we will assume an attacker scenario in which the attacker obtains a vector containing the exact plot values at locations $\boldsymbol{r}_i$, $i = 1, \ldots, N$ and left-multiplies that vector by $\boldsymbol{C}_h^{-1}$ to obtain estimates of the measurement values g_i, $i = 1, \ldots, N$.

4 Proposed Method and Main Result

Our method to prevent the disclosure of sensitive information consists of disturbing the plot of (1), by adding random noise to the numerator $g(\boldsymbol{r})$, $\boldsymbol{r} \in \mathcal{D}$, so that an attacker observes

$$\tilde{m}_h(\boldsymbol{r}) = \frac{\sum_{i=1}^N g_i k\left((\boldsymbol{r} - \boldsymbol{r}_i)/h\right) + \epsilon(\boldsymbol{r})}{\sum_{i=1}^N k\left((\boldsymbol{r} - \boldsymbol{r}_i)/h\right)}, \quad \boldsymbol{r} \in \mathcal{D}, \tag{3}$$

instead of (1), where we define $\tilde{m}_h(\boldsymbol{r}) = 0$ if $f_h(\boldsymbol{r}) = 0$. The random noise $\epsilon(\boldsymbol{r})$ will be generated as a Gaussian random field, with mean 0 and covariance function

$$\mathrm{Cov}\left(\epsilon(\boldsymbol{r}), \epsilon(\boldsymbol{s})\right) = \sigma^2 k\left(\frac{\boldsymbol{r}-\boldsymbol{s}}{h}\right), \quad \boldsymbol{r}, \boldsymbol{s} \in \mathcal{D},$$

where σ is the standard deviation of the magnitude of the added noise. The kernel k should be a proper covariance function, which is the case when for all $h > 0$, $m \in \mathbb{N}$ and $\boldsymbol{s}_i \in \mathbb{R}^2$, $i = 1, \ldots, m$, the corresponding matrix $\boldsymbol{K}_h$ is positive definite, see Chapt. 1 of [1]. In this way, (3) will be continuous, just as (1), whenever a continuous kernel function is used and f_h vanishes nowhere.

Adding random noise to the plot implies that the attacker's estimates will be stochastic as well. This fact should be captured in a rule that describes whether it is safe to publish the noised kernel weighted average. It brings us to the following sensitivity rule, that states that a plot is considered unsafe to publish when any measurement value estimate that the attacker makes lies with probability greater than α within p percent of the true value. Such a sensitivity rule can be seen as a stochastic counterpart of the well known $p\%$ rule for tabular data, which is elaborated on in [10].

Definition 1. *For* $0 < p \leq 100$ *and* $0 \leq \alpha < 1$*, a plot is said to be* unsafe according to the $(p\%, \alpha)$ rule for an attacker scenario *whenever the estimates* $\hat{g}_i$ *of* g_i*,* $i = 1, \ldots, N$*, computed according to the scenario, satisfy*

$$\max_{i=1,\ldots,N} P\left\{\left|\frac{\hat{g}_i - g_i}{g_i}\right| < \frac{p}{100}\right\} > \alpha, \tag{4}$$

where we take $|(\hat{g}_i - g_i)/g_i| = |\hat{g}_i|$ *if* $g_i = 0$.

When applying the $(p\%, \alpha)$ rule, we normally choose p and α to be small, so that a plot is safe when small relative errors in the recalculation happen with small probability. Theorem 1 implies that the plot of (1) cannot be safe for any $(p\%, \alpha)$ rule. Furthermore, we note that high values of p and low values of α correspond to a stricter rule: If a plot is safe according the $(p\%, \alpha)$ rule, then for any $\tilde{p} \leq p$ and $\tilde{\alpha} \geq \alpha$, the plot is also safe according to the $(\tilde{p}\%, \tilde{\alpha})$ rule.

Our main result is the following theorem, that gives the standard deviation of the magnitude of the noise ϵ in (3) needed to ensure that the plot is safe according to the $(p\%, \alpha)$ rule. In Sect. 5, we will prove the theorem.

Theorem 2. *Suppose that the kernel* $k\colon \mathbb{R}^2 \to \mathbb{R}$ *is a proper covariance function and* $g_i > 0$*,* $i = 1, \ldots, N$*. Then the plot of (3) is safe according to the* $(p\%, \alpha)$ *rule for our attacker scenario of Sect. 3 if*

$$\sigma \geq \frac{p}{100\,\Phi^{-1}\left((1+\alpha)/2\right)} \max_{i=1,\ldots,N} \left\{\frac{g_i}{\sqrt{\left(\boldsymbol{K}_h^{-1}\right)_{ii}}}\right\}. \tag{5}$$

5 Proof of Theorem 2

Recall that the attacker observes (3). In matrix notation, (3) reads

$$\boldsymbol{m}_h + \tilde{\boldsymbol{\epsilon}} = \boldsymbol{C}_h \boldsymbol{g} + \tilde{\boldsymbol{\epsilon}},$$

where

$$\tilde{\boldsymbol{\epsilon}} = (\tilde{\epsilon}_i)_{i=1}^N = \left(\frac{\epsilon(\boldsymbol{r}_i)}{\sum_{j=1}^N k\left((\boldsymbol{r}_i - \boldsymbol{r}_j)/h\right)} \right)_{i=1}^N . \tag{6}$$

If the attacker left-multiplies the vector of observed plot values by $\boldsymbol{C}_h^{-1}$ to recalculate $\boldsymbol{g}$, just as he could do in (2), he will now make an error, because the observed values are $\boldsymbol{m}_h + \tilde{\boldsymbol{\epsilon}}$ instead of $\boldsymbol{m}_h$. When we write $\hat{\boldsymbol{g}} = (\hat{g}_i)_{i=1}^N$ for the vector of recalculated measurement values, we obtain

$$\hat{\boldsymbol{g}} = \boldsymbol{C}_h^{-1}(\boldsymbol{m}_h + \tilde{\boldsymbol{\epsilon}}) = \boldsymbol{g} + \boldsymbol{C}_h^{-1}\tilde{\boldsymbol{\epsilon}}. \tag{7}$$

Recall that $\boldsymbol{C}_h$ is invertible because $\boldsymbol{K}_h$ is positive definite since k is a proper covariance function.

By the next lemma, that is the result of basic probability theory, it suffices, in order to prove Theorem 2, to show that for our attacker scenario of Sect. 3 and using the plot of (3), for $i = 1, \ldots, N$, the recalculated value $\hat{g}_i$ follows a normal distribution with mean g_i and variance $\sigma^2 \left(\boldsymbol{K}_h^{-1}\right)_{ii}$.

Lemma 1. *Whenever $\hat{g}_i$ follows a normal distribution with mean g_i, (4) is equivalent with*

$$\max_{i=1,\ldots,N} \frac{p\ g_i}{100\ \Phi^{-1}\left(\frac{1+\alpha}{2}\right)\sqrt{\mathrm{Var}(\hat{g}_i)}} > 1.$$

Now, let us compute the variance of the recalculated measurement values. For all $i = 1, \ldots, N$, combining (7) with the fact that $\epsilon(\boldsymbol{r}_i)$, $i = 1, \ldots, N$, follows a multivariate normal distribution with zero mean and covariance matrix $\sigma^2 \boldsymbol{K}_h$, the i-th recalculated value $\hat{g}_i$ will follow a normal distribution with mean g_i and variance

$$\mathrm{Var}(\hat{g}_i) = \sum_{j=1}^N \sum_{k=1}^N \mathrm{Cov}\left(\left(\boldsymbol{C}_h^{-1}\right)_{ij} \tilde{\epsilon}_j, \left(\boldsymbol{C}_h^{-1}\right)_{ik} \tilde{\epsilon}_k \right).$$

Rewriting $\tilde{\epsilon}_j$ and $\tilde{\epsilon}_k$ according to (6), taking factors outside the covariance term and substituting $\sigma^2 \left(\boldsymbol{K}_h\right)_{jk} = \sigma^2 \left(\boldsymbol{C}_h\right)_{kj} \sum_{m=1}^N \left(\boldsymbol{K}_h\right)_{km}$ for $\mathrm{Cov}(\epsilon_j, \epsilon_k)$, we obtain

$$\mathrm{Var}(\hat{g}_i) = \sigma^2 \sum_{j=1}^N \sum_{k=1}^N \frac{\left(\boldsymbol{C}_h^{-1}\right)_{ij} \left(\boldsymbol{C}_h^{-1}\right)_{ik}}{\sum_{m=1}^N \left(\boldsymbol{K}_h\right)_{jm}} \left(\boldsymbol{C}_h\right)_{kj}.$$

Now, we can work out the multiplications of inverse matrices and use

$$\boldsymbol{K}_h^{-1} = \left(\frac{\left(\boldsymbol{C}_h^{-1}\right)_{ij}}{\sum_{m=1}^N \left(\boldsymbol{K}_h\right)_{jm}} \right)_{i,j=1}^N$$

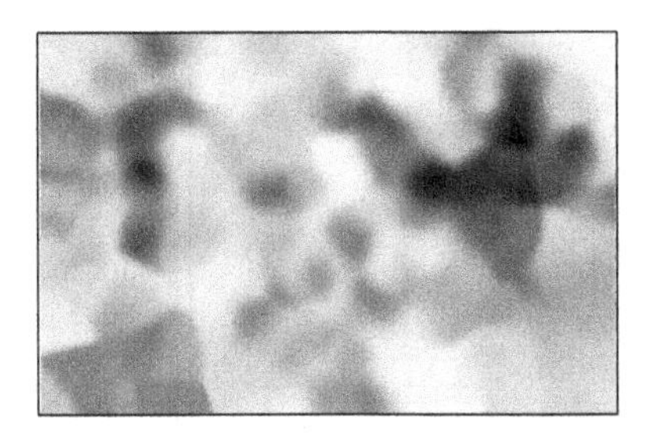

Fig. 1. Unprotected (left panel) and protected (right panel) kernel weighted average of our entire synthetic dataset, according to a (10%, 0.1) rule for a Gaussian kernel with bandwidth $h = 250\,\mathrm{m}$

to get the result

$$\mathrm{Var}(\hat{q}_i) = \sigma^2 \left(\boldsymbol{K}_h^{-1}\right)_{ii},$$

which, together with Lemma 1, proves Theorem 2.

6 Case Study

We want to be able to compare unprotected plots with protected plots, so we cannot use original, confidential data. Hence we used a synthetic dataset, based on real data of energy consumption by enterprises. The original data contained enterprises in the region 'Westland' of The Netherlands. This region is known for its commercial greenhouses as well as enterprises from the Rotterdam industrial area. We perturbed the locations of the enterprises and we assigned random values for the energy consumption drawn from a log-normal distribution with parameters estimated from the original data. We introduced some spatial dependency in the energy consumption to mimic the compact industrial area and the densely packed greenhouses. The final dataset consists of some 8348 locations and is also included in the sdcSpatial R-package that can be found on CRAN [12].

Figure 1 shows the unprotected kernel weighted average (1) and the protected kernel weighted average (3) that satisfies the (10%, 0.1) rule. A Gaussian kernel with a bandwidth of 250 m was used. We computed a safe lower bound for the standard deviation σ of the random noise by (5). The plot of (3) resulting from that computation looks almost exactly identical to the plot of (1). Only at parts of the boundary where the population density is very small, the added disturbance is perceptible by the eye.

When the bandwidth would be taken smaller, the standard deviation of the noise would become large enough for the disturbance to be visually apparent. However, working on this scale, it would be hard to see the details in that situation. Thus, we plotted a subset of the data, restricting ourselves to a square of 2 km × 2 km and all 918 enterprises contained in that square. The results of our method on the data subset are visible in Fig. 2 for $h = 100\,\mathrm{m}$ and in Fig. 3

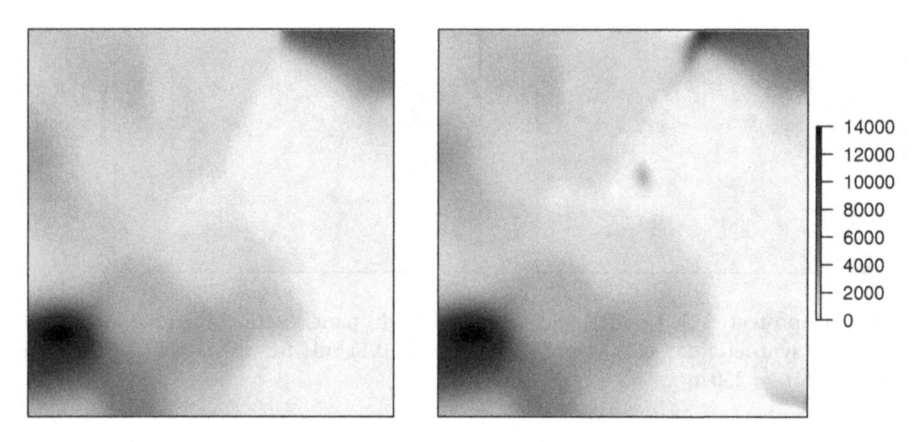

Fig. 2. Unprotected (left panel) and protected (right panel) kernel weighted average of a part of our synthetic dataset, according to a $(10\%, 0.1)$ rule for a Gaussian kernel with bandwidth $h = 100\,\text{m}$

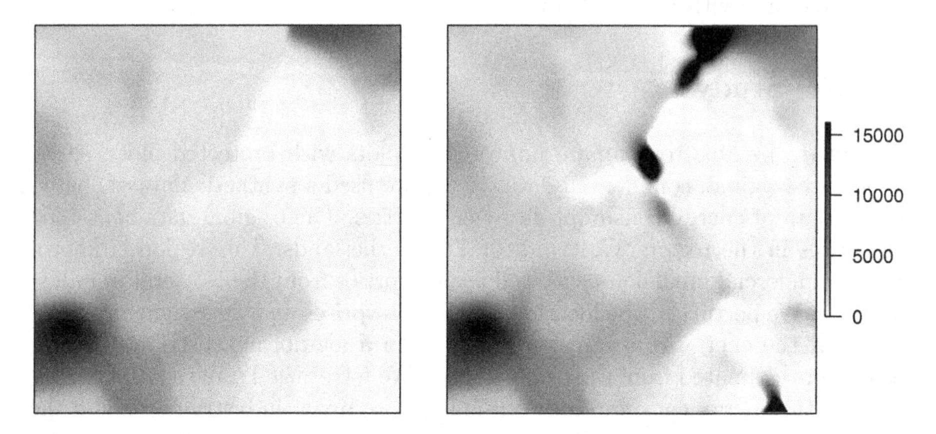

Fig. 3. Unprotected (left panel) and protected (right panel) kernel weighted average of a part of our synthetic dataset, according to a $(10\%, 0.1)$ rule for a Gaussian kernel with bandwidth $h = 80\,\text{m}$

for $h = 80\,\text{m}$, while Fig. 4 displays the spatial structure of the locations in our entire synthetic dataset and the subset thereof.

We see that the necessary disturbance to the plot is smaller in Fig. 3 than in Fig. 2. In order to be able to compare the results for different bandwidths, Fig. 5 contains two graphs that show the influence of the bandwidth on σ for our synthetic data set. Note that the total disturbance of the plot is also influenced by the denominator of (3), that increases with increasing bandwidth if the used kernel is decreasing in $||\boldsymbol{r}||$. The graph of the entire dataset shows a steep decrease of σ around $h = 5$. This is caused by the quick increase of the diagonal elements of $\boldsymbol{K}_h^{-1}$ due to $\boldsymbol{K}_h$ becoming less similar to the identity matrix. For $h \leq 5$ a single company with a very large energy consumption dominates the value of σ.

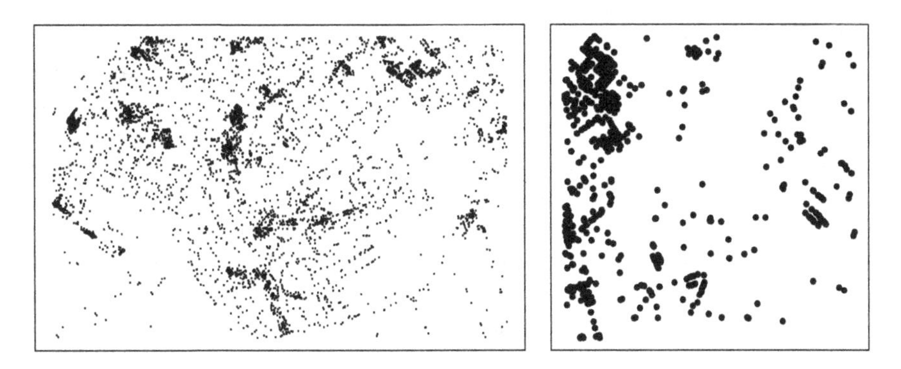

Fig. 4. Map of enterprise locations in our entire dataset (left panel) and in the data subset (right panel)

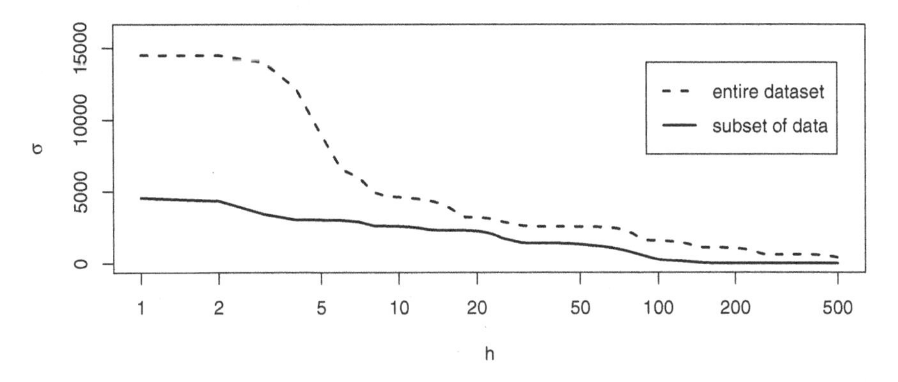

Fig. 5. Standard deviation σ of added noise for different bandwidths

Since this company is not present in the subset that we work with, a smaller σ may be used for the subset, also for $h \leq 5$.

7 Discussion

In this paper we introduced a new sensitivity rule that is applicable in the scenario that an attacker knows both the kernel and the bandwidth used to produce the map, reads off the plotted values at the population elements and estimates the measurement values by solving a system of linear equations. To protect the plot, we proposed to disturb the data by adding noise and derived a rule on how large the disturbance to the plot should be before publishing it.

To investigate the efficacy of the proposed method a case study was carried out. It indicated that for a bandwidth that is large relative to the population density, the disturbance needed was very small. When zooming in, however, the disturbance to the plot was visually apparent.

During this research, some other interesting results were found that fall outside the scope of this paper. For details we refer to [11]. For instance, in our attacker scenario we assumed that the bandwidth is known to the attacker. If the bandwidth were unknown to the attacker, simulations indicate that in many cases, the bandwidth can be retrieved from the plot of (1) by repeatedly guessing a bandwidth, solving the linear system for that bandwidth, making a plot using the recalculated values and the guessed bandwidth and calculating the similarity between the original and the recovered plot.

Secondly, many kernels with a compact support, including the uniform and Epanechnikov kernel, are discontinuous or not infinitely differentiable at the boundary of their support. An attacker can often use such information to obtain the bandwidth or a single measurement value by considering plot values close to that boundary.

We close with some final remarks and perspectives. At first glance, it might seem more natural to add noise to the kernel weighted average itself rather than to the numerator of (1). However, typically more noise should then be added, resulting in a less visually attractive map. Furthermore, the proposed method agrees with the intuition that densely populated areas need less protection, since the standard deviation of the noise is inversely proportional to the kernel smoothed population density. Note that the addition of noise in our method might lead to negative or extremely large values of (3) at locations where the population density is very small. In our figures, these locations were given the minimal or maximal colour scale values, to result in a realistic map for the user.

It would be interesting to look at the utility of our plot for different bandwidths. Figure 5 is a first step in this direction but more research is needed.

Our method requires that all $\boldsymbol{r}_i$, $i = 1, \ldots, N$ are distinct. It would be interesting to look into a scenario in which population elements can have the same location, since these might partly protect each other for disclosure. If one would introduce grid cells and use a single location for elements in the same cell, a similar analysis could lead to explicitly taking the resolution of the plot into account. Alternatively, rounding the plot values or using a discrete color scale may be a useful approach to obtaining some level of disclosure control.

Finally, we restricted ourselves to a single simple attacker scenario. It would be interesting to investigate alternative scenarios in which the attacker is particularly interested in a single value, uses other locations to read off the plot or tries to eliminate the added noise.

References

1. Abrahamsen, P.: A review of Gaussian random fields and correlation functions. Technical report 917, Norwegian Computing Center (1997)
2. Berman, M., Diggle, P.: Estimating weighted integrals of the second-order intensity of a spatial point process. J. R. Stat. Soc. **51**, 81–92 (1989)
3. Borruso, G.: Network density and the delimitation of urban areas. Trans. GIS **7**(2), 177–191 (2003)
4. Bowman, A.W., Azzalini, A.: Applied Smoothing Techniques for Data Analysis. Oxford University Press, Oxford (1997)

5. Chacón, J.E., Duong, T.: Multivariate Kernel Smoothing and its Applications. CRC Press, Boca Raton (2018)
6. Chainey, S., Reid, S., Stuart, N.: When is a hotspot a hotspot? A procedure for creating statistically robust hotspot maps of crime. In: Kidner, D., Higgs, G., White, S. (eds.) Innovations in GIS 9: Socio-economic Applications of Geographic Information Science, pp. 21–36. Taylor and Francis (2002)
7. Danese, M., Lazzari, M., Murgante, B.: Kernel density estimation methods for a geostatistical approach in seismic risk analysis: the case study of Potenza Hilltop Town (Southern Italy). In: Gervasi, O., Murgante, B., Laganà, A., Taniar, D., Mun, Y., Gavrilova, M.L. (eds.) ICCSA 2008. LNCS, vol. 5072, pp. 415–429. Springer, Heidelberg (2008). https://doi.org/10.1007/978-3-540-69839-5_31
8. Davies, T.M., Hazelton, M.L.: Adaptive kernel estimation of spatial relative risk. Stat. Med. **29**(23), 2423–2437 (2010)
9. Diggle, P.J.: A kernel method for smoothing point process data. J. R. Stat. Soc. **34**, 138–147 (1985)
10. Hundepool, A., et al.: Statistical Disclosure Control. Wiley Series in Survey Methodology. Wiley (2012). ISBN 978-1-119-97815-2
11. Hut, D.A.: Statistical disclosure control when publishing on thematic maps. Master's thesis, University of Twente, Enschede, The Netherlands (2020)
12. de Jonge, E., de Wolf, P.P.: sdcSpatial: Statistical Disclosure Control for Spatial Data. https://CRAN.R-project.org/package=sdcSpatial, r package version 0.2.0.9000
13. de Jonge, E., de Wolf, P.-P.: Spatial smoothing and statistical disclosure control. In: Domingo-Ferrer, J., Pejić-Bach, M. (eds.) PSD 2016. LNCS, vol. 9867, pp. 107–117. Springer, Cham (2016). https://doi.org/10.1007/978-3-319-45381-1_9
14. Lee, M., Chun, Y., Griffith, D.A.: An evaluation of kernel smoothing to protect the confidentiality of individual locations. Int. J. Urban Sci. **23**(3), 335–351 (2019). https://doi.org/10.1080/12265934.2018.1482778
15. van Lieshout, M.N.M.: On estimation of the intensity function of a point process. Methodol. Comput. Appl. Probab. **14**, 567–578 (2012)
16. O'Keefe, C.M.: Confidentialising maps of mixed point and diffuse spatial data. In: Domingo-Ferrer, J., Tinnirello, I. (eds.) PSD 2012. LNCS, vol. 7556, pp. 226–240. Springer, Heidelberg (2012). https://doi.org/10.1007/978-3-642-33627-0_18
17. Silverman, B.W.: Density Estimation for Statistics and Data Analysis. Chapman & Hall, London (1986)
18. Suñé, E., Rovira, C., Ibáñez, D., Farré, M.: Statistical disclosure control on visualising geocoded population data using quadtrees. In: Extended Abstract at NTTS 2017 (2017). http://nt17.pg2.at/data/x_abstracts/x_abstract_286.docx
19. Wand, M.P., Jones, M.C.: Kernel Smoothing. CRC Press, Boca Raton (1994)
20. Wang, Z., Liu, L., Zhou, H., Lan, M.: How is the confidentiality of crime locations affected by parameters in kernel density estimation? Int. J. Geo-Inf. **8**(12), 544–556 (2019). https://doi.org/10.3390/ijgi8120544
21. Watson, G.S.: Smooth regression analysis. Sankhya: Indian J. Stat. **26**(4), 359–372 (1964)
22. de Wolf, P.P., de Jonge, E.: Location related risk and utility. Presented at UNECE/Eurostat Worksession Statistical Data Confidentiality, Skopje, 20–22 September (2017). https://www.unece.org/fileadmin/DAM/stats/documents/ece/ces/ge.46/2017/3_LocationRiskUtility.pdf
23. de Wolf, P.-P., de Jonge, E.: Safely plotting continuous variables on a map. In: Domingo-Ferrer, J., Montes, F. (eds.) PSD 2018. LNCS, vol. 11126, pp. 347–359. Springer, Cham (2018). https://doi.org/10.1007/978-3-319-99771-1_23

Record Linkage and Alternative Methods

Bayesian Modeling for Simultaneous Regression and Record Linkage

Jiurui Tang(✉), Jerome P. Reiter, and Rebecca C. Steorts

Department of Statistical Science, Duke University, Durham, USA
{jiurui.tang,jreiter}@duke.edu, beka@stat.duke.edu

Abstract. Often data analysts use probabilistic record linkage techniques to match records across two data sets. Such matching can be the primary goal, or it can be a necessary step to analyze relationships among the variables in the data sets. We propose a Bayesian hierarchical model that allows data analysts to perform simultaneous linear regression and probabilistic record linkage. This allows analysts to leverage relationships among the variables to improve linkage quality. Further, it enables analysts to propagate uncertainty in a principled way, while also potentially offering more accurate estimates of regression parameters compared to approaches that use a two-step process, i.e., link the records first, then estimate the linear regression on the linked data. We propose and evaluate three Markov chain Monte Carlo algorithms for implementing the Bayesian model, which we compare against a two-step process.

1 Introduction

Increasingly, data analysts seek to link records across data sets to facilitate statistical analyses. As a prototypical example, a health researcher seeks to link data from a previously completed study to patients' electronic medical records to collect long-term outcomes, with the ultimate goal of estimating relationships between the long-term outcomes and baseline covariates. Such linkages are performed readily when unique identifiers, such as social security numbers, are available for all records in all data sets.

Often, however, one or more of the data sets do not have unique identifiers, perhaps because they were never collected or are not made available due to privacy concerns. In such cases, analysts have to link records based on indirect identifiers, such as name, date of birth, and demographic variables [1,2]. Generally, such indirect identifiers contain distortions and errors. As a result, they can differ across the data sets, which can make it difficult to determine the correct record linkages. This uncertainty should be quantified and propagated to statistical inferences, although typically this is not done.

In the statistics literature, the most popular method for linking records via indirect identifiers is based on the probabilistic record linkage (RL) approach

R. C. Steorts—This research was partially supported by the National Science Foundation through grants SES1131897, SES1733835, SES1652431 and SES1534412.

J. Domingo-Ferrer and K. Muralidhar (Eds.): PSD 2020, LNCS 12276, pp. 209–223, 2020.
https://doi.org/10.1007/978-3-030-57521-2_15

of Newcombe et al. [15], which was later extended and formalized by Fellegi and Sunter [4]. Many extensions to the Fellegi-Sunter (FS) model have been proposed [e.g., 18,23,24]. A common drawback of these and other probabilistic RL methods [e.g., 7,12] is the difficulty in quantifying linkage uncertainty, and propagating that uncertainty to statistical inferences. These limitations have led to developments of RL approaches from Bayesian perspectives [e.g., 3,5,6,9–11,14,16,17,19–21,25,26].

In this article, we propose a Bayesian model for performing probabilistic RL and linear regression simultaneously. The proposed model quantifies uncertainty about the linkages and propagates this uncertainty to inferences about the regression parameters. We focus on bipartite RL—that is, the analyst seeks to merge two data sets—assuming that individuals appear at most once in each data set. As we illustrate, the model can leverage relationships among the dependent and independent variables in the regression to potentially improve the quality of the linkages. This also can increase the accuracy of resulting inferences about the regression parameters.

We use a Bayesian hierarchical model that builds on prior work by Sadinle [17], who proposed a Bayesian version of the FS model for merging two data sets. In fact, one of our primary contributions is to turn the model in [17] into a procedure for jointly performing probabilistic RL and fully Bayesian inference for regression parameters. We also propose and evaluate the effectiveness of three algorithms for fitting the Bayesian hierarchical model, focusing on both the quality of the linkages and on the accuracy of the parameter estimates.

2 Review of Bayesian Probabilistic Record Linkage

In this section, we review the Bayesian bipartite RL model of [17]. Consider two data sets $\mathbf{A}_1$ and $\mathbf{A}_2$, containing n_1 and n_2 records, respectively. Without loss of generality, assume $n_1 \geq n_2$. Our goal is to link records in $\mathbf{A}_1$ to records in $\mathbf{A}_2$. We further assume that $\mathbf{A}_1$ and $\mathbf{A}_2$ do not contain duplicate records; that is, each record in $\mathbf{A}_1$ corresponds to a single individual, as is the case for $\mathbf{A}_2$. We assume that some of the same individuals are in $\mathbf{A}_1$ and $\mathbf{A}_2$.

To characterize this, we define the random variable $\mathbf{Z} = (Z_1, \ldots, Z_{n_2})$ as the vector of matching labels for the records in $\mathbf{A}_2$. For $j = 1, \ldots, n_2$, let

$$Z_j = \begin{cases} i, & \text{if record } i \in \mathbf{A}_1 \text{ and } j \in \mathbf{A}_2 \text{ refer to the same entity;} \\ n_1 + j, & \text{if record } j \in \mathbf{A}_2 \text{ does not have a match in } \mathbf{A}_1. \end{cases}$$

Analysts determine whether a pair of records (i, j) is a link, i.e., whether or not $Z_j = i$, by comparing values of variables that are common to $\mathbf{A}_1$ and $\mathbf{A}_2$. Suppose we have F common variables, also known as *linking variables* or *fields*. For $f = 1, \ldots, F$, let γ_{ij}^f represent a score that reflects the similarity of field f for records i and j. For example, when field f is a binary variable, we can set $\gamma_{ij}^f = 1$ when record i agrees with record j on field f, and $\gamma_{ij}^f = 0$ otherwise. When field f is a string variable like name, we can calculate a similarity metric like the Jaro-Winkler distance [8,22] or the Levenshtein edit distance [13]. We can convert these string metrics to γ_{ij}^f by categorizing the scores into a multinomial variable, where the categories represent the strength of agreement. We illustrate this approach in Sect. 4.

For each record (i, j) in $\mathbf{A}_1 \times \mathbf{A}_2$, let $\boldsymbol{\gamma}_{ij} = (\gamma_{ij}^1, \ldots, \gamma_{ij}^F)$. We assume $\boldsymbol{\gamma}_{ij}$ is a realization of a random vector Γ_{ij} distributed as

$$\Gamma_{ij}|Z_j = i \overset{iid}{\sim} \mathcal{M}(\mathbf{m}), \qquad \Gamma_{ij}|Z_j \neq i \overset{iid}{\sim} \mathcal{U}(\mathbf{u}), \qquad \text{where}$$

$\mathcal{M}(\mathbf{m})$ represents the model for comparison vectors among matches, and $\mathcal{U}(\mathbf{u})$ represents the model for comparison vectors among non-matches. For each field f, we let $m_{f\ell} = \mathbb{P}(\Gamma_{ij}^f = \ell|Z_j = i)$ be the probability of a match having level ℓ of agreement in field f, and let $u_{f\ell} = \mathbb{P}(\Gamma_{ij}^f = \ell|Z_j \neq i)$ be the probability of a non-match having level ℓ of agreement in field f. Let $\mathbf{m}_f = (m_{f1}, \ldots, m_{fL_f})$ and $\mathbf{u}_f = (u_{f1}, \ldots, u_{fL_f})$; let $\mathbf{m} = (\mathbf{m}_1, \ldots, \mathbf{m}_F)$ and $\mathbf{u} = (\mathbf{u}_1, \ldots, \mathbf{u}_F)$.

For computational convenience, it is typical to assume the comparison fields are conditionally independent given the matching status of the record pairs. Let $\boldsymbol{\Theta} = (\mathbf{m}, \mathbf{u})$. The likelihood of the comparison data can be written as

$$\mathcal{L}(\mathbf{Z}|\boldsymbol{\Theta}, \boldsymbol{\gamma}) = \prod_{i=1}^{n_1} \prod_{j=1}^{n_2} \prod_{f=1}^{F} \prod_{\ell=0}^{L_f} \left[m_{f\ell}^{I(Z_j=i)} u_{f\ell}^{I(Z_j \neq i)} \right]^{I(\gamma_{ij}^f = \ell)}, \tag{1}$$

where $I(\cdot) = 1$ when its argument is true and $I(\cdot) = 0$ otherwise.

Sometimes, there are missing values in the linking variables. Although we do not consider this possibility in our simulations, we summarize how to handle missing values in the model, assuming ignorable missing data. With conditional independence and ignorability, we can marginalize over the missing comparison variables. The likelihood of the observed comparison data can be written as

$$\mathcal{L}(\mathbf{Z}|\boldsymbol{\Theta}, \boldsymbol{\gamma}^{\text{obs}}) = \prod_{f=1}^{F} \prod_{\ell=0}^{L_f} m_{f\ell}^{a_{f\ell}(\mathbf{Z})} u_{f\ell}^{b_{f\ell}(\mathbf{Z})}, \qquad \text{where} \tag{2}$$

$a_{f\ell}(\mathbf{Z}) = \sum_{i,j} I_{\text{obs}}(\gamma_{ij}^f) I(\gamma_{ij}^f = \ell) I(Z_j = i)$ and $b_{f\ell}(\mathbf{Z}) = \sum_{i,j} I_{\text{obs}}(\gamma_{ij}^f) I(\gamma_{ij}^f = \ell) I(Z_j \neq i)$. For a given $\mathbf{Z}$, these represent the number of matches and non-matches with observed disagreement level ℓ in field f. Here, $I_{\text{obs}}(\cdot) = 1$ when its argument is observed, and $I_{\text{obs}}(\cdot) = 0$ when its argument is missing.

To define the prior distributions for $\mathbf{m}$ and $\mathbf{u}$, for all fields f, let $\boldsymbol{\alpha}_f = (\alpha_{f0}, \ldots, \alpha_{fL_f})$ and $\boldsymbol{\beta}_f = (\beta_{f0}, \ldots, \beta_{fL_f})$. We assume that $\mathbf{m}_f \sim \text{Dirichlet}(\boldsymbol{\alpha}_f)$ and $\mathbf{u}_f \sim \text{Dirichlet}(\boldsymbol{\beta}_f)$, where $\boldsymbol{\alpha}_f$ and $\boldsymbol{\beta}_f$ are known parameters. In our simulation studies, we set all entries of $\boldsymbol{\alpha}_f$ and $\boldsymbol{\beta}_f$ equal to 1 for every field.

We use the prior distribution for $\mathbf{Z}$ from [17]. For $j = 1, \ldots, n_2$, let the indicator variable $I(Z_j \leq n_1) \mid \pi \overset{iid}{\sim} \text{Bernoulli}(\pi)$, where π is the proportion of matches expected a priori. Let $\pi \sim \text{Beta}(\alpha_\pi, \beta_\pi)$, where the prior mean $\alpha_\pi/(\alpha_\pi + \beta_\pi)$ represents the expected percentage of overlap. Let $n_{12}(\mathbf{Z}) = \sum_j I(Z_j \leq n_1)$

be the number of matches according to $\mathbf{Z}$. The prior specification implies that $n_{12}(\mathbf{Z}) \sim$ Beta-Binomial$(n_2, \alpha_\pi, \beta_\pi)$ after marginalizing over π. That is,

$$\mathbb{P}(n_{12}(\mathbf{Z})) = \binom{n_2}{n_{12}(\mathbf{Z})} \frac{\mathrm{B}(n_{12}(\mathbf{Z}) + \alpha_\pi, n_2 - n_{12}(\mathbf{Z}) + \beta_\pi)}{\mathrm{B}(\alpha_\pi, \beta_\pi)}. \tag{3}$$

We assume that, conditional on the value of $n_{12}(\mathbf{Z})$, all the possible bipartite matchings are equally likely a priori. There are $n_1!/(n_1 - n_{12}(\mathbf{Z}))!$ such bipartite matchings. Thus, the prior distribution for $\mathbf{Z}$ is

$$\mathbb{P}(\mathbf{Z}|\alpha_\pi, \beta_\pi) = \mathbb{P}(n_{12}(\mathbf{Z})|\alpha_\pi, \beta_\pi)\mathbb{P}(\mathbf{Z}|n_{12}(\mathbf{Z}), \alpha_\pi, \beta_\pi) \tag{4}$$

$$= \frac{(n_1 - n_{12}(\mathbf{Z}))!}{n_1!} \frac{\mathrm{B}(n_{12}(\mathbf{Z}) + \alpha_\pi, n_2 - n_{12}(\mathbf{Z}) + \beta_\pi)}{\mathrm{B}(\alpha_\pi, \beta_\pi)}. \tag{5}$$

3 The Bayesian Hierarchical Model for Simultaneous Regression and Record Linkage

In this section, we present the Bayesian hierarchical model for regression and RL, and propose three Markov chain Monte Carlo (MCMC) algorithms for fitting the model in practice. Throughout, we assume the explanatory variables $\mathbf{X}$ are in $\mathbf{A}_1$, and the response variable $\mathbf{Y}$ is in $\mathbf{A}_2$.

3.1 Model Specification

We assume the standard linear regression, $\mathbf{Y}|\mathbf{X}, \mathbf{V}, \mathbf{Z} \sim N(\mathbf{X}\beta, \sigma^2\mathbf{I})$. Here, $\mathbf{V}$ are linking variables used in the RL model but not in the regression model. Analysts can specify prior distributions on (β, σ^2) that represent their beliefs. A full specification of the joint distribution of $(\mathbf{Y}, \mathbf{X}|\mathbf{V})$ requires analysts to specify some marginal model for $\mathbf{X}$, written generically as $f(\mathbf{X}|\mathbf{V})$. In some contexts, however, it is not necessary to specify $f(\mathbf{X}|\mathbf{V})$, as we explain in Sect. 3.2. Critically, this model assumes that the distribution of $(\mathbf{Y}, \mathbf{X}|\mathbf{V})$ is the same for matches and non-matches. Finally, for the RL component of the model, we model $\mathbf{Z}$ using the Bayesian FS approach in Sect. 2.

For the simulation studies, we illustrate computations with the Bayesian hierarchical model using univariate Y and univariate X. Assume $X \sim N(\mu, \tau^2)$. As a result, in the simulations, the random variable (X, Y) follows a bivariate normal distribution with

$$\begin{bmatrix} X \\ Y \end{bmatrix} \sim N\left(\begin{bmatrix} \mu \\ \beta_0 + \beta_1\mu \end{bmatrix}, \begin{bmatrix} \tau^2 & \beta_1\tau^2 \\ \beta_1\tau^2 & \sigma^2 + \beta_1^2\tau^2 \end{bmatrix} \right).$$

We assume a normal-Gamma prior on the regression parameters. Letting $\phi = 1/\sigma^2$, we have $\phi \sim G(.5, .5)$ and $\beta|\phi \sim N(b_0, \Phi_0\phi^{-1})$ where $b_0 = [3, 1]^T$ and Φ_0 is a 2×2 identity matrix. When needed, we assume Jeffrey's prior $p(\mu, \tau^2) \propto 1/\tau^2$.

3.2 Estimation Strategies

Even in the relatively uncomplicated set-up of the simulation study, it is not possible to compute the posterior distribution of the model parameters in closed form. Therefore, we consider three general strategies for MCMC sampling in order to approximate the posterior distribution.

First, we propose an MCMC sampler that uses only the linked records when estimating the full conditional distribution of the regression parameters in each iteration of the sampler. This method generates potentially different samples of linked records at each iteration. A key advantage of this method is that it does not require imputation of $\mathbf{X}$; hence, analysts need not specify $f(\mathbf{X}|\mathbf{V})$. We call this the joint model without imputation, abbreviated as JM. Second, we propose an MCMC sampler that imputes the missing values of $\mathbf{X}$ for those records in $\mathbf{A}_2$ without a link, and updates the regression parameters in each iteration using the linked pairs as well the imputed data. We call this the joint model with imputation, abbreviated as JMI. Third, we propose an MCMC sampler that is similar to JMI but uses an extra step when imputing the missing values of $\mathbf{X}$. Specifically, at each iteration, we (i) sample values of the regression parameters from a conditional distribution based on only the linked records, (ii) use the sampled parameters to impute missing values in $\mathbf{X}$, and (iii) update regression coefficients based on linked as well as imputed pairs. By adding step (i), we aim to reduce potential effects of a feedback loop in which less accurate estimates of regression parameters result in less accurate estimates of the conditional distribution of $\mathbf{X}$, and so on through the MCMC iterations. We call this the joint model with imputation and reduced feedback (JMIF).

For JMI and JMIF, inferences are based on every entity in $\mathbf{A}_2$, whereas for JM inferences are based on the subsets of linked pairs, which can differ across MCMC iterations. Analysts should keep these differences in mind when selecting an algorithm that suits their goals.

3.3 Details of the MCMC Samplers

In this section, we present the mathematical details for implementing the three proposed MCMC samplers. Before doing so, we present an algorithm for a two-step approach, where we perform RL and then use the linked data for regression. The three proposed MCMC samplers for the Bayesian hierarchical model utilize parts of the algorithm for the two-step approach.

3.3.1 Two Step Approach (TS)

Given the parameter values at iteration t of the sampler, we need to sample new values $\mathbf{m}_f^{[t+1]} = (m_{f0}^{[t+1]}, \ldots, m_{fL_f}^{[t+1]})$ and $\mathbf{u}_f^{[t+1]} = (u_{f0}^{[t+1]}, \ldots, u_{fL_f}^{[t+1]})$, where $f = 1, \ldots, F$. We then sample a new value $\mathbf{Z}^{[t+1]} = (Z_1^{[t+1]}, \ldots, Z_{n_2}^{[t+1]})$. The steps are as follows.

T.1 For $f = 1, \ldots, F$, sample

$$\mathbf{m}_f^{[t+1]}|\gamma^{obs}, \mathbf{Z}^{[t]} \sim \text{Dirichlet}(a_{f0}(\mathbf{Z}^{[t]}) + \alpha_{f0}, \ldots, a_{fL_f}(\mathbf{Z}^{[t]}) + \alpha_{fL_f}), \quad (6)$$

$$\mathbf{u}_f^{[t+1]}|\boldsymbol{\gamma}^{obs}, \mathbf{Z}^{[t]} \sim \text{Dirichlet}(b_{f0}(\mathbf{Z}^{[t]}) + \beta_{f0}, \ldots, b_{fL_f}(\mathbf{Z}^{[t]}) + \beta_{fL_f}). \tag{7}$$

Collect these new draws into $\boldsymbol{\Theta}^{[t+1]}$. Here, each

$$a_{fl}(\mathbf{Z}) = \sum_{i,j} I_{obs}(\boldsymbol{\gamma}_{ij}^f) I(\gamma_{ij}^f = l) I(Z_j = i), \tag{8}$$

$$b_{fl}(\mathbf{Z}) = \sum_{i,j} I_{obs}(\boldsymbol{\gamma}_{ij}^f) I(\gamma_{ij}^f = l) I(Z_j \neq i). \tag{9}$$

T.2 Sample the entries of $\mathbf{Z}^{[t+1]}$ sequentially. Having sampled the first $j-1$ entries of $\mathbf{Z}^{[t+1]}$, we define $\mathbf{Z}_{-j}^{[t+(j-1)/n_2]} = (Z_1^{[t+1]}, \ldots, Z_{j-1}^{[t+1]}, Z_{j+1}^{[t]}, \ldots, Z_{n_2}^{[t]})$. We sample a new label $Z_j^{[t+1]}$, with the probability of selecting label $q \in \{1, \ldots, n_1, n_1 + j\}$ given by $p_{qj}(\mathbf{Z}_{-j}^{[t+(j-1)/n_2]} \mid \boldsymbol{\Theta}^{[t+1]})$. This can be expressed for generic $\mathbf{Z}_{-j}$ and $\boldsymbol{\Theta}$ as

$$p_{qj}(\mathbf{Z}_{-j}|\boldsymbol{\Theta}) \propto \begin{cases} \exp[w_{qj}] I(Z_{j'} \neq q, \forall j' \neq j), & \text{if } q \leq n_1 \\ [n_1 - n_{12}(\mathbf{Z}_{-j})] \dfrac{n_2 - n_{12}(\mathbf{Z}_{-j}) - 1 + \beta_\pi}{n_{12}(\mathbf{Z}_{-j}) + \alpha_\pi}, & \text{if } q = n_1 + j; \end{cases} \tag{10}$$

where $w_{qj} = \log[\mathbb{P}(\boldsymbol{\gamma}_{qj}^{obs}|Z_j = q, \mathbf{m})/\mathbb{P}(\boldsymbol{\gamma}_{qj}^{obs}|Z_j \neq q, \mathbf{u})]$ is equivalently

$$w_{qj} = \sum_{f=1}^{F} I_{obs}(\boldsymbol{\gamma}_{qj}^f) \sum_{l=0}^{L_f} \log\left(\frac{m_{fl}}{u_{fl}}\right) I(\gamma_{qj}^f = l). \tag{11}$$

The normalizing constant for $p_{qj}(\mathbf{Z}_{-j}|\boldsymbol{\Theta})$ is

$$\prod_{i=1}^{n_1} \prod_{k=1}^{n_2} \prod_{f=1}^{F} \prod_{\ell=0}^{L_f} \left[m_{f\ell}^{I(Z_k=i)} u_{f\ell}^{I(Z_k \neq i)} \right]^{I(\gamma_{ij}^f = \ell)} u_{fl}^{I(Z_j \neq i)}$$
$$\times \frac{(n_1 - (n_{12}(\mathbf{Z}_{-j}) + 1))!}{n_1!} \frac{(n_{12}(\mathbf{Z}_{-j}) + \alpha_\pi)!(n_2 - (n_{12}(\mathbf{Z}_{-j}) + 1) + \beta_\pi - 1)!}{(n_2 + \alpha_\pi + \beta_\pi - 1)} \tag{12}$$

where $k \neq j$.

T.3 We now add the regression parameters to the sampler. For any draw of $\mathbf{Z}^{[t+1]}$, we sample $\beta^{[t+1]}$ and $(\sigma^2)^{[t+1]} = \phi^{-1}$ from

$$\beta^{[t+1]}|\phi, \mathbf{Y}, \mathbf{Z}^{[t+1]} \sim N(b_n, (\phi \Phi_n)^{-1}) \tag{13}$$

$$\phi|\mathbf{Y}, \mathbf{Z}^{[t]} \sim G\Big(\tfrac{n+1}{2}, \tfrac{1}{2}(SSE + 1 + \hat{\beta}^T \widetilde{\mathbf{X}}^T \widetilde{\mathbf{X}} \hat{\beta} + b_0^T \phi_0 b_0 - b_n^T \Phi_n b_n)\Big) \tag{14}$$

where $SSE = \widetilde{\mathbf{Y}}^T [\mathbf{I} - \widetilde{\mathbf{X}}(\widetilde{\mathbf{X}}^T \widetilde{\mathbf{X}})^{-1} \widetilde{\mathbf{X}}^T] \widetilde{\mathbf{Y}}$,$\Phi_n = \widetilde{\mathbf{X}}^T \widetilde{\mathbf{X}} + \Phi_0$, $b_n = \Phi_n^{-1}(\widetilde{\mathbf{X}}^T \widetilde{\mathbf{X}} \hat{\beta} + \phi_0 b_0)$, and $\hat{\beta} = (\widetilde{\mathbf{X}}^T \widetilde{\mathbf{X}})^{-1} \widetilde{\mathbf{X}}^T \widetilde{\mathbf{Y}}$. Here, $\widetilde{\mathbf{X}}$ and $\widetilde{\mathbf{Y}}$ are the subsets of $\mathbf{X}$ and $\mathbf{Y}$ belonging to only the linked cases at iteration t.

Steps **T.1** and **T.2** are the same as those used in [17]; we add **T.3** to sample the regression parameters. Alternatively, and equivalently, analysts can run **T.1** and **T.2** until MCMC convergence, then apply **T.3** to each of the resulting draws of $\mathbf{Z}^{[t]}$ to obtain the draws of the regression parameters.

3.3.2 Joint Method Without Imputation (JM)

The sampler for the JM method uses **T.1**, but it departs from **T.2** and **T.3**. As we shall see, in JM we need the marginal density $f(\mathbf{Y})$. This can be approximated with a standard univariate density estimator. Alternatively, one can derive it from $f(\mathbf{Y}|\mathbf{X})$ and extra assumptions about $f(\mathbf{X})$, although these extra assumptions obviate one of the advantages of JM compared to JMI and JMIF. In the simulations, for convenience we use the fact that (Y, X) are bivariate normal when computing the marginal density of $\mathbf{Y}$, as evident in step **J.2**. Step **J.2** can be omitted when using means to compute $f(\mathbf{Y})$ that do not leverage a joint model for $(\mathbf{Y}, \mathbf{X})$.

J.1 Sample $\mathbf{m}_f^{[t+1]}$ and $\mathbf{u}_f^{[t+1]}$ using **T.1**.

J.2 Sample $\mu^{[t+1]}$ and $(\tau^2)^{[t+1]}$ using $1/\tau^2 \sim G((n-1)/2, \sum(X_i - \bar{X})^2)$ and $\mu|\tau^2 \sim N(\bar{X}, \tau^2)$. We use all of $\mathbf{X}$ in this step.

J.3 Given $\mathbf{Z}^{[t]}$, sample $\beta^{[t+1]}$ and $(\sigma^2)^{[t+1]}$ from (13) and (14).

J.4 Sample $\mathbf{Z}^{[t+1]}$ sequentially. Having sampled the first $j-1$ entries of $\mathbf{Z}^{[t+1]}$, we define $\mathbf{Z}_{-j}^{[t+(j-1)/n_2]} = (Z_1^{[t+1]}, \ldots, Z_{j-1}^{[t+1]}, Z_{j+1}^{[t]}, \ldots, Z_{n_2}^{[t]})$. Then we sample a new label $Z_j^{[t+1]}$ with probability $p_{qj}(\mathbf{Z}_{-j}^{[t+(j-1)/n_2]} \mid \mathbf{\Theta}^{[t+1]}, \mathbf{X}, \mathbf{Y})$ of selecting label $q \in \{1, \ldots, n_1, n_1 + j\}$. For generic $(\mathbf{Z}_{-j}, \mathbf{\Theta}, \mathbf{X}, \mathbf{Y})$, we have $f(\mathbf{Z}_{-j}|\mathbf{\Theta}, \mathbf{X}, \mathbf{Y}) \propto f(\mathbf{Y}, \mathbf{X}|\mathbf{\Theta}, \mathbf{Z}_{-j}) f(\mathbf{Z}_{-j}|\mathbf{\Theta})$. For $q \leq n_1$, and using the definition for w_{qj} in (11), we thus have

$$p_{qj}(\mathbf{Z}_{-j}|\mathbf{\Theta}, \mathbf{X}, \mathbf{Y}) \propto \exp[w_{qj}] I(Z_{j'} \neq q, \forall j' \neq j) \prod_{i \neq q, i \in \mathbf{A}_{12}} f(X_i, Y_i|\mathbf{Z}_{-j})$$

$$\times \prod_{i \neq q, i \in \mathbf{A}_{1-}} f(X_i) \prod_{i \neq q, i \in \mathbf{A}_{2-}} f(Y_i) f(Y_j, X_q) \tag{15}$$

$$\propto \exp[w_{qj}] I(Z_{j'} \neq q, \forall j' \neq j) \frac{f(Y_j, X_q)}{f(Y_j) f(X_q)} \tag{16}$$

$$= \exp[w_{qj}] I(Z_{j'} \neq q, \forall j' \neq j) \frac{f(Y_j|X_q)}{f(Y_j)}. \tag{17}$$

Here, $\mathbf{A}_{12}$ is the set of matched records, $\mathbf{A}_{1-}$ is the set of records in $\mathbf{A}_1$ without a match in $\mathbf{A}_2$, and $\mathbf{A}_{2-}$ is the set of records in $\mathbf{A}_2$ without a match in $\mathbf{A}_1$.

For $q = n_1 + j$, after some algebra to collect constants, we have

$$p_{qj}(\mathbf{Z}_{-j}|\mathbf{\Theta}, \mathbf{X}, \mathbf{Y}) \propto [n_1 - n_{12}(\mathbf{Z}_{-j})] \frac{n_2 - n_{12}(\mathbf{Z}_{-j}) - 1 + \beta_\pi}{n_{12}(\mathbf{Z}_{-j}) + \alpha_\pi}.$$

3.3.3 Joint Method with Imputation (JMI)

The sampler for JMI is similar to the sampler for JM, except we impute $\mathbf{X}$ for non-matches in $\mathbf{A}_2$. Thus, we require a model for $\mathbf{X}$, which we also use to compute $f(\mathbf{Y})$. In accordance with the simulation set-up, we present the sampler with $X \sim N(\mu, \tau^2)$.

I.1 Sample $\mathbf{m}_f^{[t+1]}$ and $\mathbf{u}_f^{[t+1]}$ using **J.1**.

I.2 Sample $\mu^{[t+1]}$ and $(\tau^2)^{[t+1]}$ using **J.2**.

I.3 Impute $\mathbf{X}_{mis}$ for those records in $\mathbf{A}_2$ without a matched X. For the sampler in the simulation study, the predictive distribution is

$$X_{mis} \sim N\Big(\mu + \frac{\beta_1\tau^2}{\sigma^2 + \beta_1^2\tau^2}(Y - \beta_0 - \beta_1\mu), \tau^2 - \frac{\beta_1^2\tau^4}{\sigma^2 + \beta_1^2\tau^2}\Big). \tag{18}$$

In JMI, we use the values of $(\beta^{[t]}, (\sigma^2)^{[t]}, (\tau^2)^{[t+1]})$ in (18). Once we have $\mathbf{X}_{mis}^{[t+1]}$, in the full conditional distributions for $(\beta^{[t+1]}, (\sigma^2)^{[t+1]})$ we use both the matched and imputed data for all records in $\mathbf{A}_2$, with the priors in Sect. 3.3.1. As a result, we draw $\beta^{[t+1]}$ and $(\sigma^2)^{[t+1]}$ based on (13) and (14), but let $(\widetilde{\mathbf{X}}, \widetilde{\mathbf{Y}})$ include both the linked pairs and imputed pairs in $\mathbf{A}_2$.

I.4 Sample $\mathbf{Z}^{[t+1]}$ sequentially using **J.4**.

3.3.4 Joint Method with Imputation and Reduced Feedback (JMIF)

The sampler for JMIF is like the sampler for JMI, but we use a different predictive model for $\mathbf{X}_{mis}$. We again present the sampler with $X \sim N(\mu, \tau^2)$.

F.1 Sample $\mathbf{m}_f^{[t+1]}$ and $\mathbf{u}_f^{[t+1]}$ using **J.1**.

F.2 Sample $\mu^{[t+1]}$ and $(\tau^2)^{[t+1]}$ using **J.2**.

F.3 Given $\mathbf{Z}^{[t]}$, take a draw (β^*, σ^{2*}) from the full conditional distributions in (13) and (14), using only the linked cases at iteration t. We impute $\mathbf{X}_{mis}$ for those records in $\mathbf{A}_2$ without a matched X using (β^*, σ^{2*}). For the sampler in the simulation study, we use (18) with (β^*, σ^{2*}) and $(\tau^2)^{[t+1]}$. Once we have $\mathbf{X}_{mis}^{[t+1]}$, in the full conditional distributions for $(\beta^{[t+1]}, \sigma^{[t+1]})$ we use both the matched and imputed data for all records in $\mathbf{A}_2$, with the priors in Sect. 3.3.1. As a result, we draw $\beta^{[t+1]}$ and $(\sigma^2)^{[t+1]}$ based on (13) and (14), but let $(\widetilde{\mathbf{X}}, \widetilde{\mathbf{Y}})$ include both the linked pairs and imputed pairs in $\mathbf{A}_2$.

F.4 Sample $\mathbf{Z}^{[t+1]}$ sequentially using (**J.4**).

3.4 MCMC Starting Values

Sadinle [17] starts the MCMC sampler by assuming none of the records in file $\mathbf{A}_2$ have a link in file $\mathbf{A}_1$. We do not recommend this starting point for the hierarchical model, as it is beneficial to specify an initial set of links to determine sensible starting values for the linear regression parameters. Instead, we employ a standard FS algorithm—implemented using the `RecordLinkage` package in `R`—to determine a set of links to use as a starting point.

4 Simulation Studies

We generate simulated data sets using the `RLdata10000` data set from the `RecordLinkage` package in `R`. The `RLdata10000` contains 10,000 records; 10% of these records are duplicates belonging to 1,000 individuals. The `RLdata10000` includes linking variables, which the developers of the data set have distorted to create uncertainty in the RL task. To create $\mathbf{A}_2$, we first randomly sample $n_{12} = 750$ individuals from the 1,000 individuals with duplicates. We then sample 250 individuals from the remaining 8,000 individuals in `RLdata10000`. This ensures that each record in $\mathbf{A}_2$ belongs to one and only one individual. To create $\mathbf{A}_1$, we first take the duplicates for the 750 records in $\mathbf{A}_2$; these are true matches. Next, we sample another 250 records from the individuals in `RLdata10000` but not in $\mathbf{A}_2$. Thus, we have 750 records that are true links, and 250 records in each file that do not have a match in the other file. We repeat this process independently in each simulation run.

In both files, in each simulation run, we generate the response and explanatory variables, as none are available in the `RLdata10000`. For each sampled record i, we generate $x_i \sim N(0,1)$ and $y_i \sim N(\beta_0 + \beta_1 x_i, \sigma^2)$ in each simulation run. We set $\beta_0 = 3$ and $\sigma^2 = 1$. We consider $\beta_1 \in \{.4, .65, .9\}$, to study how the correlation between X and Y affects performance of the methods.

We use four linking variables: the first name and last name, and two constructed binary variables based on birth year and birth day. For the constructed variables, we create an indicator of whether the individual in the record was born before or after 1974, and another indicator of whether the individual in the record was born before or after the 16th day of the month.

To compare first and last name, we use the Levenshtein edit distance (LD), defined as the minimum number of insertions, deletions, or substitutions required to change one string into the other. We divide this distance by the length of the longest string to standardize it. The final measure is in the range of $[0, 1]$, where 0 represents total agreement and 1 total disagreement. Following [17], we categorize the LD into four levels of agreement. We set $f = 1$ and $\gamma_{ij}^f = 3$ when the first names for record i and j match perfectly ($LD = 0$); $\gamma_{ij}^f = 2$ when these names show mild disagreement ($0 < LD \leq .25$); $\gamma_{ij}^f = 1$ when these names show moderate disagreement ($.25 < LD \leq .5$); and, $\gamma_{ij}^f = 0$ when these names show extreme disagreement ($LD \geq .5$). The same is true for last names with $f = 2$. For the constructed binary variables based on birth day and year, we set $\gamma_{ij}^f = 1$ when the values for record i and j agree with each other, and $\gamma_{ij}^f = 0$ otherwise.

We create two scenarios to represent different strengths of the information available for linking. The strong linkage scenario uses all four linking variables, and the weak linkage scenario uses first and last name only.

4.1 Results

Table 1 displays averages of the estimated regression coefficients over 100 simulation runs. Across all scenarios, on average the point estimates of β_1 from the

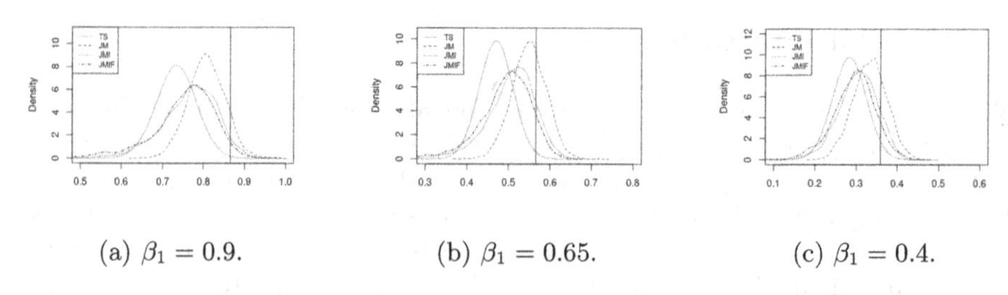

(a) $\beta_1 = 0.9$. (b) $\beta_1 = 0.65$. (c) $\beta_1 = 0.4$.

Fig. 1. Posterior density of β_1 in one arbitrarily chosen data set for each value of β_1 under the strong linking information scenario. Posterior distribution for the two-step approach is in solid green, for JM is in dashed red, for JMI is in dotted blue, and for JMIF is in dotdash black. Vertical lines are estimates of β_1 when using all the correct links.

Bayesian hierarchical model, regardless of the MCMC algorithm, are at least as close to the true β_1 as the point estimates from the two-step approach. The hierarchical model offers the largest improvements in accuracy over the two-step approach when the correlation between X and Y is strongest and the information in the linking variables is weakest. In this scenario, the model takes advantage of information in the relationship between the study variables that the two-step approach cannot. In contrast, when the correlation between X and Y is weakest and the information in the linking variables is strongest, there is little difference in the performances of the hierarchical model and two-step approach. These patterns are illustrated in Fig. 1.

Generally, all the algorithms tend to underestimate β_1 in these simulations. It is practically impossible to identify all the true links. Therefore, the regression is estimated with some invalid pairs of (x_i, y_i). This attenuates the estimates of β_1. The hierarchical model tends to overestimate β_0 slightly. The difference is most noticeable when the correlation between X and Y is strong. Generally, on average the two-step approach offers more accurate estimates of β_0, although the differences are practically irrelevant.

Among the hierarchical models, JM outperforms JMI and JMIF, with JMIF slightly better than JMI. This is because inaccuracies in the estimated distribution of $\mathbf{X}_{mis}$ in JMI and JMIF are propagated to the estimated distributions of (β_0, β_1). To illustrate this, suppose in a particular MCMC iteration the value of β_1 is somewhat attenuated, which in turn leads to inaccuracy in the parameters of the imputation model for $X|Y$. As a result, the imputed values of $\mathbf{X}_{mis}$ are not samples from an accurate representation of $f(X|Y)$. Thus, the full conditional distribution of (β_0, β_1) is estimated from completed-data that do not follow the relationship between X and Y. The inaccurate samples of (β_0, β_1) then create inaccurate imputations, and the cycle continues. In contrast, in any iteration, JM samples coefficients using only the records deemed to be links (in that iteration), thereby reducing the effects of feedback from imprecise imputations. This also explains why JMIF yields slightly more accurate estimates of β_1 than JMI when the correlation between X and Y is strong.

Table 1. Summary of simulation results for regression coefficients. Results based on 100 runs per scenario. The true $\beta_0 = 3$ in all scenarios. For each reported average, the Monte Carlo standard errors are smaller than .01. "Strong" refers to scenarios where we use all four comparison fields, and "Weak" refers to scenarios where we use only two comparison field's.

	Results for β_1				Results for β_0			
	TS	JM	JMIF	JMI	TS	JM	JMIF	JMI
Strong								
$\beta_1 = .90$	.73	.85	.80	.79	3.01	3.05	3.04	3.04
$\beta_1 = .65$	.52	.60	.56	.55	3.00	3.02	3.01	3.01
$\beta_1 = .40$	.32	.36	.33	.32	2.99	3.00	3.00	3.00
Weak								
$\beta_1 = .90$	.60	.82	.78	.76	3.00	3.05	3.04	3.03
$\beta_1 = .65$	.42	.57	.53	.53	3.00	3.03	3.02	3.02
$\beta_1 = .40$	.27	.34	.32	.32	3.00	3.01	3.01	3.01

Table 2 displays averages across the 100 simulation runs of six standard metrics for the quality of the record linkages. These include the average numbers of correct links (CL), correct non-links (CNL), false negatives (FN), and false positives (FP), as well as the false negative rate (FNR) and false discovery rate (FDR). These are formalized in Appendix A. The results in Table 2 indicate that the hierarchical model offers improved linkage quality over the two-step approach, regardless of the estimation algorithm. In particular, the hierarchical model tends to have smaller FP and larger CNL than the two-step approach. The difference in CNL is most apparent when the information in the linking variables is weak and when the correlation between X and Y is strong. The hierarchical model tends to have higher CL than the two-step approach, but the difference is practically important only when the linkage information is weak and the correlation is relatively strong ($\beta_1 = .9, \beta_1 = .65$). Overall, the hierarchical model has lower FDR compared to the two step approach.

5 Discussion

The simulation results suggest that the Bayesian hierarchical model for simultaneous regression and RL can offer more accurate coefficient estimates than the two-step approach in which one first performs RL then runs regression on linked data. The hierarchical model is most effective when the correlation between the response and explanatory variable is strong. The hierarchical model also can improve linkage quality, in particular by identifying more non-links. This is especially the case when the information in the linking variables is not strong. In all scenarios, the relationship between the response and explanatory variable complements the information from the comparison vectors, which helps us declare record pairs more accurately.

Table 2. Summary of linkage quality across 100 simulation runs. Averages in first four columns have standard errors less than 3. Averages in the last two columns have Monte Carlo standard errors less than .002.

		CL	CNL	FN	FP	FNR	FDeR
Strong							
$\beta_1 = .90$	JM	702	152	47	128	.06	.15
	JMIF	702	155	48	125	.06	.15
	JMI	702	155	48	125	.06	.15
	TS	697	123	53	167	.07	.19
$\beta_1 = .65$	JM	702	139	48	144	.06	.17
	JMIF	701	143	49	140	.06	.16
	JMI	701	143	49	141	.06	.17
	TS	698	119	52	172	.07	.20
$\beta_1 = .40$	JM	698	131	52	158	.07	.18
	JMIF	698	133	52	155	.07	.18
	JMI	698	133	52	155	.07	.18
	TS	697	121	53	170	.07	.20
Weak							
$\beta_1 = .90$	JM	636	114	114	223	.15	.26
	JMIF	636	116	114	222	.15	.26
	JMI	636	115	114	223	.15	.26
	TS	620	53	130	315	.17	.34
$\beta_1 = .65$	JM	632	93	118	256	.16	.29
	JMIF	631	94	119	255	.16	.29
	JMI	631	92	119	256	.16	.30
	TS	621	51	129	317	.17	.34
$\beta_1 = .40$	JM	625	69	125	291	.17	.32
	JMIF	624	69	126	291	.17	.32
	JMI	625	69	125	291	.17	.32
	TS	620	50	130	319	.17	.34

As with any simulation study, we investigate only a limited set of scenarios. Our simulations have 75% of the individuals in the target file as true matches. Future studies could test whether the hierarchical model continues to offer gains with lower overlap rates, as well as different values of other simulation parameters. We used a correctly specified linear regression with only one predictor. Appendix B presents a simulation where the linear regression is mis-specified; the model continues to perform well. We note that the hierarchical model without imputation for missing $\mathbf{X}$ extends readily to multivariate $\mathbf{X}$. When outcomes are binary, analysts can use probit regression in the hierarchical model. The model also can be modified for scenarios where one links to a smaller file containing explanatory variables. In this case, we use the marginal distribution of $\mathbf{X}$ and conditional distribution of $\mathbf{X}|\mathbf{Y}$ rather than those for $\mathbf{Y}$ and $\mathbf{Y}|\mathbf{X}$ in (17).

A Record Linkage Evaluation Metrics

Here, we review the definitions of the average numbers of correct links (CL), correct non-links (CNL), false negatives (FN), and false positives (FP). These allow one to calculate the false negative rate (FNR) and false discovery rate (FDR) [19]. For any MCMC iteration t, we define $\text{CL}^{[t]}$ as the number of record pairs with $Z_j \leq n_1$ and that are true links. We define $\text{CNL}^{[t]}$ as the number of record pairs with $Z_j > n_1$ that also are not true links. We define $\text{FN}^{[t]}$ as the number of record pairs that are true links but have $Z_j > n_1$. We define $\text{FP}^{[t]}$ as the number of record pairs that are not true links but have $Z_j \leq n_1$. In the simulations, the true number of true links is $\text{CL}^{[t]}+\text{FN}^{[t]} = 750$, and the estimated number of links is $\text{CL}^{[t]}+\text{FP}^{[t]}$. Thus, $\text{FNR}^{[t]} =$ is $\text{FN}^{[t]}/(\text{CL}^{[t]}+\text{FN}^{[t]})$. The $\text{FDR}^{[t]} = \text{FP}^{[t]}/(\text{CL}^{[t]}+\text{FP}^{[t]})$, where by convention we take $\text{FDR}^{[t]} = 0$ when both the numerator and denominator are 0. We report the FDR instead of the FPR, as an algorithm that does not link any records has a small FPR, but this does not mean that it is a good algorithm. Finally, for each metric, we compute the posterior means across all MCMC iterations, which we average across all simulations.

B Additional Simulations with a Mis-specified Regression

As an additional simulation, we examine the performance of the hierarchical model in terms of linkage quality when we use a mis-specified regression. The true data generating model is $\log(\mathbf{Y})|\mathbf{X},\mathbf{V},\mathbf{Z} \sim N(\mathbf{X}\beta, \sigma^2\mathbf{I})$, but we incorrectly assume $\mathbf{Y}|\mathbf{X},\mathbf{V},\mathbf{Z} \sim N(\mathbf{X}\beta, \sigma^2\mathbf{I})$ in the hierarchical model. Table 3 summarizes the measures of linkage quality when the linkage variables have weak information. Even though the regression component of the hierarchical model is mis-specified, the hierarchical model still identifies more correct non-matches than the two-step approach identifies, although the difference is less obvious than when we use the correctly specified regression. We see a similar trend when the information in the linking variables is strong, albeit with smaller differences between the two-step approach and the hierarchical model.

Table 3. Results for simulation with mis-specified regression component in the hierarchical model. Entries summarize the linkage quality across 100 simulation runs. Averages in first four columns have standard errors less than 3. Averages in the last two columns have Monte Carlo standard errors less than .002.

		CL	CNL	FN	FP	FNR	FDeR
$\beta_1 = .90$	JM	3625	69	125	292	.17	.32
	JMIF	624	70	126	291	.17	.32
	JMI	624	69	126	292	.17	.32
	TS	619	51	131	318	.17	.34
$\beta_1 = .65$	JM	626	62	124	299	.17	.32
	JMIF	626	62	124	299	.17	.32
	JMI	626	62	124	299	.17	.32
	TS	622	49	128	319	.17	.34
$\beta_1 = .40$	JM	623	56	127	309	.17	.33
	JMIF	623	56	127	309	.17	.33
	JMI	623	57	127	309	.17	.33
	TS	622	50	128	318	.17	.34

References

1. Christen, P.: A survey of indexing techniques for scalable record linkage and deduplication. IEEE Trans. Knowl. Data Eng. **24**(9), 1537–1555 (2012)
2. Christen, P.: Data linkage: the big picture. Harv. Data Sci. Rev. **1**(2) (2019)
3. Dalzell, N.M., Reiter, J.P.: Regression modeling and file matching using possibly erroneous matching variables. J. Comput. Graph. Stat. **27**, 728–738 (2018)
4. Fellegi, I.P., Sunter, A.B.: A theory for record linkage. J. Am. Stat. Assoc. **64**(328), 1183–1210 (1969)
5. Fortini, M., Liseo, B., Nuccitelli, A., Scanu, M.: On Bayesian record linkage. Res. Off. Stat. **4**, 185–198 (2001)
6. Gutman, R., Afendulis, C.C., Zaslavsky, A.M.: A Bayesian procedure for file linking to analyze end-of-life medical costs. J. Am. Stat. Assoc. **108**(501), 34–47 (2013)
7. Hof, M.H., Ravelli, A.C., To, A.H.Z.: A probabilistic record linkage model for survival data. J. Am. Stat. Assoc. **112**(520), 1504–1515 (2017)
8. Jaro, M.A.: Advances in record-linkage methodology as applied to matching the 1985 census of Tampa, Florida. J. Am. Stat. Assoc. **84**, 414–420 (1989)
9. Larsen, M.D.: Iterative automated record linkage using mixture models. J. Am. Stat. Assoc. **96**, 32–41 (2001)
10. Larsen, M.D.: Comments on hierarchical Bayesian record linkage. In: Proceedings of the Section on Survey Research Methods. ASA, Alexandria, VA, 1995–2000 (2002)
11. Larsen, M.D.: Advances in record linkage theory: hierarchical Bayesian record linkage theory. In: Proceedings of the Section on Survey Research Methods. ASA, Alexandria, VA, pp. 3277–3284 (2005)
12. Larsen, M.D., Rubin, D.B.: Iterative automated record linkage using mixture models. J. Am. Stat. Assoc. **96**(453), 32–41 (2001)

13. Levenshtein, V.I.: Binary codes capable of correcting deletions, insertions, and reversals. Soviet Phys. Doklady **10**, 707–710 (1965)
14. Marchant, N.G., Steorts, R.C., Kaplan, A., Rubinstein, B.I., Elazar, D.N.: d-blink: distributed end-to-end Bayesian entity resolution (2019). arXiv preprint arXiv:1909.06039
15. Newcombe, H.B., Kennedy, J.M., Axford, S.J., James, A.P.: Automatic linkage of vital records: computers can be used to extract "follow-up" statistics of families from files of routine records. Science **130**(3381), 954–959 (1959)
16. Sadinle, M.: Detecting duplicates in a homicide registry using a Bayesian partitioning approach. Ann. Appl. Stat. **8**(4), 2404–2434 (2014). MR3292503
17. Sadinle, M.: Bayesian estimation of bipartite matchings for record linkage. J. Am. Stat. Assoc. **112**, 600–612 (2017)
18. Sadinle, M., Fienberg, S.E.: A generalized Fellegi-Sunter framework for multiple record linkage with application to homicide record systems. J. Am. Stat. Assoc. **108**(502), 385–397 (2013)
19. Steorts, R.C.: Entity resolution with empirically motivated priors. Bayesian Anal. **10**(4), 849–875 (2015). MR3432242
20. Steorts, R.C., Hall, R., Fienberg, S.E.: A Bayesian approach to graphical record linkage and deduplication. J. Am. Stat. Assoc. **111**(516), 1660–1672 (2016)
21. Tancredi, A., Liseo, B.: A hierarchical Bayesian approach to record linkage and population size problems. Ann. Appl. Stat. **5**(2B), 1553–1585 (2011). MR2849786
22. Winkler, W.E.: String comparator metrics and enhanced decision rules in the Fellegi-Sunter model of record linkage. In: Proceedings of the Section on Survey Research Methods. ASA, Alexandria, VA, pp. 354–359 (1990)
23. Winkler, W.E.: Overview of record linkage and current research directions. Technical report. Statistics #2006-2, U.S. Bureau of the Census (2006)
24. Winkler, W.E.: Matching and record linkage. Wiley Interdisc. Rev.: Comput. Stat. **6**(5), 313–325 (2014)
25. Zanella, G., Betancourt, B., Wallach, H., Miller, J., Zaidi, A., Steorts, R.C.: Flexible models for microclustering with application to entity resolution. In: Proceedings of the 30th International Conference on Neural Information Processing Systems, NIPS 2016, NY, USA. Curran Associates Inc., pp. 1425–1433 (2016)
26. Zhao, B., Rubinstein, B.I.P., Gemmell, J., Han, J.: A Bayesian approach to discovering truth from conflicting sources for data integration. Proc. VLDB Endow. **5**(6), 550–561 (2012)

Probabilistic Blocking and Distributed Bayesian Entity Resolution

Ted Enamorado[1,3] and Rebecca C. Steorts[2,3](✉)

[1] Department of Political Science, Washington University in St. Louis, St. Louis, USA
ted@wustl.edu
[2] Department of Statistical Science and Computer Science, Duke University, Durham, USA
beka@stat.duke.edu
[3] Principal Mathematical Statistician, United States Census Bureau, Suitland, USA

Abstract. Entity resolution (ER) is becoming an increasingly important task across many domains (e.g., official statistics, human rights, medicine, etc.), where databases contain duplications of entities that need to be removed for later inferential and prediction tasks. Motivated by scaling to large data sets and providing uncertainty propagation, we propose a generalized approach to the blocking and ER pipeline which consists of two steps. First, a probabilistic blocking step, where we consider that of [5], which is ER record in its own right. Its usage for blocking allows one to reduce the comparison space greatly, providing overlapping blocks for any ER method in the literature. Second, the probabilistic blocking step is passed to any ER method, where one can evaluate uncertainty propagation depending on the ER task. We consider that of [12], which is a joint Bayesian method of both blocking and ER, that provides a joint posterior distribution regarding both the blocking and ER, and scales to large datasets, however, it does it a slower rate than when used in tandem with [5]. Through simulation and empirical studies, we show that our proposed methodology outperforms [5,12] when used in isolation of each other. It produces reliable estimates of the underlying linkage structure and the number of true entities in each dataset. Furthermore, it produces an approximate posterior distribution and preserves transitive closures of the linkages.

1 Introduction

One of the most increasingly common problems in computer science and machine learning is merging databases that contain duplicate entities, often without a unique identifier due to privacy and confidentiality reasons. This task is referred as entity resolution (ER) or record linkage, and it is known to be difficult due to the noise and distortions as well as the size of the databases [2]. The scalability of

R. C. Steorts—This research was partially supported by the Alfred P. Sloan Foundation.

J. Domingo-Ferrer and K. Muralidhar (Eds.): PSD 2020, LNCS 12276, pp. 224–239, 2020.
https://doi.org/10.1007/978-3-030-57521-2_16

ER approaches has been recently studied. Considering T tables[1] (or databases) with N records each, one must make $O(N^T)$ comparisons using brute force approaches. In order to avoid such computational constraints, the number of comparisons needs to be reduced without comprising the ER accuracy. One typically achieves this via blocking, which seeks to place similar records into partitions (or blocks), and treating records in different blocks as non-co-referent. ER techniques are then applied *within* blocks.

The simplest blocking method is known as *traditional blocking*, which picks certain fields (e.g., gender and year of birth) and places records in the same block if and only if they agree on all fields. This amounts to a deterministic *a priori* judgment that these fields are error-free. Other *probabilistic blocking* methods use probability, likelihood functions, or scoring functions to place records in similar partitions. For example, techniques based upon locality sensitive hashing (LSH) utilize all the features of a record, and can be adjusted to ensure that all the records are manageably small. Recent work, such as, [21] introduced novel data structures for sorting and fast approximate nearest-neighbor look-up within blocks produced by LSH. This approach gave a good balance between speed and recall, but their technique was specific to nearest neighbor search with similarity defined by the hash function. Such methods are fast and have high recall (true positive rate), but suffer from low precision, rather, too many false positives.[2]

Turning to the ER literature, the most popular and widely used approach is that of [6]. While there exists a well-known literature on probabilistic record linkage, there are few available implementations that scale to large data sets commonly used in research or industry. Recently, [5] addressed this by developing a scalable implementation, called `fastLink`, of the seminal record linkage model of [6]. In addition, the authors extended work of [8] in order to incorporate auxiliary information such as population name frequency and migration rates. The authors used parallelization and hashing to merge millions of records in a near real-time on a laptop computer, and provided open-source software of their proposed methodology. [5] compared their open-source software to that of [16], which utilizes that of [6]. For example, `fastLink` provided a 22 times speed up compared to the `RecordLinkage` package when merging datasets of 50,000 records each. Further, [5] showed that these speed gains grow orders of magnitude larger as the size of the datasets goes beyond 50,000 records. To our knowledge, this method has never been proposed as a blocking step.

While there have been a large number of new proposals in the Bayesian literature, many of these have difficulty scaling due to the curse of dimensionality with Markov chain Monte Carlo (MCMC) [14,15,17–19,22].[3] Specifically, even with the use of blocking, these methods still scale quadratically with the number of

[1] A *table* is an ordered (indexed) collection of records, which may contain duplicates (records for which all attributes are identical).

[2] This approach is called *private* if, after the blocking is performed, all candidate records pairs are compared and classified into coreferent/non-confererent using computationally intensive "private" comparison and classification techniques [4].

[3] For a review of Bayesian methods, we refer the interested reader to [11].

records. Existing attempts to manage scalability apply to deterministic blocking outside the Bayesian framework, thereby sacrificing accuracy and proper treatment of uncertainty [7,9,10,13,14,18–20]. The only work among these which deviates from deterministic blocking, is a post-hoc blocking approach proposed by [13]. In addition to coarse-grained deterministic blocking, they apply data-dependent post-hoc blocking formulated as restricted MCMC. This can yield significant efficiency gains, but may induce approximation error in the posterior, since pairs outside the post-hoc blocks cannot be linked.

We are only aware of one *joint* Bayesian and blocking method that scales to realistically sized datasets, where [12] have proposed a joint model for blocking and ER that is fully Bayesian, where they propose scaling via distributed computing using a partially collapsed Gibbs sampler. Given that the model is joint, the error of the blocking and the entity resolution is propagated exactly, which contrasts the approach of [5,13]. In both of these approaches, the blocking task cannot propagated exactly into the ER task given that these two approaches do not propose a joint model. Thus, any errors that are made in the blocking task are passed to the entity resolution task and cannot be corrected. Accuracy may suffer such that the methods can be more computationally scalable.

In this paper, we make the following contributions. Motivated by the need to both scale to large data sets as well as provide uncertainty propagation, first we propose utilizing a probabilistic blocking step. Specifically, we consider that of [5], which, as noted above, is an ER record in its own right. We propose its usage for blocking as it allows one to reduce the comparison space, providing overlapping blocks for any ER method in the literature. Second, we propose that the probabilistic blocking step is passed to any ER method, where one can evaluate uncertainty propagation depending on the proposed ER task. Specifically, we consider that of [12], which is a joint Bayesian method of both blocking and ER, which scales to databases in the millions, and provides a joint posterior distribution regarding both the blocking and ER task at hand. Third, we consider the computational complexity of our proposed methodology. Fourth, we consider simulation studies of our proposed methodology, where we compare to baselines of `fastLink` and `dblink`. Finally, we provide a discussion of our proposed method and avenues for future work.

The remainder of the paper is as follows. Section 2 proposes a general framework for probabilistic blocking and ER. We recommend a specific proposal (see Sects. 2.1, 2.2, and 2.3). Section 2.4 provides the computational complexity. Section 3 provides the evaluation metrics. Section 4 provides simulation studies on our proposed methods with comparisons to two baselines from the literature. Section 6 concludes.

2 Proposed Methodology

In this section, we provide our proposed methodology, which is a framework for performing a pipeline of probabilistic blocking and ER. We assume that we have T aligned databases that contain missing values, and we assume that these databases may be large. First, given that all-to-all comparisons is not feasible due to the total number of records, we first perform dimension reduction using probabilistic blocking, where we assume the blocks can be overlapping. This is important as this allows mistakes to be corrected in the ER task. Second, the blocks are passed to an ER task, such that the duplications can be removed. In addition, it is important that both accuracy and inference can be assessed as these are often goals of the ER task.

Specifically, in this paper, we consider the work of [5] as our proposal for probabilistic blocking, given that it is based upon the seminal work of [6]. This method is appealing given its simplicity, popularity, and immense speed for a blocking method that can then be passed to a fully unsupervised method regarding the ER task, such as [12]. In Sects. 2.1 and 2.2, we review the methods of [5,12], where we have built an open source pipeline regarding this approach to be used by the computer science, statistics, and social science communities.

2.1 Review of `fastLink`

In this section, we review the method of [5] and then describe how it can be utilized for probabilistic blocking for the first time to our knowledge. The Fellegi-Sunter model is fit to the data based on agreement patterns (discrete comparisons) of each attribute across all pairs of records (i, j) between two data sets $\mathcal{A}$ and $\mathcal{B}$. Formally, for each attribute a, we define $\rho_a(i, j)$ to be a discrete variable e.g., to allow for three levels of similarity, we define $\rho_a(i, j)$ to be a factor variable with three levels, in which 0, 1, and 2 indicate that the values of two records for this variable are different, similar, and identical (or nearly so), respectively. Based on this definition, the model can be written as the following two-class mixture model with the latent variable C_{ij}, indicating coreference $C_{ij} = 1$ or a non-coreference $C_{ij} = 0$ for the pair (i, j),

$$\rho_a(i,j) \mid C_{ij} = c \overset{\text{indep.}}{\sim} \text{Discrete}(\pi_{ac}) \quad (1)$$

$$C_{ij} \overset{\text{i.i.d.}}{\sim} \text{Bernoulli}(\mu), \quad (2)$$

where π_{ac} is a vector of length $\mathcal{H}_a$, which is the number of possible values taken by $\rho_a(i, j)$, containing the probability of each agreement level for the ath variable given that the pair is coreferent ($c = 1$) or non-coreferent ($c = 0$), and μ represents the probability of coreference across all pairwise comparisons. Once the model is fit to the data, the coreference probability is computed via Bayes rule based on the maximum likelihood estimates of the model parameters (μ, π_{ac}),[4]

$$\begin{aligned}\xi_{ij} &= \Pr(C_{ij} = 1 \mid m(i,j), \rho(i,j), \mu, \pi_{ac}) \\ &= \frac{\mu \prod_{a=1}^{A} \left(\prod_{h=0}^{\mathcal{H}_k - 1} \pi_{a1l}^{\mathbf{1}\{\rho_a(i,j)=h\}} \right)^{1-m_a(i,j)}}{\sum_{c=0}^{1} \mu^c (1-\mu)^{1-c} \prod_{a=1}^{A} \left(\prod_{h=0}^{\mathcal{H}_k - 1} \pi_{acl}^{\mathbf{1}\{\rho_a(i,j)=l\}} \right)^{1-m_a(i,j)}}\end{aligned} \quad (3)$$

[4] The Expectation-Maximization algorithm is used to recover estimates for μ and π_{ac} (M-Step) and ξ_{ij} (E-Step).

where $m_a(i,j)$ indicates whether the value of variable a is missing for pair (i,j) (a missing value occurs if at least one record for the pair is missing the value for the variable). We say that record j is a potential link of record i based on the coreference probability ξ_{ij}. Note that the model assumes (1) independence across pairs, (2) independence across linkage fields conditional on the latent variable C_{ij}, and (3) missing at random conditional on C_{ij} [5].

2.2 Review of dblink

In this section, we review the method of **dblink**, and then describe how we integrate it with the approach of [5] for the first time to our knowledge.

[12] extended [17], where ER was framed as a bipartite matching problem between records and latent entities. Specifically, the work of [17] introduced a model called **blink** which extended [18] to include both categorical and noisy string attributes using an empirically-motivated prior. [12] have proposed a scalable, joint blocking and ER model under a Bayesian framework referred to as "distributed **blink**" or **dblink**. To our knowledge, this is the first scalable and distributed model where uncertainty both the blocking and the ER stages is able to be accounted for exactly. This contrasts other Bayesian frameworks, where the blocking uncertainty is approximated or not accounted for when passed to the Bayesian ER task. The output from the method is a joint posterior of the blocks and the linkage structure. In addition, the proposed method has a representation that enables distributed inference at the partition level, resulting in a sub-linear algorithm (as the number of records grows).

Again, assume a total of T tables (databases) indexed by t, each with R_t records (rows) indexed by r and A aligned attributes (columns) indexed by a. Assume a fixed population of entities of size E indexed by e, where entity e is described by a set of attributes $\boldsymbol{y}_e = [y_{ea}]_{a=1\ldots A}$, which are aligned with the record attributes. Within this framework, the population of entities is partitioned into P blocks using a blocking function PartFn, which maps an entity e to a block according to its attributes $\boldsymbol{y}_e$. Assume each record (t,r) belongs to a block γ_{tr} and is associated with an entity λ_{tr} within that block. Denote the value of the a-th attribute for record (t,r) by x_{tra}, which is a noisy observation of the associated entity's true attribute value $y_{\lambda_{tra}}$. Assume that some attributes x_{tra} are missing completely at random through the indicator variable o_{tra}. Denote θ_{ta} as a distortion probability with an assumed prior distribution, where α_a and β_a are fixed hyper-parameters.

The records are generated by selecting an entity uniformly at random and copying the entity's attributes subject to distortion. First, one chooses a partition assignment γ_{tr} at random in proportion to the partition sizes: $\gamma_{tr}|\boldsymbol{Y} \overset{\text{ind.}}{\sim} \text{Discrete}_{p\in\{1\ldots P\}}[|\mathcal{E}_p|/E]$. Second, one chooses an entity assignment λ_{tr} uniformly at random from partition γ_{tr}: $\lambda_{tr}|\gamma_{tr}, \boldsymbol{Y} \overset{\text{ind.}}{\sim} \text{DiscreteUniform}[\mathcal{E}_{\gamma_{tr}}]$. Third, for each attribute a, one draws a distortion indicator z_{tra}: $z_{tra}|\theta_{ta} \overset{\text{ind.}}{\sim} \text{Bernoulli}[\theta_{ta}]$.

Fourth, for each attribute a, one draws a record value x_{tra}:

$$x_{tra}|z_{tra}, y_{\lambda_{tr}a} \overset{\text{ind.}}{\sim} (1 - z_{tra})\delta(y_{\lambda_{tr}a}) + z_{tra}\,\text{Discrete}_{v \in \mathcal{V}_a}[\psi_a(v|y_{\lambda_{tr}a})] \quad (4)$$

where $\delta(\cdot)$ represents a point mass. If $z_{tra} = 0$, x_{tra} is copied directly from the entity. Otherwise, x_{tra} is drawn from the domain $\mathcal{V}_a$ according to the distortion distribution ψ_a. In the literature, this is known as a hit-miss model [3]. Finally, for each attribute a, one draws an observed indicator o_{tra}: $o_{tra} \overset{\text{ind.}}{\sim} \text{Bernoulli}[\eta_{ta}]$. If $o_{tra} = 1$, x_{tra} is observed, otherwise it is missing. Figure 1 in Appendix A provides a diagram for the `dblink` model.

Assume the following notation for convenience: A boldface lower-case variable denotes the set of *all attributes*: e.g. $\boldsymbol{x}_{tr} = [x_{tra}]_{a=1\ldots A}$. A boldface capital variable denotes the set of *all index combinations*: e.g. $\boldsymbol{X} = [x_{tra}]_{t=1\ldots T; r=1\ldots R_t; a=1\ldots A}$. In addition, let $\mathbf{X}^{(o)}$ denote the x_{tra}'s for which $o_{tra} = 1$ (observed) and let $\mathbf{X}^{(m)}$ denote the x_{tra}'s for which $o_{tra} = 0$ (missing). The authors perform ER by inferring the *joint* posterior distribution over the block assignments $\boldsymbol{\Gamma} = [\gamma_{tr}]_{t=1\ldots T; r=1\ldots R_t}$, the linkage structure $\boldsymbol{\Lambda} = [\lambda_{tr}]_{t=1\ldots T; r=1\ldots R_t}$, and the true entity attribute values $\boldsymbol{Y} = [y_{ea}]_{e=1\ldots E; a=1\ldots A}$, conditional on the observed record attribute values $\mathbf{X}^{(o)}$.

Remark 1. The authors operate in a fully unsupervised setting, as they do not condition on ground truth data for the links or entities. Inferring $\boldsymbol{\Gamma}$ is equivalent to the *blocking* stage of ER, where the records are partitioning into blocks to limit the comparison space. Inferring $\boldsymbol{\Lambda}$ is equivalent to the *matching/linking* stage of ER, where records that refer to the same entities are linked together. Inferring $\boldsymbol{Y}$ is equivalent to the *merging* stage, where linked records are combined to produce a single representative record. By inferring $\boldsymbol{\Gamma}$, $\boldsymbol{\Lambda}$ and $\boldsymbol{Y}$ jointly, they are able to propagate uncertainty between the three stages.

2.3 Putting It Together: `fastLink-dblink`

In this paper, we propose a two-stage approach that combines the strength of two probabilistic entity resolution approaches, where crucially in the first stage, we do not rely on traditional blocking strategies. As already mentioned, the first stage makes use of `fastLink` as a pre-processing step designed to identify a set of possible matches. Then, based on the set of possible matches produced by `fastLink`, we further refine the entity resolution process via `dblink`, where we determine the co-reference structure of the linkages, where transitivity is satisfied, and we are able to have approximate measures of uncertainty, which include the posterior distribution of the linkage structure, precision, and recall. Our proposed method strikes a balance between the two methods. `fastLink` provides very fast linking, but no uncertainty propagation, whereas, `dblink` provides exact uncertainty propagation. Thus, with `fastLink-dblink` we settle for fast approximate uncertainty quantification. As our results show, such a balance pays off in practice.

2.4 Computational Complexity

In this section, we provide the computational complexity of `fastLink-dblink`. Let N_{max} represent the total number of records in all databases. Assume without loss of generality that for `fastLink` we de-duplicate records, i.e., $N_{\mathcal{A}} = N_{\mathcal{B}} = N_{max}$, and that each table is of equal size i.e., $N_t = N$ for all $t = 1, \ldots, T$. Further, assume that for `dblink` we perform ER, meaning we find duplicate records within and across the N^*_{max} records in all databases, where $N^*_{max} \ll N_{max}$ is the number of records classified as possible matches by `fastLink`. Finally, assume that S_G is the total number of MCMC iterations.

Theorem 1. *The computational complexity of* `fastLink-dblink` *is*

$$O(\Upsilon N_{max}(N_{max} - 1)) + O(S_G N^*_{max}), \quad \text{where} \quad \Upsilon = \frac{\omega}{2Q}.$$

Proof. First, we prove the computational complexity of `fastLink`. Assume that A represents the total number of attributes. The computational complexity of all-to-all record comparison of the FS method is $O(\frac{A}{2}N_{max}(N_{max} - 1))$, which is this is computationally intractable as N_{max} grows. `fastLink` compares only the unique instance of a value per linkage field, leading to a time complexity equal to $O(\frac{\omega}{2}N_{max}(N_{max} - 1))$, where $\omega = \sum_{a=1}^{A} \kappa_{a,\mathcal{A}}\kappa_{a,\mathcal{B}}$ and $\kappa_{a,t} \in [0, 1]$ represents the share of unique values per linkage field a per dataset $t \in \{\mathcal{A}, \mathcal{B}\}$. Unless $\kappa_{a,t} = 1$ for all a and t, we have that $O(\frac{\omega}{2}N_{max}(N_{max}-1)) \ll O(\frac{A}{2}N_{max}(N_{max}-1))$. In addition, `fastLink` takes advantage of multithreading/parallelization and divides the problem into B equally-sized partitions which are $O(\frac{\omega}{2B}N_{max}(N_{max}-1))$ each. While these partitions are quadratic, if either N_{max} or ω grows, one can always choose B to be large and guarantee computational feasibility. Therefore, with Q threads available and $Q \ll B$, the final complexity is $O(\Upsilon N_{max}(N_{max} - 1))$ with $\Upsilon = \frac{\omega}{2Q}$.[5]

Second, we prove the computational complexity of `dblink`. Recall S_G is the number of MCMC iterations. Let $M = \frac{1}{A}\sum_{a=1}^{A} M_a$ be the average number of possible values per field ($M \geq 1$). The computational complexity of each Gibbs iteration is dominated by the conditional distributions of `dblink`. The update for the conditional distribution of λ_{tr} is $O(|\mathcal{L}_{tr}|)$, where $\mathcal{L}_{tr}$ is the set of candidate links for record (t, r). This is a linear time operation given that it is done by inverted index and scales linearly using hashed based representations. The update for θ_{ta} is $O(aN^*_{max})$, where N^*_{max} is the total number of records in all the tables (data sets). The update for z_{tra} is $O(aMN^*)$, where N^* is the number of records per table that have been classified as possible matches by `fastLink`. Therefore, the total order of the algorithm is $O(|\mathcal{L}_{tr}|) + O(AN^*_{max}) + O(AMN^*)$ for each Gibbs iteration. For S_G iterations, the algorithm is $O(S_G|\mathcal{L}_{tr}|) + O(S_G AN^*_{max}) + O(S_G AMN^*)$. Recall that

[5] After the comparisons are made, to retrieve a match or a pairwise comparison, the lookup time is $O(H)$ where H is the number of unique agreement patterns that are observed. Note that in practice, H is often times smaller than the number of possible agreement patterns ($\prod_{a=1}^{A} \mathcal{H}_a$).

$N^* < N^*_{max}$. If $|\mathcal{L}_{tr}|$, A, and M are much smaller than N^*_{max}, then the computational complexity is $O(S_G N^*_{max})$. Thus, the total computational complexity of `fastLink-dblink` is $O(\Upsilon N^*_{max}(N^*_{max} - 1)) + O(S_G N^*_{max})$, where $\Upsilon = \frac{\omega}{2Q}$.

3 Evaluation Metrics

In this section, we describe evaluation metrics used in the paper. In order to evaluate ER performance, we compute the pairwise precision, recall, and F-measure [1]. In addition, we evaluate each method by standard cluster metrics in the literature (if possible), based upon the adjusted Rand index (ARI) and the percentage error in the number of clusters. In addition, we also evaluate any Bayesian method regarding inference based upon its posterior mean and standard error.

4 Simulation Studies

In this section, we describe our simulation studies, which are based upon two synthetic datasets (`RLdata500` and `RLdata10000`) from the `RecordLinkage` package in `R` with a total of 500 and 10,000 total records and 10 percent duplication. Feature information available is first and last name and full date of birth. There is a unique identifier available such that one can ascertain the accuracy of proposed methods. In our simulation study for `RLdata500`, we consider two blocking criterions (loose and strict) based upon `fastLink`. We compare our baselines of `fastLink` (no blocking) and `dblink` (no blocking) to our proposed method. In our simulation study for `RLdata10000` , we consider the same type of criterion — loose and strict and the same comparisons as done for `RLdata500`. We do not utilize any partitioning (blocking) of `dblink` given the small size of the data sets.

In the case of the `fastLink` step, we use three categories for the string valued variables (first and last names), i.e., agreement (or nearly agreement), partial agreement, and disagreement, based on the Jaro-Winkler string distance with 0.94 and 0.84 as the cutpoints. For the numeric valued fields (day, month, and year of birth), we use a binary comparison, based on exact matches. For each data set, and based on the match probability ξ_{ij} from `fastLink`, we use 0.60 (loose) and 0.90 (strict) as the thresholds to produce two sets of likely duplicates.[6]

Next, we feed the set of possible duplicates per data set to `dblink`. For both simulation studies, we run the `dblink` sampler for 100,000 Gibbs iterations (using a thin of 10). For each simulation study, we utilize the Edit distance for string distances, with a threshold of 7.0, and a maximum similarity of 10.0. First and last name are compa red using Edit distance, while birth year, birth month, and birth day are assumed to be categorical variables with low distortion. We assume a maximum cluster size of 10. For `RLdata500`, $\alpha = 0.5, \beta = 50$ and `RLdata10000`, $\alpha = 10, \beta = 10000$. Convergence diagnostics can be found in Appendix B.

[6] These thresholds are selected such that the estimated rate of false positives is less than 1% (strict) and 10% (loose).

4.1 Simulation Study Results

In this section, we describe the results from our simulation studies. Table 1 reports the pairwise precision, recall, F-measure, the adjusted Rand index (ARI), and the percentage error in the number of clusters for the `RLdata500` and `RLdata10000` simulation studies, where we provide a baseline comparisons to `dblink`, `fastLink-L`, and `fastLink-S`. For `RLdata500`, when we consider `dblink` and `fastLink-L`, we find that the precision and recall are the same at 0.9091 and 1.0000. This intuitively makes sense given `dblink` has no blocking and the blocking criterion for `fastLink` is allowed to be non-conservative (or loose). When one considers, `fastLink-S`, the precision and recall are 0.9800 and 1.0000, respectively. We find a substantial improvement regarding the precision due to the pre-processing step of `fastLink` combined with `dblink`. For example, with `fastLink-dblink-L`, the precision improves from 0.9091 to 0.9434; and with `fastLink-dblink-S`, the precision goes from 0.9091 to 1.0000. Finally, we see a computational speed up in the method of `dblink` regarding runtime from 1 h (no probabilistic blocking) to 10 min (`fastLink-dblink`) on a standard Linux server with one core. Note that the computational run time of `fastLink` is 1.2 s.

Turning to `RLdata10000`, when we consider `dblink`, we find that the precision is 0.6334 and recall is 0.9970, illustrating there is room for improvement in the precision. When one considers `fastLink-L`, we observe the precision is 0.9371 and recall is 0.9840, and when turning to `fastLink-S`, we observe evaluation metrics of 0.9589 and 0.9820, respectively. We see improvements in the precision due to the pre-processing step of `fastLink` combined with `dblink`. For `fastLink-dblink-L`, the precision found using `dblink` improves from 0.6334 to 0.9428 and the recall goes from 0.9970 to 1. In the case of `fastLink-dblink-S`, it has a precision of 0.9563 and recall of 0.9989. Finally, we see a computational speed up in the method of `dblink` regarding runtime from 3 h (no probabilistic blocking) to 40 min (`fastLink-dblink`) on a standard Linux server with one core. The computational run time of `fastLink` is 12 s.

Table 2 reports the posterior mean for the number of entities and the corresponding standard error (if available). It is not possible to calculate these values for `fastLink`. For the `RLdata500` simulation study, we find that `dblink` provides a posterior mean of 443.68 with standard error of 2.02. `fastLink-dblink-L` provides a posterior mean of 452.82 with standard error of 0.60 and `fastLink-dblink-S` provides a posterior mean of 451.62 with standard error of 0.55. Observe that `dblink` underestimates the true population size of 450, whereas our proposed method using both loose and strict blocking provides a slight overestimate.

In the case of `RLdata10000`, `dblink` provides a posterior mean of 8012.83 with standard error of 26.26. Observe that this estimate is far off from the true population value of 9000. `fastLink-dblink-L` provides a posterior mean of 8968.95 with a posterior standard error of 2.60 and `fastLink-dblink-S` provides a posterior mean of 8985.33 with posterior standard error of 2.55. Here, we see that the loose and strict blocking provide a slight underestimate of the true population size.

Table 1. Comparison of Matching Quality. "ARI" stands for adjusted Rand index and "Err. # clust." is the percentage error in the number of clusters.

Dataset	Method	Pairwise measures			Cluster measures	
		Precision	Recall	F1-score	ARI	Err. # clust
RLdata500	dblink	0.9091	1.0000	0.9524	0.9523	–1.55%
	fastLink-L	1.0000	0.9800	0.9899	—	—
	fastLink-S	1.0000	0.9800	0.9899	—	—
	fastLink-dblink-L	1.0000	0.9591	0.9791	0.9789	+0.67%
	fastLink-dblink-S	1.0000	1.0000	1.0000	1.0000	+0.44%
RLdata10000	dblink	0.6334	0.9970	0.7747	0.7747	–10.97%
	fastLink-L	0.9371	0.9840	0.9600	—	—
	fastLink-S	0.9589	0.9820	0.9704	—	—
	fastLink-dblink-L	0.9428	1.0000	0.9705	0.9705	–0.34%
	fastLink-dblink-S	0.9563	0.9989	0.9772	0.9771	–0.17%

Table 2. Comparison of Posterior Mean and Standard Error for the Number of Unique Entities.

Dataset	Method	Posterior Estimates	
		Mean	Std. Error
RLdata500	dblink	443.68	2.02
	fastLink-L	–	–
	fastLink-S	–	–
	fastLink-dblink-L	452.82	0.60
	fastLink-dblink-S	451.62	0.55
RLdata10000	dblink	8012.83	26.26
	fastLink-L	–	–
	fastLink-S	–	–
	fastLink-dblink-L	8968.95	2.60
	fastLink-dblink-S	8985.33	2.55

5 Application to the National Long Term Care Study

In this section, we apply our proposed methodology to the National Long Care Term Study (NLTCS), which is a longitudinal study on health and well-being of those in the United States that are older than sixty-five years. This data set consists of three waves, with duplications across but not within each data set. The entire number of individuals in the study is 57,077. Due to privacy and confidentiality, the data was anonymized. One goal of this study is to reveal how one can easily link such data (which is often public) based upon the following features: full DOB, state, location of doctor's office, and gender. The true number of unique individuals in this data set is 34,945. We illustrate that via `fastLink-dblink`, we are able to link records a higher level of accuracy than previously proposed methods, which raises important questions regarding linking anonymized health data given that an increasing amount of it is publicly available or shared with industry given the rise of technology applications. Specifically, with the rise of virus spread (e.g. SARS-CoV-2 (COVID-19)) and increased contact tracing, one would expect the accuracy of these methods to only increase with additional features. This raises important questions regarding privacy tradeoffs regarding the collections of such data that can be matched with extremely sensitive information.

5.1 Results to the National Long Term Care Study

In this section, we provide results to the `NLTCS`. Table 3 reports the pairwise precision, recall, F-measure, the adjusted Rand index (ARI), and the percentage error in the number of clusters for the `NLTCS`, where we provide a baseline comparison to `dblink`, `fastLink-L`, and `fastLink-S`. For the `NLTCS`, when we consider `dblink`, we find that the precision and recall are 0.8319 and 0.9103, respectively – slightly better than of `fastLink-L`, which has a precision of 0.7977 and recall of 0.9101. In the case of `fastLink-S`, the precision and recall are 0.9094 and 0.9087, respectively. Similar to the simulations studies, we see a significant improvement when combining `fastLink` with `dblink`. Specifically, when utilizing `fastLink-dblink-L`, both the precision and recall improve to 0.8712 and 0.9363, respectively. In the case of `fastLink-dblink-S`, we also see a remarkable improvement in terms of recall, moving from 0.9103 (`dblink`) to 0.9971. Finally, we see a computational speed up of `dblink` regarding runtime from 4 h (no probabilistic blocking) to 3 h (`fastLink-dblink`) on a standard Linux server with 1 core. The computational run time of `fastLink` is 16 min.[7]

Table 3. Comparison of matching quality. "ARI" stands for adjusted Rand index and "Err. # clust." is the percentage error in the number of clusters.

Dataset	Method	Pairwise measures			Cluster measures	
		Precision	Recall	F1-score	ARI	Err. # clust.
`NLTCS`	`dblink`	0.8319	0.9103	0.8693	0.8693	-22.09%
	`fastLink-L`	0.7977	0.9101	0.7963	—	—
	`fastLink-S`	0.9094	0.9087	0.9090	—	—
	`fastLink-dblink-L`	0.8712	0.9363	0.9025	0.9025	-13.01%
	`fastLink-dblink-S`	0.9094	0.9971	0.9512	0.9512	2.79%

Table 4 reports the posterior mean and standard error (if available). For the `NLTCS`, we find that `dblink` provides a posterior mean of 29,990 with standard

[7] Parameter settings for `fastLink-dblink` can be found in Appendix C.

Table 4. Comparison of Posterior Mean and Standard Error for the Number of Unique Entities.

Dataset	Method	Posterior Estimates	
		Mean	Std. Error
`NLTCS`	`dblink`	29,990	63.48
	`fastLink-L`	—	—
	`fastLink-S`	—	—
	`fastLink-dblink-L`	30,397	55.46
	`fastLink-dblink-S`	35,922	13.07

error of 63.48. `fastLink-dblink-L` provides a posterior mean of 30,397 with standard error of 55.46 and `fastLink-dblink-S` provides a posterior mean of 35,922 with standard error of 13.07. Observe that `dblink` underestimates the true population size of 34,945, whereas our proposed method using the strong blocking provides a slight overestimate and the loose blocking provides an underestimate. Convergence diagnostics can be found in Appendix B.

6 Discussion

In this paper, motivated by the need to scale ER to large data sets in fast and accurate ways, we have proposed a principled two-stage approach to blocking and ER, where our recommendation is based upon first performing the blocking scheme based upon `fastLink`, where one can perform this step for around 50 K records in just minutes. Our recommendation for the ER stage, is using a joint blocking and Bayesian method such that one obtains a posterior distribution, and thus, uncertainty propagation. Of course, there are many parameter choices that must be selected for our proposed framework, however, extensive simulation and sensitivity analyses have guided these choices [12,17]. One area for future work would be making these methods more automated, however this would come at the cost of being more computationally demanding. Another area for future work would be automating more string distance functions in our pipeline.

We believe that this open-source pipeline could be useful for official statistics agencies looking for guidance regarding how to improve their own. Finally, as illustrated the data from the `NLTCS`, we recommend caution with methods given the high accuracy for categorical data that is anonymized. This tends to lead to guidance to the community regarding sharing sensitive medical data that could be linked or tracking is used, and one's personal privacy could be in risk without any prior consent or knowledge.

A Plate Diagram for `dblink`

In this section, we provide a plate diagram for the `dblink` model (Fig. 1).

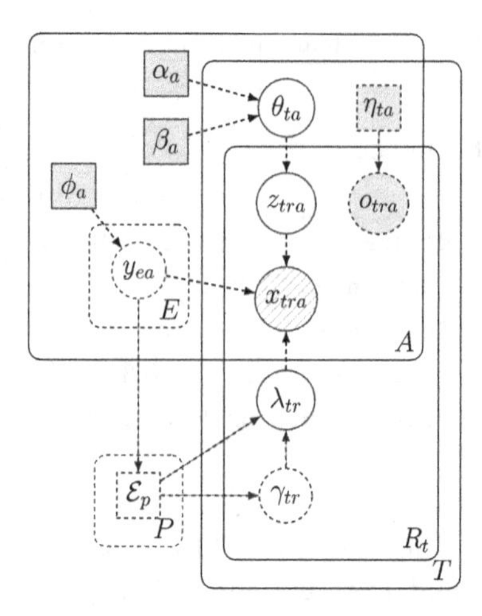

Fig. 1. Plate diagram for `dblink`. Extensions to `blink` are highlighted dashed blue lines. Circular nodes represent random variables, whereas square nodes represent deterministic variables. (Un)shaded nodes represent (un)observed variables. Arrows represent conditional dependence, and plates represent replication over an index.

B Comparison of `dblink` and `fastLink-dblink-S` and Convergence Diagnostics

In this section, we provide convergence diagnostics after passing the preprocessing step of `fastLink`to `dblink`. For all experiments, we utilized 100,000 Gibbs iterations (with a thin of 10). Figures 2 and 3 contain trace plots for two summary statistics as a function of running time, where the first plot illustrates this for `RLdata10000` and the second illustrates this for the `NLTCS`. The convergence diagnostics plot for `RLdata500` is very similar to that of `RLdata10000`. The left most plot provides the number of observed entities versus time. The right most plot provides the distortion for each feature versus time. For both `RLdata10000` and `NLTCS`, we consider `dblink` as a baseline and compare this to `fastLink-dblink`.

Figure 2 (left plot) illustrates that the convergence for `fastLink-dblink` is faster slower than that of `dblink`, however, the inferential posterior mean is much more accurate under `fastLink-dblink-S` (see teal line). Figure 2 (right plot) illustrates that the the convergence for `fastLink-dblink`is faster than that of `dblink` for each feature. Figure 3 (left plot) illustrates that the convergence for `fastLink-dblink` is much faster than that of `dblink`. Again, the posterior mean is much more accurate under `fastLink-dblink-S` (see teal line). Figure 2 (right plot) illustrates that the the convergence for `fastLink-dblink` is much faster than that of `dblink` for each feature.

C Parameter Settings for NLTCS Experiments

In this section, we describe the parameter settings for `fastLink-dblink`. In the case of the `fastLink` step, we use a binary comparison, based on exact matches. For each data set, and based on the match probability ξ_{ij} from `fastLink`, we use 0.98 (loose) and 0.90 (strict) as the thresholds to produce two sets of likely duplicates. As before, these threshold are selected such that the estimated rate of false positives is less than 1% (strict) and 10% (loose). In the case of the `dblink` step of `fastLink-dblink`, we run 150,000 iterations of the Gibbs sampler (using a thin value of 10). In addition, we used a KD-tree with two-levels of partitioning, which result in a total of four partitions, which were the following: DOB month, DOB day, DOB year, and state. All six features were treated as categorical variables. For all experiments, $\alpha = 448.134, \beta = 44813.4$. For all experiments, the threshold is 7.0, the max-similarity is 10.0, and the expected cluster max size is 10.

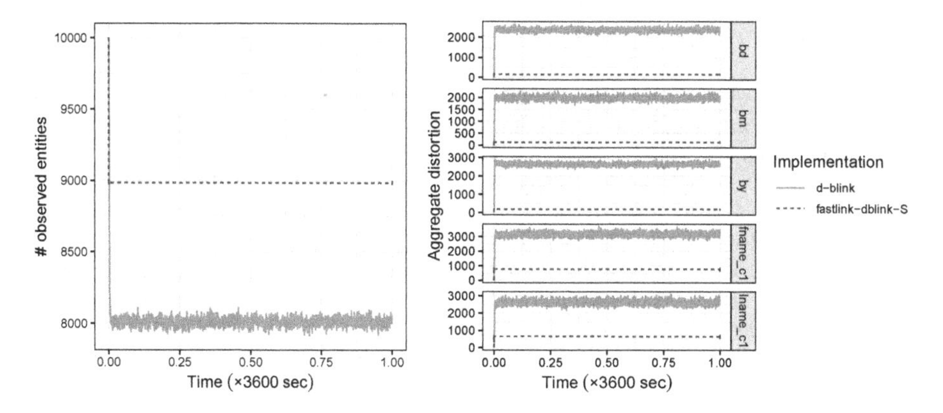

Fig. 2. Comparison of convergence rates for `dblink` and `fastLink-dblink-S` for `RLdata10000`. The summary statistics for `dblink` and `fastLink-dblink-S` (number of observed entities on the left and attribute distortions on the right) converge within 1 h (3600 s).

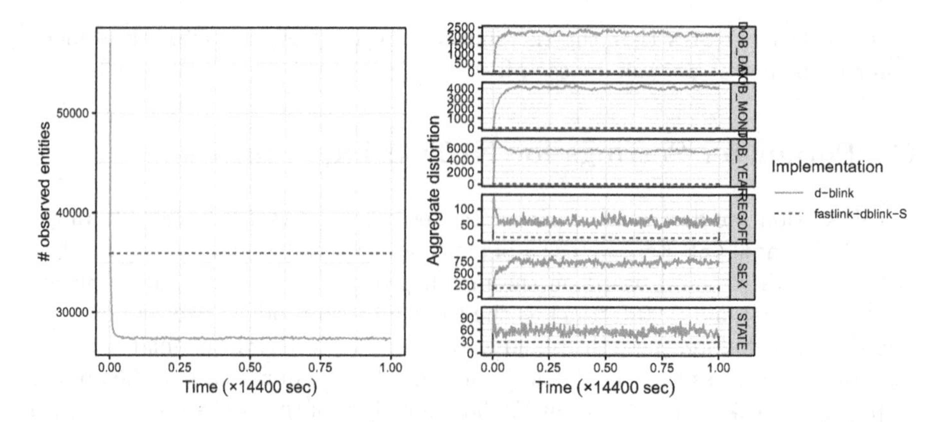

Fig. 3. Comparison of convergence rates for `dblink` and `fastLink-dblink-S` for `NLTCS`. The summary statistics for `dblink` and `fastLink-dblink-S` (number of observed entities on the left and attribute distortions on the right) converge within 4 h (14400 s).

References

1. Christen, P.: Data Matching: Concepts and Techniques for Record Linkage, Entity Resolution, and Duplicate Detection. Springer, Berlin, Data-Centric Systems and Applications (2012)
2. Christen, P.: Data linkage: the big picture. Harvard Data Sci. Rev. (2019)
3. Copas, J.B., Hilton, F.J.: Record linkage: statistical models for matching computer records. J. R. Stat. Soc. Ser. (Stat. Soc.) **153**, 287–320 (1990)
4. De Vries, T., et al.: Robust record linkage blocking using suffix arrays. In: Proceedings of the 18th ACM Conference on Information and Knowledge Management, pp. 305–314 (2009)
5. Enamorado, T., Fifield, B., Imai, K.: Using a probabilistic model to assist merging of large-scale administrative records. Am. Polit. Sci. Rev. **113**, 353–371 (2019)
6. Fellegi, I.P., Sunter, A.B.: A theory for record linkage. J. Am. Stat. Assoc. **64**, 1183–1210 (1969)
7. Gutman, R., Afendulis, C.C., Zaslavsky, A.M.: A Bayesian procedure for file linking to analyze end-of-life medical costs. J. Am. Stat. Assoc. **108**, 34–47 (2013)
8. Lahiri, P., Larsen, M.D.: Regression analysis with linked data. J. Am. Stat. Assoc. **100**, 222–230 (2005)
9. Larsen, M.D.: Advances in record linkage theory: hierarchical Bayesian record linkage theory. In: Proceedings of the Survey Research Methods Section. American Statistical Association, pp. 3277–3284 (2005)
10. Larsen, M.D.: An experiment with hierarchical Bayesian record linkage. arXiv preprint arXiv:1212.5203 (2012)
11. Liseo, B., Tancredi, A.: Some advances on Bayesian record linkage and inference for linked data (2013). URL http://www. ine. 418/es/e/essnetdi ws2011/ppts/Liseo Tancredi. pdf

12. Marchant, N.G., et al.: d-blink: Distributed end-to-end Bayesian entity resolution. arXiv preprint arXiv:1909.06039 (2019)
13. McVeigh, B.S., Spahn, B.T., Murray, J.S.: Scaling Bayesian probabilistic record linkage with post-hoc blocking: an application to the california great registers. Tech. rep (2020)
14. Sadinle, M.: Detecting duplicates in a homicide registry using a Bayesian partitioning approach. Ann. Appl. Stat. **8**, 2404–2434 (2014)
15. Sadinle, M.: Bayesian estimation of bipartite matchings for record linkage. J. Am. Stat. Assoc. **112**, 600–612 (2017)
16. Sariyar, M., Borg, A.: Record linkage in R. R package. version 0.4-10 (2016). http://cran.r-project.org/package=RecordLinkage
17. Steorts, R.C.: Entity resolution with empirically motivated priors. Bayesian Anal. **10**, 849–875 (2015)
18. Steorts, R.C., Hall, R., Fienberg, S.E.: A Bayesian approach to graphical record linkage and deduplication. J. Am. Stat. Assoc. **111**, 1660–1672 (2016)
19. Tancredi, A., Liseo, B.: A hierarchical Bayesian approach to record linkage and population size problems. Ann. Appl. Stat. **5**, 1553–1585 (2011)
20. Tancredi, A., Steorts, R., Liseo, B., et al.: A unified framework for de-duplication and population size estimation. Bayesian Anal. (2020)
21. Vatsalan, D., Christen, P., O'Keefe, C.M., Verykios, V.S.: An evaluation framework for privacy-preserving record linkage. J. Priv. Confidentiality **6**, 3 (2014)
22. Zanella, G., et al.: Flexible models for microclustering with application to entity resolution. In: Proceedings of the 30th International Conference on Neural Information Processing Systems. NIPS 2016, Curran Associates Inc., NY, USA, pp. 1425–1433 (2016)

Secure Matrix Computation: A Viable Alternative to Record Linkage?

Jörg Drechsler[1](✉) and Benjamin Klein[2](✉)

[1] Institute for Employment Research, Regensburger Str. 104, 90478 Nuremberg, Germany
joerg.drechsler@iab.de
[2] Ludwig-Maximilians-Universität, Ludwigstr. 33, 80539 Munich, Germany
B.Klein@campus.lmu.de

Abstract. Linking data from different sources can enrich the research opportunities in the Social Sciences. However, datasets can typically only be linked if the respondents consent to the linkage. Strategies from the secure multi-party computation literature, which do not require linkage on the record level, might be a viable alternative in this context to avoid biases due to selective non-consent. In this paper, we evaluate whether such a strategy could actually be successfully applied in practice by replicating a study based on linked data available at the Institute for Employment Research. We find that almost identical results could be obtained without the requirement to link the data. However, we also identify several problems suggesting that the proposed strategy might not be feasible in many practical contexts without further modification.

Keywords: Confidentiality · Record linkage · Secure multi-party computation · Privacy · Oaxaca-Blinder decomposition

1 Introduction

Linking data collected through surveys with administrative data is a popular and cost-efficient strategy for enriching the information available for research. However, privacy regulations typically require that survey respondents consent to the linkage. Non-consent and other reasons for impossible or unsuccessful record linkage lead to incomplete data and therefore a reduction in statistical power. However, the main concern regarding non-consent is that non-consenters could be systematically different from consenters introducing bias in subsequent analysis based on the linked data. These biasing effects have been confirmed in several studies (see for example, [18,19]).

To avoid the risk of non-consent bias, we propose using strategies from the secure multi-party computation literature. Secure multi-party computation addresses the problem, how to jointly analyze databases distributed among multiple entities that are either unwilling or unable to share their databases directly. We will focus on strategies for analyzing vertically partitioned data. Vertically

J. Domingo-Ferrer and K. Muralidhar (Eds.): PSD 2020, LNCS 12276, pp. 240–254, 2020.
https://doi.org/10.1007/978-3-030-57521-2_17

partitioned databases are data sources that contain the same data subjects but provide different information on those subjects (in contrast to horizontally partitioned databases which cover the same attributes but for different subjects). To give an example, assume that agency A collects information on income and spending, whereas agency B collects health information for the same individuals. Both agencies might be interested in analyzing the relationships between income and health status but neither of them is willing to share their data, either because of distrust or because confidentiality regulations do not permit the dissemination of the data. The literature for analyzing vertically partitioned data deals with strategies to obtain information regarding the joint distribution of the two data sources without the need for linking the data at the record level. For example, [12] provide protocols to obtain coefficients and standard errors of linear regression models in this situation. The method is privacy preserving, since only aggregate information—sample means and covariances—of the two datasets need to be exchanged. Other protocols allow for privacy preserving logistic regression [7], linear discriminant analysis [5], and maximum likelihood estimation for exponential family models [14].

While all protocols have been developed for contexts in which the data agencies are unwilling or unable to share their data, the approach can also be used to assess non-consent bias: We can use these protocols to obtain the estimates that would have been observed, if all survey respondents had agreed to linkage (note that no consent is required for this approach since no information is linked on the individual level). A prerequisite of these techniques is that both data sources contain the same individuals and need to be ordered in the same way. In our context this means that even though actual data linkage is not necessary, survey respondents need to be identified in the administrative dataset. If sharing identifiers, such as social security number or address information is problematic, privacy preserving record linkage techniques (see for example [6,17,21] or [22]) using encrypted identifiers can be utilized for identification. Alternatively, if the administrative data act as a sampling frame for the survey data, the sample IDs could be used to identify and sort the relevant records from the administrative data. In this case, the department responsible for drawing the initial sample could act as a trusted third party, which uses the sample IDs to select the appropriate cases from the administrative data and ensures that both data files are sorted in the same order. It would be crucial under this scenario that researchers working with the data would not be able to simply merge the two data files, i.e., the datasets would need to be hosted on two different platforms with strict regulations which information could be exchanged between the two platforms.

In this paper we evaluate whether vertical partitioning can be a realistic alternative to data linkage in an applied setting. To do so, we try to replicate results from a report by [10], which uses data from the National Educational Panel linked to the Integrated Employment Biographies, a large administrative database gathered by the Institute for Employment Research to study the relationship between participation in further education programs and the extend to which the job consists of high routine tasks. Note that most research requires substantial preprocessing of the data such as variable recoding, imputation, or outlier removal before the actual model of interest can be fitted. Thus, the question of whether vertical partitioning can be a substitute for record linkage to avoid consent bias does not only depend on whether the model of interest can be computed in a secure way. It also requires that all preprocessing steps can be performed independently on the two data sources.

The remainder of the paper is organized as follows. In Sect. 2 we review the concept of vertical partitioning with a focus on the linear regression model, as this will be the model we need for our application. In Sect. 3 we introduce the two data sources, summarize the research of [10], and present the results of our replication study. The paper concludes with a discussion of practical problems that we encountered when trying to replicate the study without linking the data sources.

2 Secure Multi-party Computation

Fienberg [7] defines secure multi-party computation "as efficient algorithms, to securely evaluate a function whose inputs are distributed among several parties". Most of the secure multi-party computation literature focuses on analyzing horizontally partitioned databases (see, for example, [9,13]). Since we are searching for alternatives to record linkage, we will only focus on strategies for analyzing vertically partitioned databases in the remainder of the paper. Algorithms for different types of analyses have been proposed in the literature. Since our application discussed in Sect. 3 uses an Oaxaca-Blinder decomposition, which is based on a standard linear regression model, we will describe the protocol of [12] in more detail below. Other approaches for analyzing vertically partitioned datasets include [7] for the logistic model and [5] for linear discriminant analysis.

2.1 Estimating Linear Regression Models Using Vertically Partitioned Databases

Strategies for computing secure linear regression models have been proposed for example by [20] and [12]. However, the approach by [20] assumes that the dependent variable of the model is available to all parties involved in the secure analysis. Obviously, this is an unrealistic scenario for our application as it would require to link the dependent variable to one of the data sources, which would not be possible for those units that did not consent to the linkage. The approach of [12] only requires that covariance matrices and sample means can be shared between the parties involved. While this can also be problematic – for example, such a strategy will never offer formal privacy guarantees such as differential privacy and we will discuss some further caveats in the conclusions–, disseminating sample means and covariance matrices is considered acceptable at most statistical agencies. In the following, we discuss, how the approach of [12] can be used to obtain regression coefficients and their standard errors for the standard linear model.

Karr et al. consider the general setting, in which the full database X is distributed among K agencies $A_1, A_2, \ldots, A_K$. Let $X_1, X_2, \ldots, X_K$ denote the subsets of variables from the full database held by the different agencies and let $p_1, p_2, \ldots, p_K$ denote the number of attributes contained in each of the subsets. Some further assumptions are necessary, for the approach to be feasible: None of the attributes is included in more than one database, or if it is, the agencies coordinate to remove the duplicated attribute. There are no missing values in any of the databases, and the agencies collude in the sense that they follow the protocol and share the true values obtained from their computations.

A key element in linear regression analysis is the (scaled) covariance matrix $X'X$. Once the matrix is available, other statistics such as the regression coefficients and their standard errors can be computed easily using the sweep operator as we will illustrate below. Thus, [12] mostly focus on how to compute the matrix in a secure way. We will review their proposed methods below assuming that only two parties are involved, as this mimics the linkage scenario that we are concerned about. The methods can be extended easily to allow for more than two parties.

2.2 Secure Computation of $X'X$

Considering only two parties the cross product $X'X$ can be written as

$$X'X = \begin{bmatrix} X_1'X_1 & X_1'X_2 \\ X_2'X_1 & X_2'X_2 \end{bmatrix}$$

The on-diagonal elements can be computed, using only one dataset at a time. To compute the off-diagonal elements, [12] propose using a secure multiplication protocol. Specifically, assuming X comprises n records, they propose the following steps:

1. A_1 generates g n-dimensional vectors $Z = \{Z_1, \ldots, Z_g\}$, which are orthogonal to X_1, that is, it holds that $Z'^{(i)}X_1^{(j)} = 0$ for $i = 1, \ldots, p_1$ and $j = 1, \ldots, p_1$. Z can be computed using a QR-decomposition of the model-matrix X_1 (see [12] for details).
2. A_1 sends Z to A_2.
3. A_2 calculates $W = (I - ZZ')X_2$, where I is an $(n \times n)$ identity matrix. W is sent back to A_1
4. A_1 can generate $X_1'X_2$ by calculating: $X_1'W = X_1'(I - ZZ')X_2 = X_1'X_2$

The variable g controls the degree of data protection of the protocol. For example, if $g = 0$, $W = X_2$ exposes the complete X_2 matrix to A_1. [12] derive an optimal value for g, which minimizes the loss of protection for both agencies, in the sense that the amount of information shared is the same for both data owners. The optimal value is given by $\frac{p_2}{p_1+p_2} * n$ (this fixes a typo in [12]). Using g can be problematic for large sample sizes. As Z is a matrix of dimension $n \times g$, and

g corresponds to n times the proportion of attributes owned by A_1, computing ZZ' can become unwieldy. To avoid excessive computational burden, pragmatic choices such as setting a maximum threshold for g might have to be made in practice. Given that all data sources are owned by the same agency anyway in our context, we feel that choosing an appropriate value for g is less important as long as it is bounded away from 0 and $n - p$.

Assuming that $X_1'X_1$ and $X_2'X_2$ can also be shared without violating confidentiality, the protocol enables the secure computation of the full matrix $X'X$. We note that this assumption can be problematic in some situations. For example, if X_1 contains binary attributes, releasing $X_1'X_1$ will reveal how many cases in the database share these attributes. Furthermore, the average value for the subset of records sharing the attribute will be revealed for all variables contained in X_1. To illustrate, assume that X_1 contains information about HIV status and income. Releasing $X_1'X_1$ will reveal how many units have HIV and their average income. Developing strategies for dealing with this problem is beyond the scope of this paper.

2.3 Obtaining Regression Coefficients and Standard Errors

Once $X'X$ is available, regression results for linear regression models can be computed easily using the sweep operator. The sweep operator [1] can be used to compute maximum likelihood estimates for linear regression models. Following the notation in [15], let G be a $p \times p$ symmetric matrix. G is said to be swept on row and column k if it is replaced by a matrix H consisting of the following elements:

$$\begin{aligned} h_{kk} &= -1/g_{kk}, \\ h_{jk} &= h_{kj} = g_{jk}/g_{kk}, \quad j \neq k, \\ h_{jl} &= g_{jl} - g_{jk}g_{kl}/g_{kk}, \quad j \neq k, l \neq k. \end{aligned}$$

The sweep operator can be used to obtain the desired regression results for our context by applying the following procedure. Let X be a $n \times p$ matrix containing the p predictor variables in the regression of interest. Likewise, let Y denote a $n \times 1$ vector containing the data for the dependent variable. Note that the variables in X will contain variables from both data sources. Let $Z = \{X, Y\}$ and let $A = \{\bar{X}_1, \ldots, \bar{X}_p, \bar{Y}\}$ be a $1 \times (p+1)$-vector containing the means of all variables. Finally, let $B = (Z'Z)/n$, where the matrix $Z'Z$ is computed following the protocol given above. Define the $(p+2) \times (p+2)$ symmetric matrix G as

$$G = \begin{pmatrix} -1 & A \\ A' & B \end{pmatrix}$$

Regression results for the linear regression of Y on X can be obtained by sweeping the matrix G on the rows and columns 1 to $p+1$. Specifically, let the resulting matrix H be defined as

$$H = \text{SWP}[1, 2, \ldots, p+1]G = \begin{pmatrix} -D & E \\ E' & F \end{pmatrix},$$

where D is $(p+1) \times (p+1)$, E is $(p+1) \times 1$, and F is 1×1. Then F holds the residual variance of the linear regression model, E provides the regression coefficients and D can be used to calculate the covariance matrix of the estimated regression coefficients. Specifically, it holds that $D = n \cdot [(1,X)'(1,X)]^{-1}$, that is D is a scaled version of inverse of the covariance $X'X$ of the predictor variables including the intercept. Hence, an unbiased estimate of the covariance matrix of the regression coefficients $\hat{\beta}$ can be computed as

$$V\hat{a}r(\hat{\beta}) = \hat{\sigma}^2 \cdot [(1,X)'(1,X)]^{-1} = n/(n-p)F \cdot D/n,$$

where the adjustment factor $n/(n-p)$ is required to ensure that F–the maximum likelihood estimate of the residual variance–is unbiased.

3 Application

In the application section, we evaluate whether the vertical partitioning approach can be successfully applied in practice to avoid the requirement to link the data sources on the record level. To do so we try to replicate a study by [10], which investigates the relationship between participation in further education programs and the extend to which the job consists of high routine tasks. The authors find that while modern technologies mainly replace employees in professions with a high share of routine tasks, people working in these professions participate less in further training. For their analysis the authors use data from the adult cohort of the National Education Panel linked to the Integrated Employment Biographies of the Institute for Employment Research. Before describing the two data sources and their linkage below, we emphasize again that the question of whether the vertical partitioning approach can be a useful alternative to record linkage does not only depend on whether it would produce the same estimates for the final model. A substantial amount of work in applied research is typically devoted to data preparation and exploratory data analysis. Variables might be regrouped, because of problems with sparse categories, missing values might be imputed, outliers might have to be removed, and the fit of the final model needs to be evaluated. It is not clear ex ante whether all the steps that the authors took before running their final model would be reproducible if the data cannot be linked on the record level. The authors of the paper kindly agreed to share their entire script with us so we are able to evaluate whether all data preparation steps could also have been performed without linking the data. We will discuss our findings in Sect. 3.2 before presenting the results for the final model.

3.1 The Data

This section provides an overview of the two data sources and discusses the linkage of both sources. It borrows heavily from [8].

National Educational Panel Study. The National Educational Panel Study (NEPS) is carried out by the Leibniz Institute for Educational Trajectories at the University of Bamberg. The NEPS collects longitudinal data on competency development, educational processes, educational decisions, and returns to education in Germany. Panel surveys on different age cohorts are conducted which provide data throughout the life course. The NEPS Starting Cohort 6 collects data specifically on the adult cohort. After a longer period between the first and second waves, which were carried out in 2007/2008 and 2009/2010, respectively, surveys for the adult cohort have been conducted yearly since 2011. The sample is drawn from municipality registration records of residents using a two-stage cluster sampling design with communities defining the primary sampling units and simple random sampling without replacement of individuals at the second stage. The target population of the adult cohort comprises residents in Germany who were born between 1944 and 1986 [4].

Integrated Employment Biographies. The Integrated Employment Biographies (IEB) consist of administrative data obtained from social security notifications and different business processes of the German Federal Employment Agency. The different data sources are integrated for and by the Institute for Employment Research.

Employment information is provided for every employee covered by social security. Note that this excludes individuals who did not enter the labour market and individuals who were self employed since these groups are not subject to mandatory social security contributions. For individuals who received benefits in accordance with the Social Code Books (SGB) II & III, benefit recipient data are generated respectively. Further data are generated for individuals who were registered as job seekers with the Federal Employment Agency or who participated in an employment or training program. We refer to [11] for a detailed description of the different data sources and of the IEB.

Linkage of the Two Data Sources. Record linkage of NEPS and IEB data was carried out based on the following non-unique identifiers: first and last name, date of birth, sex, and address information – postal code, city, street name, and house number. There are 17,140 respondents in the NEPS adult cohort. The number of linkage consenters among the NEPS adult cohort was 15,215. Thus, the linkage consent rate was 88.8%. Of those units that consented 77.4%, that is, 11,778 units, could be linked deterministically and 8.9% (1,053 units) probabilistically, and the remainder could not be linked at all. In the NEPS linkage, a link is called deterministic if the identifiers either match exactly or differ only in a way that the probability for false positive links is still extremely low.

3.2 Review of the Data Preparation Steps

The authors apply several preprocessing steps before running their final model of interest. In this section we give a brief summary of these steps focusing on the question whether they are reproducible without linking the data.

Step 1: Preparing the data sources: In this step, the relevant subset of the data (in terms of time and content) is identified, some variables are recoded and new variables are generated based on existing variables. Furthermore, additional information from other data sources is added. For the IEB data, which is organized as (un-)employment spell data, that is, each row in the data represents an (un-)employment spell, this step mostly deals with generating aggregated person level information, such as work experience or tenure at the current employer. Furthermore, aggregated information such as whether the individual is currently employed or participates in vocational training is generated based on the detailed information about the current work status recorded in the IEB. Finally, additional information about the employer is added from another administrative data source of the IAB, the Establishment History Panel (BHP).

The data on the adult cohort of the NEPS contain more than 1,000 variables stored in several subfiles grouped by content. The majority of the data preparation steps focus on selecting, labeling, and merging the relevant variables from the different subfiles. Some subfiles have to be reshaped into wide format to allow the inclusion of lagged variables in later analysis steps. New variables are also derived based on the available information. To give an example, the authors generate an index variable, which measures the amount of the workload spent on routine tasks. The variable, which is bounded between 6 and 30 is generated by aggregating over several variables from the NEPS survey. Using this variable, the authors generate an indicator variable to identify respondents working in jobs with a high level of routine tasks.

All processing steps are performed independently in both data sources. Thus, this step can be fully replicated even if the databases are never linked.

Step 2: Imputation: Since the information in the IEB is based on administrative processes, the number of missing cases is very small for most variables. Thus, the authors only impute two variables in the IEB: education and income. Information on education is often missing or inconsistent as reporting the education level is not mandatory. For the imputation of the education variable, the authors use a simple deterministic approach, which is based on the assumption that the level of education can only increase over time, but never decrease. In a first step, all reported (un-)employment spells for each individual are sorted chronologically. Then, starting with the earliest reported education level, the observed level of education is always carried over to the next spell unless the reported level of education is higher for this spell. In this case, the higher education level is accepted and this higher level will be used when moving on to the next spell. This strategy fulfills two goals: it imputes previously reported levels of education if the education information is missing in certain spells, but it also fixes inconsistencies in the data, by editing reported education levels, which are lower than those previously reported. Since values for education are imputed using only its own history, this step can also be replicated without linking the data. The income variable does not contain any missing values since all employers are required to report the income for all their employees to the social security administration. However, income is only reported up to the contribution limit of the social security system and thus higher incomes are top coded. To deal with the problem, imputed values above the contribution limit are provided in the IEB and the authors use these imputed values. Imputations are generated using a tobit regression model. For the tobit model all predictors are taken from the IEB. This can be seen critical as it implicitly assumes that conditional on the variables from the IEB the income information is independent from the variables included in the NEPS data. On the other hand it implies that this imputation step can also be replicated without linking the data.

To impute missing values in the NEPS data, the authors rely on a set of STATA routines provided by the research data center of the NEPS. These routines are meant as a convenience tool for the applied researcher as they automatically impute missing values in the data where possible. However, given that the routines are developed only for the survey data and not for the linked dataset, they make a similar independence assumption as the imputation routines for the education variable described above. While methodologically questionable, this implies again that these routines can be run without actually linking the data.

Step 3: Identifying the appropriate employment spells: Since the final model uses information from both data sources to measure the impact of various individual and job characteristics on the probability to participate in further training, it is important that the linked data provide information on the same job. To ensure this, the authors always link the employment spell of each individual that covers the NEPS interview date (if more than one spell covers the date, the spell with the higher income is selected). Obviously, this strategy would require linking the interview date to the IEB and thus it cannot be fully replicated. We implemented the following alternative, which serves as a proxy if linkage is not possible: In a first step, we identified the field period of the NEPS survey by searching for the earliest and latest interview date. Then we identified all employment spells overlapping with the field period. In a final step we selected the spell of each individual with the highest overlap with the field period. The rational behind the approach is that assuming a uniform distribution for the number of interviews at each day of the field period, the strategy will select the spell having the highest probability that the interview took place during this spell. As we show in the next section, this strategy only has modest effects on the final results.

3.3 Results for the Model of Interest

As discussed previously, the authors are interested in measuring the impact of various individual and job characteristics on the probability to participate in further training. To estimate the impacts, the authors use an Oaxaca-Blinder decomposition [3,16] taking into account information about the respondents, their education, socio-economic, firm and job characteristics. The Oaxaca-Blinder decomposition is a popular strategy in economics to decompose the observed difference between two groups into a component that can be explained by observed characteristics and an unexplained component. For example, most studies on the gender wage gap rely on the Oaxaca-Blinder decomposition treating the unexplained component as an estimate for discrimination (see for example [2]). The approach proceeds by fitting the same regression model independently for both subgroups. Assuming a linear regression model and using the indices A and B to distinguish the two subgroups, the two models are $Y_A = X_A\beta_A + u_A$ and $Y_B = X_B\beta_B + u_B$, where Y is the variable of interest, X is the matrix of explanatory variables and u denotes the error term. Based on these two models, we can write the difference in the means of Y as

$$\bar{Y}_A - \bar{Y}_B = \bar{X}_A\beta_A - \bar{X}_B\beta_B = \beta_A(\bar{X}_A - \bar{X}_B) + \bar{X}_B(\beta_A - \beta_B).$$

The first term $\beta_A(\bar{X}_A - \bar{X}_B)$ is the part of the observed difference in Y which can be explained by observed characteristics. The second term $\bar{X}_B(\beta_A - \beta_B)$ contains the unexplained difference.

Since [10] also use a linear regression model in the decomposition, we can rely on the secure protocol described in Sect. 2 to replicate the results of the authors. Specifically, defining the two groups as those working in highly routine based jobs and those that do not, the authors run a linear regression in which the dependent variable is an indicator, whether the unit participated in further training, and the set of 46 predictor variables can be grouped into seven categories: education, health status, further individual characteristics that serve as control variables, job characteristics, firm characteristics, information about the management style, and information whether the firm generally supports further training.

Note that using a secure protocol for the two regression models poses an additional problem: The party, which holds the indicator variable for the two groups cannot simply share this information with the other party. The other party would learn which unit belongs to group A or group B, although she would not know which group is which. In our example this would imply that the owner of the administrative database would still not know, whether group A or group B contains the individuals that classify their job as mostly routine based. Still, the amount of protection from this might be limited in practice. For example, the owner of the administrative database could use the information on the job classification to identify an individual (for example a university professor) in the database, for which it is prudent to assume that she classified herself as working in a job with few routine tasks. This knowledge would then automatically reveal the value of the group indicator for all other units in the database.

However, sharing the group indicator is not required. Estimates for the two subgroups can also be obtained by running one regression model on the entire dataset, but interacting all the variables with the group indicator. The point estimates from this model will be exactly the same as the point estimates when running two separate models. Still, the variance estimates will generally differ, since one constant residual variance is assumed for the entire dataset instead of allowing for two independent variances for the two subgroups. It might seem that nothing is gained by using this strategy as one of the parties will not be able to compute the interaction terms without sharing of the group indicator. However, this is not required. As long as the party that owns the group indicator includes all interactions with the group indicator and the indicator itself, it suffices if the other party includes all two-way interactions and squared terms of its own variables. Still, some bookkeeping is required to obtain the full covariance matrix from the output of the secure protocol in this case.

Table 1 contains the results for the Oaxaca-Blinder decomposition. The results are presented for three different estimation strategies. The third column replicates the results from [10] using their code. We note that these results differ slightly (only in the third digit) from the results reported in the original paper. We can only speculate about the reasons. Most likely some bugs in the reported data have been fixed in the meantime. The forth column reports the results if the data are linked at the record level, but the alternative strategy which does not rely on the interview date to identify the correct employment spell is used. These results serve as a benchmark for the last column, which contains the results based on the vertical partitioning approach.

The upper part of the table reports summary statistics. The first two rows contain the share of training participants in the two groups (in percent), while the third row repots the difference in percentage points between the two groups. Based on the original data, we find that more than 40% of the individuals working in jobs with low rates of routine tasks participate in further training, while the uptake in the other group is only 27% resulting in a (rounded) difference of 14%. Rows four and five show to what extend the difference between the two groups can be explained by the observed characteristics. From the table it is evident that about half of the difference can be explained. The lower part of the table shows how much of the explained difference can be attributed to the different components. The results indicate that individual and job characteristics only play a minor role in explaining the difference in the training uptake. Most of the difference is explained by firm characteristics with a supportive attitude towards further training explaining most of the difference.

Comparing the results in column three and column four, we find that using the alternative strategy to identify appropriate employment spells only has minor effects on the findings. The absolute difference between the estimates is never more than 0.4% points and the substantive conclusions would not be affected.

Results in the fifth column are based on the vertical partitioning approach, that is, the data are never linked to compute the results. Comparing the results to the fourth column, we find that the point estimates are identical. The difference in the estimated standard errors can be attributed to three factors: First, as explained above, the residual variance is assumed to be the same in both groups. Second, the variance estimates in the Stata implementation of the Oaxaca-Blinder decomposition, which was used to compute the results in the third and forth column adds a variance component to account for the uncer-

Table 1. Results of the Oaxaca-Blinder decomposition from [10]. All point estimates are reported in percent. The third column contains the original results. The forth column still uses the linked data, but an alternative strategy is used to identify appropriate employment spells. The last column contains the results based on the vertical partitioning protocol.

		Original results	Alternative spell ident.	Vertical partitioning
Overall	$\bar{Y}_{R=0}$	41.08	41.11	41.11
		(0.98)	(0.97)	(0.97)
	$\bar{Y}_{R=1}$	26.71	26.88	26.88
		(1.42)	(1.42)	(1.42)
	Difference	14.36	14.23	14.23
		(1.72)	(1.72)	(1.72)
	Explained	7.15	7.31	7.31
		(1.03)	(1.02)	(0.68)
	Unexplained	7.21	6.92	6.92
		(1.70)	(1.70)	(2,40)
Explained	Education	0.61	0.97	0.97
		(0.53)	(0.54)	(0.52)
	Health	0.25	0.25	0.25
		(0.22)	(0.21)	(0.18)
	Individual characteristics	−0.94	−0.91	−0.91
		(0.38)	(0.38)	(0.35)
	Job characteristics	0.10	0.18	0.18
		(0.37)	(0.35)	(0.33)
	Firm characteristics	1.14	0.98	0.98
		(0.57)	(0.56)	(0.49)
	Management	1.85	1.80	1.80
		(0.46)	(0.45)	(0.35)
	Firm support	4.13	4.05	4.05
		(0.61)	(0.60)	(0.43)

tainty in the predictors. Third, the authors accounted for the potential clustering within firms by specifying the cluster option in Stata for the variance estimation procedure. Investigating whether it is possible to also account for the sampling design features when using a vertical partitioning approach is an interesting area for future research. However, we found that the difference in the estimated standard errors was generally small for this application.

4 Conclusion

Linking data from different sources is an increasingly popular and cost efficient strategy to facilitate research in the Social Sciences and in other fields. However, informed consent is typically required prior to the linkage. This might introduce bias in the analyses based on the linked cases if not all participants consent. To avoid linkage bias, strategies from the secure multi-party computation literature might offer an interesting alternative since they do not require to actually link the data. In this paper we evaluated whether such a strategy might be feasible in practice by replicating results from a research project based on a linked dataset. We feel that such a replication study is important since fitting the final model of interest – for which multi-party computation strategies might exist – is typically only the final step in a research project. Most of the time (and computer code) is spent on cleaning and preparing the data. To illustrate, specifying the final model required less than 20 lines of computer code, while preparing the data required almost 1200 lines of code in the replicated project.

We found that it was indeed possible to obtain very similar results for the final model. However, there are several caveats regarding these findings. From a utility perspective, the replication study was successful, because almost all of the data preparation steps happened before linking the two data sources. This may or may not be realistic for other research projects. In fact, from a methodological point of view, it would always be recommended to implement any imputation steps only after linking the data to avoid the conditional independence assumption. Generating imputations using the information from both sources without actually linking the data would be complicated. Even more critical are the required assumptions from the privacy perspective. The proposed strategy can only be successfully implemented if the following conditions hold:

- Those units that are included in both databases can be identified in both databases.
- Both databases can be sorted in the same order.
- Sharing covariances and sample means is not considered to be a violation of the confidentiality standards.

Identifying the units is straightforward if the same unique identifiers exist or if the administrative data act as a sampling frame for the survey. In other cases, privacy preserving record linkage techniques might be required. However, identifying the units is not only a methodological problem, it also raises privacy concerns. Even if it can be guaranteed that the researchers involved in the project will only have access to one of the data sources, the information is spilled, which of the units is also included in the other source. In some contexts this might not be acceptable. The strategy also poses legal questions that go beyond the standard questions regarding the risks of disclosure. Is it legally (and ethically) justifiable to compute results integrating information from different data sources even if the data subjects explicitly refused to give consent that the data sources can be linked directly? Another interesting question for future research would be to evaluate if the strategy could indeed help to avoid non-consent bias by comparing results obtained using a secure protocol and using all records in the survey, with results obtained using only the data from those units that consented to the linkage of the data.

References

1. Beaton, A.E.: The use of special matrix operators in statistical calculus. ETS Res. Bull. Ser **1964**(2), i–222 (1964)
2. Blau, F.D., Kahn, L.M.: The gender wage gap: extent, trends, and explanations. J. Econ. Lit. **55**(3), 789–865 (2017)
3. Blinder, A.S.: Wage discrimination: reduced form and structural estimates. J. Hum. Resour. **8**(4), 436–455 (1973). http://www.jstor.org/stable/144855
4. Blossfeld, H.P., Von Maurice, J.: Education as a lifelong process. Zeitschrift für Erziehungswissenschaft Sonderheft **14**, 19–34 (2011)
5. Du, W., Han, Y., Chen, S.: Privacy-preserving multivariate statistical analysis: linear regression and classification. In: Proceedings of the 2004 SIAM International Conference on Data Mining, pp. 222–233 (2004)
6. Durham, E., Xue, Y., Kantarcioglu, M., Malin, B.: Quantifying the correctness, computational complexity, and security of privacy-preserving string comparators for record linkage. Inf. Fusion **13**(4), 245–259 (2012)
7. Fienberg, S.E., Nardi, Y., Slavković, A.B.: Valid statistical analysis for logistic regression with multiple sources. In: Gal, C.S., Kantor, P.B., Lesk, M.E. (eds.) ISIPS 2008. LNCS, vol. 5661, pp. 82–94. Springer, Heidelberg (2009). https://doi.org/10.1007/978-3-642-10233-2_8
8. Gessendorfer, J., Beste, J., Drechsler, J., Sakshaug, J.W.: Statistical matching as a supplement to record linkage: a valuable method to tackle nonconsent bias? J. Official Stat. **34**(4), 909–933 (2018)
9. Ghosh, J., Reiter, J.P., Karr, A.F.: Secure computation with horizontally partitioned data using adaptive regression splines. Comput. Stat. Data Anal. **51**(12), 5813–5820 (2007)
10. Heß, P., Jannsen, S., Leber, U.: Digitalisierung und berufliche Weiterbildung: Beschäftigte, deren Tätigkeiten durch Technologien ersetzbar sind, bilden sich seltener weiter. IAB-Kurzbericht 16/2019 (2019)
11. Jacobebbinghaus, P., Seth, S.: Linked-Employer-Employee-Daten des IAB: LIAB - Querschnittmodell 2, 1993–2008. FDZ-Datenreport, Institute for Employment Research, Nuremberg, Germany (2010)
12. Karr, A., Lin, X., Reiter, J., Sanil, A.: Privacy preserving analysis of vertically partitioned data using secure matrix products. J. Official Stat. **25**, 125–138 (2009)
13. Karr, A.F., Lin, X., Sanil, A.P., Reiter, J.P.: Secure regression on distributed databases. J. Comput. Graph. Stat. **14**(2), 263–279 (2005)
14. Lin, X., Karr, A.F.: Privacy-preserving maximum likelihood estimation for distributed data. J. Priv. Confidentiality, **1**(2) (2010)
15. Little, R.J.A., Rubin, D.B.: Statistical Analysis with Missing Data, 2nd edn. Wiley, New York (2002)
16. Oaxaca, R.: Male-female wage differentials in urban labor markets. Int. Econ. Rev. **14**(3), 693–709 (1973). http://www.jstor.org/stable/2525981
17. Rao, F.Y., Cao, J., Bertino, E., Kantarcioglu, M.: Hybrid private record linkage: separating differentially private synopses from matching records. ACM Trans. Priv. Secur. (TOPS) **22**(3), 1–36 (2019)

18. Sakshaug, J.W., Huber, M.: An evaluation of panel nonresponse and linkage consent bias in a survey of employees in Germany. J. Surv. Stat. Method. **4**(1), 71–93 (2015)
19. Sakshaug, J.W., Kreuter, F.: Assessing the magnitude of non-consent biases in linked survey and administrative data. Surv. Res. Methods **6**, 113–122 (2012)
20. Sanil, A.P., Karr, A.F., Lin, X., Reiter, J.P.: Privacy preserving regression modelling via distributed computation. In: Proceedings of the Tenth ACM SIGKDD International Conference on Knowledge Discovery and Data Mining, pp. 677–682 (2004)
21. Schnell, R.: An efficient privacy-preserving record linkage technique for administrative data and censuses. Stat. J. IAOS **30**(3), 263–270 (2014)
22. Verykios, V.S., Christen, P.: Privacy-preserving record linkage. Wiley Interdisc. Rev. Data Mining Knowl. Discov. **3**(5), 321–332 (2013)

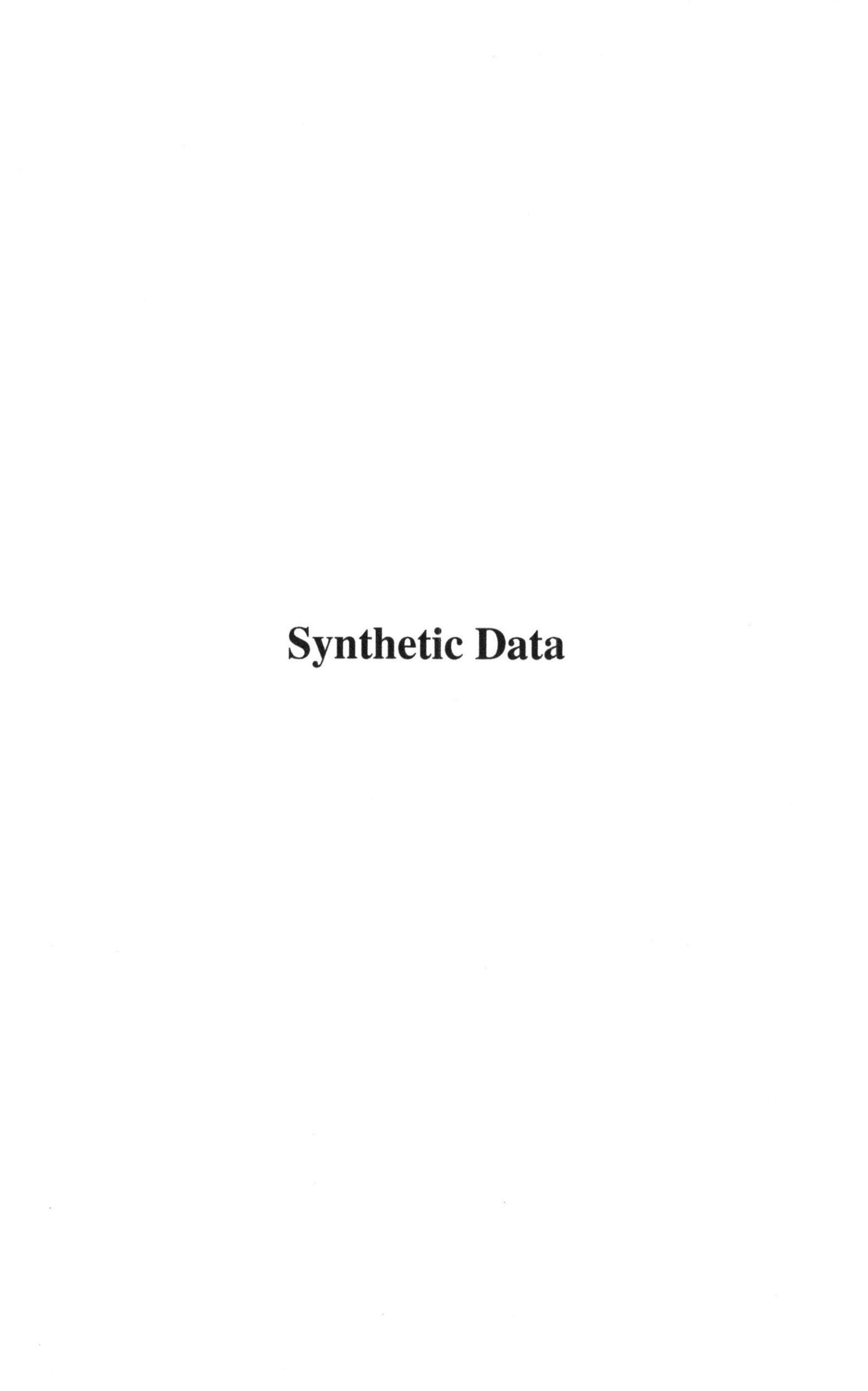

Synthetic Data

A Synthetic Supplemental Public Use File of Low-Income Information Return Data: Methodology, Utility, and Privacy Implications

Claire McKay Bowen[1(✉)], Victoria Bryant[2], Leonard Burman[1,3], Surachai Khitatrakun[1], Robert McClelland[1], Philip Stallworth[4], Kyle Ueyama[1], and Aaron R. Williams[1]

[1] Urban Institute, Washington D.C. 20024, USA
cbowen@urban.org
[2] Internal Revenue Services, Washington D.C. 20002, USA
[3] Syracuse University, Syracuse, NY 13244, USA
[4] University of Michigan, Ann Arbor, MI 48109, USA

Abstract. US government agencies possess data that could be invaluable for evaluating public policy, but often may not be released publicly due to disclosure concerns. For instance, the Statistics of Income division (SOI) of the Internal Revenue Service releases an annual public use file of individual income tax returns that is invaluable to tax analysts in government agencies, nonprofit research organizations, and the private sector. However, SOI has taken increasingly aggressive measures to protect the data in the face of growing disclosure risks, such as a data intruder matching the anonymized public data with other public information available in nontax databases. In this paper, we describe our approach to generating a fully synthetic representation of the income tax data by using sequential Classification and Regression Trees and kernel density smoothing. This synthetic data file represents previously unreleased information useful for tax policy modeling. We also tested and evaluated the tradeoffs between data utility and disclosure risks of different parameterizations using a variety of validation metrics. The resulting synthetic data set has high utility, particularly for summary statistics and microsimulation, and low disclosure risk.

Keywords: Disclosure control · Synthetic data · Utility · Classification and Regression Trees

1 Introduction

Tax data are a potentially invaluable resource for public policy analysis on a wide range of issues. The Internal Revenue Service (IRS) has for decades released a public use file (PUF) with selected information from individual income tax return

J. Domingo-Ferrer and K. Muralidhar (Eds.): PSD 2020, LNCS 12276, pp. 257–270, 2020.
https://doi.org/10.1007/978-3-030-57521-2_18

records anonymized and altered to protect against disclosure risk. Analysts in academia, nonprofit research organizations, and the private sector use the PUF to study the effects of tax policy changes on tax revenues, distribution of tax burdens, and economic incentives. For instance, the microsimulation models of various organizations, such as the American Enterprise Institute, the Urban-Brookings Tax Policy Center, and the National Bureau of Economic Research, must rely on the PUF. However, concerns about protecting data participants' privacy in the information age have required the IRS to limit the data released and distort the data in increasingly aggressive ways. As a result, the released data are becoming less useful for statistical analyses. Concerns about privacy might prevent the PUF as currently conceived from being produced.

In response to these concerns, we developed a synthetic data approach to protecting the tax data from disclosure. Although synthetic data generation has been used to protect many administrative and other sensitive data sets against disclosure, it has not previously been applied to U.S. tax return data. As a first step, we developed a synthetic process for generating the 2012 Supplemental PUF, a database of individuals who neither filed nor were claimed as a dependent on any individual income tax return in 2012, which has never been released publicly before. More information about the file is available in IRS (2019).

We organize the paper as follows. Section 2 details our synthetic data generation method on the Supplemental PUF data and outlines how our synthesis process protects privacy, including protections against disclosure of outliers and attribute disclosure. Section 3 reports and evaluates the data utility measures of the synthesized Supplemental PUF data. Further discussion of our results and our plans for future work are in Sect. 4.

2 Data Synthesis Methodology

Our ultimate objective is to synthesize records from the IRS Master File (the universe of individual income tax returns) to create a synthetic file similar to the current PUF released by the IRS Statistics of Income (SOI) Division, but with stronger privacy protections. As a proof of concept and in an effort to release useful data that had never before been made public, we have created a fully synthetic file called the Supplemental PUF based on nonfiler tax data.

Cilke (2014) defines a nonfiler as, "Any U.S. resident that does not appear on a federal income tax return filed [for] a given year." Our sample is thus comprised of people with U.S. residences who neither file nor are claimed as a dependent on a federal income tax return for a given year and do not appear to have an income tax filing requirement. Our data source is a random 0.1% sample of social security numbers appearing on information returns, excluding the people who file tax returns, for Tax Year 2012 maintained by the IRS SOI Division. Information returns are forms provided to the IRS by any business or other entity that pays income or has certain other compulsory reportable transaction with an individual. The following section provides background on SOI's disclosure methods and outlines the disclosure risks and data synthesis process for the Supplemental PUF. For interested readers, Bowen et al. (2020) provides the full report on our data synthesis methodology.

2.1 Background

Data stewards have relied on a variety of statistical disclosure control (SDC) or limitation (SDL) techniques to preserve the privacy of data while maintaining quality. However, some SDC techniques may fail to eliminate disclosure risk from data intruders armed with external data sources and powerful computers (Drechsler and Reiter 2010; Winkler 2007). These techniques may also greatly reduce the usefulness of the released data for analysis and research. For instance, currently, the SOI top codes the number of children variable in the individual income tax return included in the PUF at 3 for married filing jointly and head of household returns, 2 for single returns, and 1 for married filing separately returns (Bryant 2017). Also, the 2012 PUF aggregated 1,155 returns with extreme values into four observations, ostensibly further top coding the high/low income or extreme positive and negative values of income (Bryant 2017). These methods may degrade analyses that depend on the entire distribution, bias estimates for more complex statistical models, distort microsimulation model analyses that are sensitive to outliers, make small area estimation impossible, and hide spatial variation (Reiter et al. 2014; Fuller 1993).

2.2 Disclosure Risks

Replacing data with fully synthetic data can potentially avoid the pitfalls of previous SDC techniques. One way we can generate the synthetic data is by simulating the data generation process of the confidential data based on an underlying distribution model. This method protects against identity disclosure, because no real observations are released (Hu et al. 2014; Raab et al. 2016). Specifically, Hu et al. (2014), stated that "it is pointless to match fully synthetic records to records in other databases since each fully synthetic record does not correspond to any particular individual."

However, if not carefully designed, fully synthetic data may still risk disclosing information (Raab et al. 2016). For example, overfitting the model used to generate the synthetic data might produce a synthetic file that is "too close" to the underlying data. In the extreme case, a data synthesizer could theoretically perfectly replicate the underlying confidential data (Elliot 2015). The database reconstruction theorem proves that even noisy subset sums can be used to approximate individual records by solving a system of equations (Dinur and Nissim 2003). If too many independent statistics are published based on confidential data, then the underlying confidential data can be reconstructed with little or no error.

To protect the integrity of the tax system, the IRS has an obligation to protect confidentiality even from perceived disclosure. Yet, disclosure risks are difficult to estimate for most data sets, let alone complex synthetic data sets such as a synthetic individual income tax return database. Raab et al. (2016)

concluded that it was impractical to measure disclosure risk in the synthesized data from the UK Longitudinal Series: "Hu et al. (2014), Reiter et al. (2014), McClure and Reiter (2012) proposed other methods that can be used to identify individual records with high disclosure potential, but these methods cannot at present provide measures that can be used with (the) sort of complex data that we are synthesizing."

The underlying administrative database that is the basis for our synthetic dataset coupled with our proposed methodology affords us three critical protections: 1) Reverse engineering the SDC techniques on our high dimensional data is very computationally demanding. 2) Our synthetic data is only a fraction of the size of the underlying administrative data, sampled at a rate at most one in ten nonfilers. By limiting sample size, we reduce the potential disclosure on the properties of the underlying distribution, which is especially important for outliers. We discuss this further in Sect. 4. 3) Our synthetic data method smooths the distribution of underlying data, preserving the empirical distribution of non-sensitive observations. Specifically, the empirical distribution has a high probability density and then flattens in the tails to only reflect the general characteristics of outlier observations. This prevents data intruders from using information about outliers to infer other data about these sensitive observations.

2.3 Data Synthesis Procedure

For our synthetic data generation process, we first synthesized gender by splitting the data into two parts. One part includes the observations from the confidential data that have zeros for all 17 tax variables. The other part includes the observations with at least one non-zero tax variable. For the part with at least one non-zero value for a tax variable, we randomly assigned *gender*, X_1, values based on the underlying proportion in the confidential data.

For the part with all zeros for tax variables, we randomly assigned the *gender* value based on the proportion in the zero subsample, synthesize *age* based on *gender*, and assign zeros to all tax variables. When the tax variables had non-zero values, we used classification and regression trees (CART) to predict *age*, X_2, for each record conditional on *gender*. Since the CART method selects values at random from the final nodes, the distribution may differ slightly from the distribution of *age* by *gender* in the administrative data, but, the differences are small given the sample size.

For the remaining variables ($X_3, ..., X_{19}$), we used CART as our underlying model (Breiman et al. 1984; Reiter 2005). We estimated CART models for each variable with all observed values of the previously synthesized outcome variables as predictors. Our synthetic data generation methodology is based on the insight that a joint multivariate probability distribution can be represented as the product of sequential, conditional probability distributions (Eq. 1).

$$\begin{aligned} f(\boldsymbol{X}|\boldsymbol{\theta}) &= f(X_1, X_2, ..., X_k|\theta_1, \theta_2, ..., \theta_k) \\ &= f_1(X_1|\theta_1) \cdot f_2(X_2|X_1, \theta_2) ... f_k(X_k|X_1, ..., X_{k-1}, \theta_k) \end{aligned} \tag{1}$$

where $\boldsymbol{X}$ are the variables to be synthesized, $\boldsymbol{\theta}$ is the vector of model parameters such as regression coefficients, and k is the total number of variables (19, described in Table 1).

We chose to implement CART because the method is computationally simple and flexible (no distribution assumptions). CART also far out-performed regression-based parametric methods in several preliminary, utility tests (e.g. summary statistics) due, in part, to the large number of zero values. We used a customized version of CART from the R package `synthpop`, which contains multiple methods for creating partially-synthetic and fully-synthetic data sets and for evaluating the utility of synthetic data (Nowok et al. 2019). We used CART to partition the sample into relatively homogeneous groups, subject to the constraint that none of the partitions be too small to protect against over-fitting (Benedetto et al. 2013). In testing on the Supplemental PUF database, we found that a minimum partition size of 50 produces a good fit with adequate diversity of values within each partition.

We started with the variable with the most non-zero values–*wage* income, X_3, and then order the remaining variables, $\{X_4, X_5, ..., X_{19}\}$, in terms of their correlations with *wage*, from most to least correlated. Specifically, at each partition, the best split is the one that minimizes the error sum of squares, SSE, given that the data are partitioned into two nodes. Thus,

$$SSE = \sum_{i \in A_L} (y_i - \bar{y}_L)^2 + \sum_{i \in A_R} (y_i - \bar{y}_R)^2 \tag{2}$$

The variables $\bar{y}_L$ and $\bar{y}_R$ are the means of the left and right nodes respectively. Splits continue until there is no improvement in the splitting criteria or until the minimum size for a final node (50) is reached. Our data synthesis approach samples values from the appropriate final node and then applies our smoothing method discussed below.

To synthesize the first continuous variable X_3, (*wages* in the Supplemental PUF data), we created a kernel density function for each percentile of values predicted within groups defined by CART for this variable. While this approach seems straightforward, we had to tackle some complications. First, the variance of the Gaussian kernel must be larger when sampling outliers. If the variance is not adjusted, a data intruder who knows how the database is constructed might draw some fairly precise inferences since outlier observations in the synthetic data set would likely be relatively close to an actual observation. We used percentile smoothing, which selects the variance based on the optimal variance for a kernel density estimator estimated on observations in the percentile for each observation. As discussed later in the section, this causes the variance to grow with the value of the synthesized variable. Secondly, variables that are a deterministic function of others, such as adjusted gross income or taxable income, will be calculated as a function of the synthesized variables. We did not calculate such variables for the Supplemental PUF data.

No smoothing is applied to values of 0, which is the most common value for all continuous variables in the Supplemental PUF data. We did not consider

Table 1. Supplemental Public Use File variable description.

Variable	Description
Gender	Male or female
Age	1 to 85 (age is top-coded at 85)
Wages	Reflects the amount reported in Box 1 of Form W-2
Witholding	The combination of reported withholdings from Forms W-2, W-2G, SSA-1099, RRB-1099, 1099-B, 1099-C, 1099-DIV, 1099-INT, 1099-MISC, 1099-OID, 1099-PATR, and 1099-R
Taxable retirement income	Total taxable retirement income from the amounts reported in Box 2a of Form 1099-R for both IRA and Pension distributions
Mortgage interest	The combination of amounts reported in Boxes 1, 2, and 4 of Form 1098
Interest received	Combination of amounts reported in Interest Income from Forms 1065 Schedule K-1, 1041 Schedule K-1, 1120S Schedule K-1, and 1099-INT
Pension received	Gross pension distributions reported in Form 1099-R
Residual income	Combination of other net income/loss, cancellation of debt, gambling earnings, and net short term gains/losses from Forms 1099-Q, 1099-C, 1099-MISC, 1099-PATR, 1065 Schedule K-1, 1041 Schedule K-1, and 1120S Schedule K-1
Business income	Amounts reported in Box 7 of Form 1099-MISC
Taxable dividends	The combination of reported dividends from Forms 1065 Schedule K-1, 1041 Schedule K-1, 1120S Schedule K-1, and 1099-DIV
Qualified dividends	The amounts reported in Box 1b of Form 1099-DIV
Social security	Difference between benefits paid and benefits repaid from Form SSA-1099
Schedule E	Sum of income reported on Schedule E
Long term (LT) capital gain	Combination of capital gains from Form 1099-DIV and net long term gains/losses minus section 179 deduction from Forms 1065 Schedule K-1, 1041 Schedule K-1, and 1120S Schedule K-1
Tax-exempt interest	Amount reported in Box 8 of Form 1099-INT
Above the line	Based on the combination of amounts reported in Box 1 of Forms 5498 and 1098-E
State refund	Amounts reported in Box 2 of Form 1099-G
Taxable unemployment	Amounts reported in Box 1 of Form 1099-G

zeros to be a disclosure risk because even the variable with the most non-zero values still contains 73% zeros. In fact, many of the variables are zero for almost every record. By default, **synthpop** does not smooth values if the frequency of a single value exceeds 70%.

Subsequent variables $\{X_4, X_5, ..., X_{19}\}$ (all the tax variables) are synthesized in a similar way to X_3 by predicted values based on random draws from a kernel density estimator within groups defined by CART for observations with similar characteristics.

CART tends to over-fit data, therefore most trees are reduced based on a penalty for the number of final nodes in the tree. For the Supplemental PUF, we did not reduce trees because our minimum partition size of 50 is sufficiently large.

2.4 Additional Privacy Protection Measures

Our data synthesis methodology protects confidentiality ex ante, but we also used a set of privacy metrics to test whether the CART method might produce values that are too close to actual values or reveal too much about relationships between variables. In other words, although the synthetic data contains pseudo records, if these values are exactly the same as the confidential data, then we would be creating a large disclosure risk. We used the following metrics to adjust the precision of the data synthesis by adjusting smoothing parameters such as the minimum size of the final nodes in the CART synthesizer.

Number of Unique-Uniques: The count of unique-uniques is the number of unique rows from the confidential data that are unique in the unsmoothed synthetic data. This narrows the focus to rows that are uncommon and could carry some inferential disclosure risk.

Row-Wise Squared Inverse Frequency: For any given row in the unsmoothed synthetic data, this metric is the multiplicative inverse of the square of the number of identical rows in the confidential data. Rows that appear once are assigned a value of 1, rows that appear twice are assigned a value of 1/4, rows that appear thrice are assigned a value of 1/9, and so on.

We did not report on the results for unique-unique, and row-wise squared inverse frequency, because they were all very small.

l-diversity of Final Nodes in the CART Algorithm: We were concerned that the CART algorithm could generate final nodes that lack adequate heterogeneity to protect confidentiality. Too little heterogeneity in the final nodes could result in too much precision for the synthesizer. To ensure adequate heterogeneity, we applied l-diversity to the decision trees created by the CART algorithm (Machanavajjhala et al. 2007). Specifically, let a quasi-identifier be a collection of non-sensitive variables in a data set that could be linked to an external data source. Let a q^*-block be a unique combination of the levels of quasi-identifiers. A q^*-block is l-diverse if it contains at least l unique combinations of sensitive variables. We applied this formal measure to the CART algorithm: at each partition the split directions (left and right) are considered to be quasi-identifiers and the final nodes are considered to be q^*-blocks. The trees create the discretized space formed by quasi-identifiers, the final nodes are q^*-blocks, and the sensitive values are the values in the final nodes. We examined the minimum l-diversity in a data synthesizer and the percent of observations that came from final nodes with l-diversity less than 3. In many cases, the minimum l-diversity is 1 because some final nodes only contain zeros. We considered this to be acceptable because, as discussed earlier, zeros do not inherently present a disclosure risk.

3 Synthetic Data Quality and Evaluation

In this section, we report the synthetic Supplemental PUF utility results. Figures 2 and 3 in the Appendix compared the means and standard deviations, respectively, of the *age* and the 17 tax variables from the original and synthetic data. Overall, the original and synthetic variables had similar values, where only a few variables differed noticeably such as the standard deviation of *tax-exempt interest.*

Figure 4 in the Appendix illustrates the correlation difference between every combination of tax variables, which was 0.0013 across all variables. Most differences are close to zero. *Taxable dividends*, *qualified dividends*, *tax-exempt interest*, and *long-term capital gains* all have correlation differences that are not close to zero. This is not surprising since these variables have few non-zero values and are uncommon sources of income for nonfilers. We did not consider this a cause for concern, but it is an area for future improvement.

We applied pMSE, a propensity-score-based utility measure, that tests whether a model can distinguish between the confidential and the synthetic data. Woo et al. (2009) introduced and Snoke et al. (2018) enhanced a propensity score measure for comparing distributions and evaluating the general utility of synthetic data. We used a logistic regression with only the main effects, resulting in a p-value of the pMSE of 0.26. This suggests the model had difficulties distinguishing between the confidential and synthetic Supplemental PUF.

Ultimately, the synthetic Supplemental PUF data set will be used as an input for tax microsimulation. We built a tax calculator to compare calculations of Adjusted Gross Income (AGI), personal exemptions, deductions, regular income tax, and tax on long-term capital gains and dividends based on the confidential data and the synthetic data. The tax calculator uses a simplified version of 2012 tax law, the year of the confidential and synthetic data. The calculator assumes that all individuals are single filers and lowers the personal exemption to $500, and does not allow any tax credits and standard or itemized deductions. This unorthodox combination of rules is necessary to obtain useful calculations using the Supplemental PUF data, where we simulated a change in law that would make these individuals, who do not have a tax filing obligation under current law, into taxpayers.

Figure 1 compares results for the original and synthetic data sets across different AGI groups for count, mean tax, and total tax. Overall, the synthetic Supplemental PUF performs well on our simple tax calculator and approximates the results from the confidential data set.

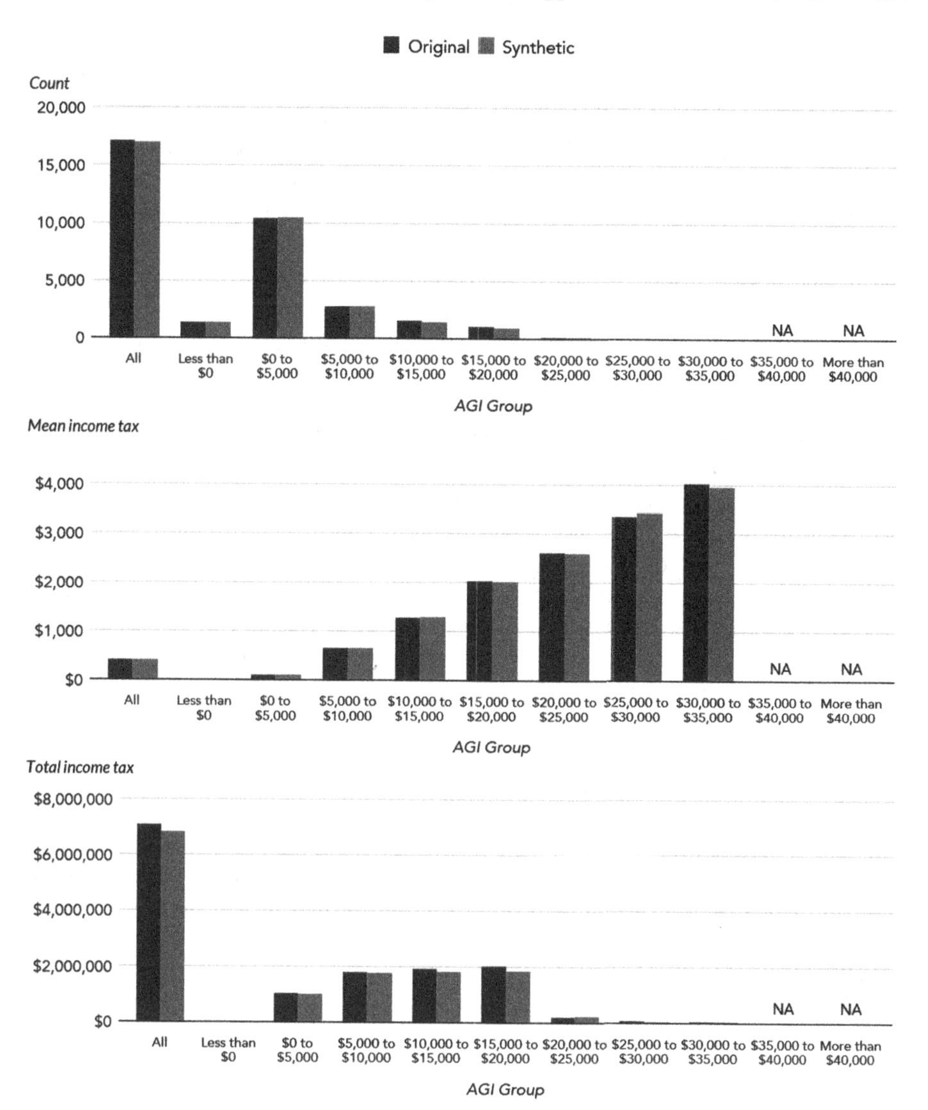

Fig. 1. Tax calculator results for the original and synthetic Supplemental PUF data.

4 Conclusions and Future Work

In this paper, we described the development and evaluation of a method to create a fully synthetic microdata nonfiler database called the Supplemental PUF. We demonstrated that the synthetic data set would not allow a data intruder to meaningfully update his or her prior distribution about any variable reported on an information return. Our method generated a synthetic data set that replicates the characteristics of the underlying administrative data while protecting individual information from disclosure. These data will be useful for policy analysts.

For future work, we will develop a synthetic data set based on the much more complex and diverse individual income tax return data, called the PUF. We do not know, a priori, how well the data synthesis methodology used for the Supplemental PUF data will replicate the underlying distributions of these data. For instance, we found that the "out-of-the-box" random forest method performed worse than CART for the Supplemental PUF data, but random forests might outperform CART for the individual income tax return data. We plan to test a range of data synthesis methods and broaden our disclosure and diagnostic analyses such as methods by Hu et al. (2014), Taub et al. (2018), and Bowen and Snoke (2020). At a minimum, our goal is to create a synthetic PUF that protects the privacy of individuals and reproduces the conditional means and variances of the administrative data. The synthetic data should also be useful for estimating the revenue and distributional impacts of tax policy proposals and for other exploratory statistical analyses.

We will still rely on the sampling rate to provide some privacy protection. Although we did not compute our test statistics for a dataset with only sampling, it is easy to see why sampling reduces disclosure risk. With a 1 in 1,000 sample, as in the synthetic Supplemental PUF, there is a 99.9% chance that any particular record is not in the sample. For the PUF, sampling will provide less protection, because we will be sampling at a higher rate (10%) due to the underlying data are more diverse. However, there is at least a 90% chance that any particular record is not in the underlying database.

Experience suggests that the synthetic data will not provide accurate estimates for complex statistical models, so a key component of this project is to create a way for researchers to run their models using the actual administrative data with parameter estimates altered to protect privacy and standard errors adjusted accordingly. Other future work includes developing and implementing a validation server, a secure process to analyze the raw confidential data. This is a natural complement to the synthetic data because researchers could use the synthetic data, which have the same record layout as the confidential data, for exploratory analysis and to test and debug complex statistical programs. Abowd and Vilhuber (2008) describe a system that provides access to the confidential version of the Survey of Income and Program Participation in which researchers receive statistical output after a privacy review by a US Census Bureau staff. Our goal is to create a similar system that would modify statistical outputs to guarantee privacy and preserve the statistical validity of estimates without requiring human review.

Acknowledgments. We are grateful to Joe Ansaldi, Don Boyd, Jim Cilke, John Czajka, Rick Evans, Dan Feenberg, Barry Johnson, Julia Lane, Graham MacDonald, Shannon Mok, Jim Nunns, James Pearce, Kevin Pierce, Alan Plumley, Daniel Silva-Inclan, Michael Strudler, Lars Vilhuber, Mike Weber, and Doug Wissoker for helpful comments and discussions.

This research is supported by a grant from Arnold Ventures. The findings and conclusions are those of the authors and do not necessarily reflect positions or policies of the Tax Policy Center or its funders.

Appendix

The appendix contains the supplementary materials to accompany the paper "A Synthetic Supplemental Public Use File of Low-Income Information Return Data: Methodology, Utility, and Privacy Implications" with additional results from the utility evaluation.

Figures 2 and 3 show the means and standard deviations, respectively, from the original and synthetic supplemental PUF data for *age* and all 17 tax variables. Figure 4 displays the correlation of the synthetic data minus the correlation of original data for all 17 tax variables.

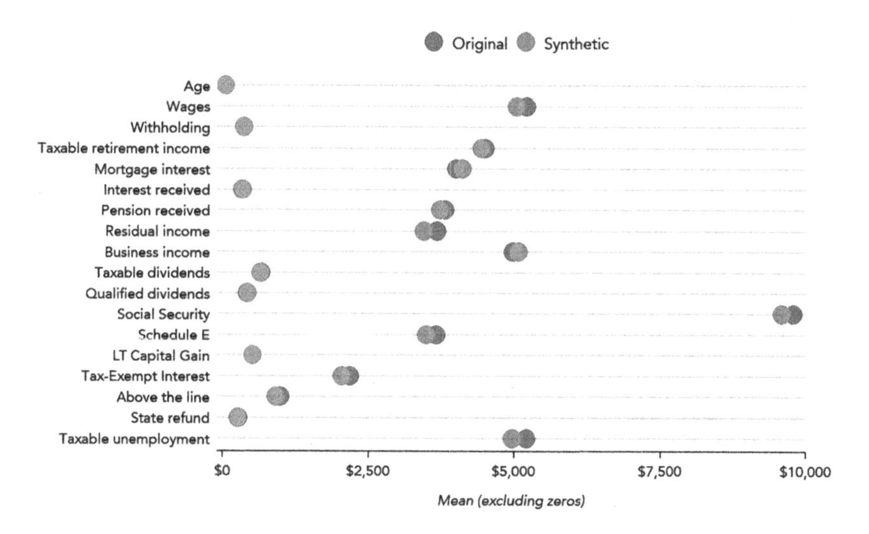

Fig. 2. Means from the original and synthetic Supplemental PUF data. *Age* is on the x-axis scale, but not in dollar amounts.

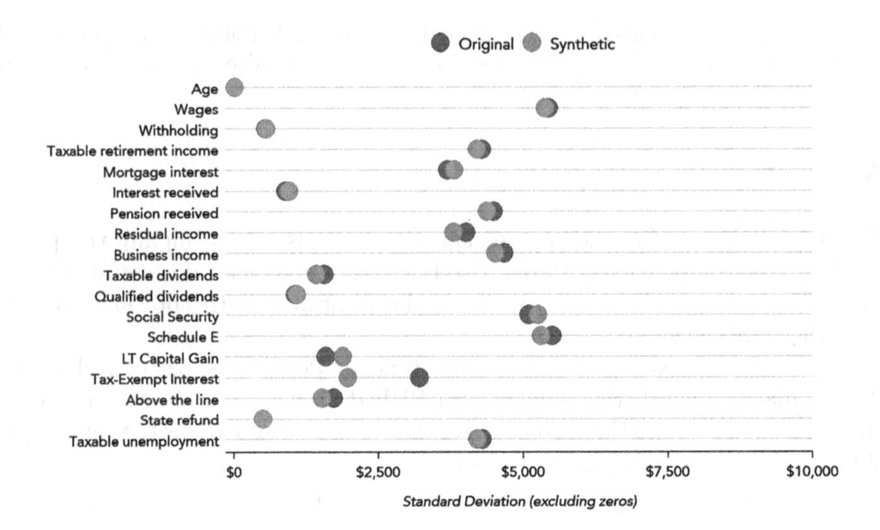

Fig. 3. Standard Deviations from the original and synthetic Supplemental PUF data. *Age* is on the x-axis scale, but not in dollar amounts.

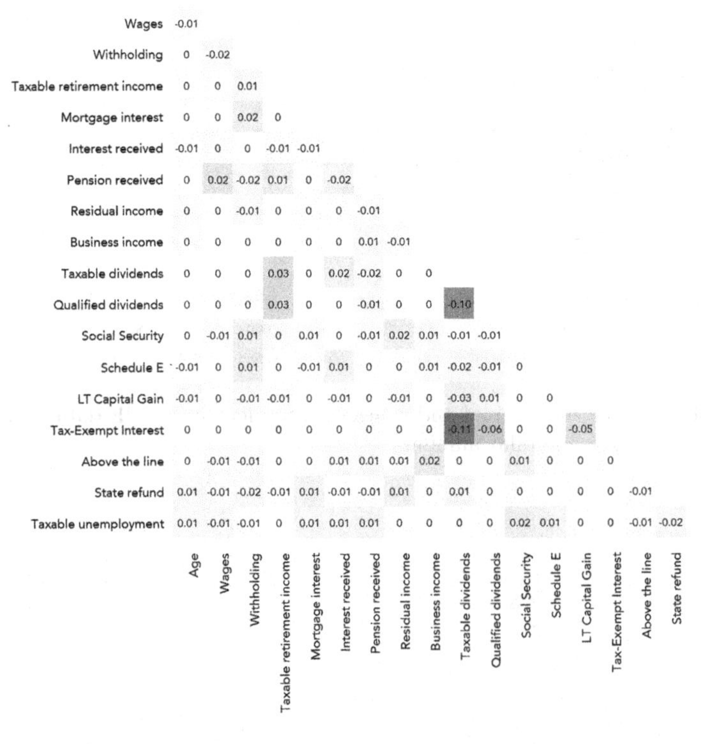

Fig. 4. Difference in Correlation (correlation of the synthetic minus the correlation of the original Supplemental PUF data).

References

Abowd, J.M., Vilhuber, L.: How protective are synthetic data? In: Domingo-Ferrer, J., Saygın, Y. (eds.) PSD 2008. LNCS, vol. 5262, pp. 239–246. Springer, Heidelberg (2008). https://doi.org/10.1007/978-3-540-87471-3_20

Benedetto, G., Stinson, M., Abowd, J.M.: The creation and use of the SIPP synthetic beta (2013)

Bowen, C.M., et al.: A synthetic supplemental public use file of low-income information return data: methodology, utility, and privacy implications. In: Domingo-Ferrer, J. Muralidhar, K. (eds.) PSD 2020. LNCS, vol. 12276, pp. 257–270. Springer, Heidelberg (2020)

Bowen, C.M., Snoke, J.: Comparative study of differentially private synthetic data algorithms from the NIST PSCR differential privacy synthetic data challenge. arXiv preprint arXiv:1911.12704 (2020)

Breiman, L., Friedman, J., Stone, C.J., Olshen, R.A.: Classification and Regression Trees. CRC Press, Boca Raton (1984)

Bryant, V.: General description booklet for the 2012 public use tax file (2017)

Cilke, J.: The case of the missing strangers: what we know and don't know about non-filers. In: 107th Annual Conference of the National Tax Association, Santa Fe, New Mexico (2014)

Dinur, I., Nissim, K.: Revealing information while preserving privacy. In: Proceedings of the Twenty-Second ACM SIGMOD-SIGACT-SIGART Symposium on Principles of Database Systems, pp. 202–210 (2003)

Drechsler, J., Reiter, J.P.: Sampling with synthesis: a new approach for releasing public use census microdata. J. Am. Stat. Assoc. **105**(492), 1347–1357 (2010)

Elliot, M.: Final report on the disclosure risk associated with the synthetic data produced by the SYLLS team. Report 2015-2 (2015)

Fuller, W.: Masking procedures for microdata disclosure. J. Off. Stat. **9**(2), 383–406 (1993)

Hu, J., Reiter, J.P., Wang, Q.: Disclosure risk evaluation for fully synthetic categorical data. In: Domingo-Ferrer, J. (ed.) PSD 2014. LNCS, vol. 8744, pp. 185–199. Springer, Cham (2014). https://doi.org/10.1007/978-3-319-11257-2_15

IRS: 2012 supplemental public use file (2019)

Machanavajjhala, A., Kifer, D., Gehrke, J., Venkitasubramaniam, M.: L-diversity: privacy beyond k-anonymity. ACM Trans. Knowl. Discov. Data (TKDD) **1**(1), 3-es (2007)

McClure, D., Reiter, J.P.: Differential privacy and statistical disclosure risk measures: an investigation with binary synthetic data. Trans. Data Priv. **5**(3), 535–552 (2012)

Nowok, B., Raab, G.M., Snoke, J., Dibben, C., Nowok, M.B.: Package 'synthpop' (2019)

Raab, G.M., Nowok, B., Dibben, C.: Practical data synthesis for large samples. J. Priv. Confid. **7**(3), 67–97 (2016)

Reiter, J.P.: Using cart to generate partially synthetic public use microdata. J. Off. Stat. **21**(3), 441 (2005)

Reiter, J.P., Wang, Q., Zhang, B.: Bayesian estimation of disclosure risks for multiply imputed, synthetic data. J. Priv. Confid. **6**(1), 17–33 (2014)

Snoke, J., Raab, G.M., Nowok, B., Dibben, C., Slavkovic, A.: General and specific utility measures for synthetic data. J. Roy. Stat. Soc. Ser. A (Stat. Soc.) **181**(3), 663–688 (2018)

Taub, J., Elliot, M., Pampaka, M., Smith, D.: Differential correct attribution probability for synthetic data: an exploration. In: Domingo-Ferrer, J., Montes, F. (eds.) PSD 2018. LNCS, vol. 11126, pp. 122–137. Springer, Cham (2018). https://doi.org/10.1007/978-3-319-99771-1_9

Winkler, W.E.: Examples of easy-to-implement, widely used methods of masking for which analytic properties are not justified. Statistics Research Division, US Bureau of the Census (2007)

Woo, M.J., Reiter, J.P., Oganian, A., Karr, A.F.: Global measures of data utility for microdata masked for disclosure limitation. J. Priv. Confid. **1**(1), 111–124 (2009)

Integrating Differential Privacy in the Statistical Disclosure Control Tool-Kit for Synthetic Data Production

Natalie Shlomo(✉)

Social Statistics Department, School of Social Sciences, University of Manchester, Humanities Bridgeford Street G17A, Manchester M13 9PL, UK
natalie.shlomo@mancester.ac.uk

Abstract. A standard approach in the statistical disclosure control tool-kit for producing synthetic data is the procedure based on multivariate sequential chained equation regression models where each successive regression includes variables from the preceding regressions. The models depend on conditional Bayesian posterior distributions and can handle continuous, binary and categorical variables. Synthetic data are generated by drawing random values from the corresponding predictive distributions. Multiple copies of the synthetic data are generated and inference carried out on each of the data sets with results combined for point and variance estimates under well-established combination rules. In this paper, we investigate whether algorithms and mechanisms found in the differential privacy literature can be added to the synthetic data production process to raise the privacy standards used at National Statistical Institutes. In particular, we focus on a differentially private functional mechanism of adding random noise to the estimating equations of the regression models. We also incorporate regularization in the OLS linear models (ridge regression) to compensate for noisy estimating equations and bound the global sensitivity. We evaluate the standard and modified multivariate sequential chained-equation regression approach for producing synthetic data in a small-scale simulation study.

Keywords: Sequential chained-equation regression models · Functional mechanism · Score functions · Ridge regression

1 Introduction

National Statistical Institutes (NSIs) have traditionally released microdata from social surveys and tabular data from censuses and business surveys. Statistical disclosure control (SDC) methods to protect the confidentiality of data subjects in these statistical data include data suppression, coarsening and perturbation. Over the years, these methods have been shown to protect against identity and attribute disclosure risks. Inferential disclosure risk arises when confidential information for data subjects can be learnt by manipulating and combining data sources and it subsumes identity and attribute disclosure risks. Up till now, inferential disclosure risk has been largely controlled by restricting

J. Domingo-Ferrer and K. Muralidhar (Eds.): PSD 2020, LNCS 12276, pp. 271–280, 2020.
https://doi.org/10.1007/978-3-030-57521-2_19

access to the data. For example, microdata from social surveys are deposited in national data archives for registered users only to access the data and census and register-based frequency tables are released as hard-copy tables after careful vetting against inferential disclosure (disclosure by differencing). However, there is growing demand for more open and easily accessible data from NSIs. Some web-based platforms are being developed to allow users to generate and download tabular data through flexible table builders. In addition, research on remote analysis servers, which go beyond generating tabular data, and synthetic microdata allow researchers access to statistical data without the need for human intervention to check for disclosures in the outputs. Synthetic data is particularly useful to researchers for learning about data structures through exploratory analysis and statistical models prior to gaining access to real data, which generally requires lengthy application processes to safe data environments. These new forms of data dissemination mean that NSIs are relinquishing some of the strict controls on released data and hence they have been focusing more on the need to protect against inferential disclosure.

Given the concerns about inferential disclosure, this has led to the statistical community actively reviewing and researching more formal privacy frameworks, in particular the differential privacy definition developed in the computer science literature. Differential privacy (DP) is a mathematical rigorous framework designed to give a well-defined quantification of the confidentiality protection guarantee. It employs a 'worst-case' scenario and avoids assumptions about which variables are sensitive, the intruder's prior knowledge and attack scenarios. DP assumes that very little can be learnt about a single data subject based on the observed output and any post-processing of the output and therefore protects against inferential disclosure. The DP mechanisms are perturbative and include additive noise and/or randomization. See Dwork et al. (2006) and Dwork and Roth (2014) and references therein for more information.

The basic form of a privacy guarantee presented by Dwork et al. (2006) is the ε - Differential Privacy defined for a perturbation mechanism M on database a as follows:

$$P(M(a) \in S) \leq e^{\varepsilon} P\left(M\left(a'\right) \in S\right) \tag{1}$$

for all output subsets S of the range of M and neighboring databases a and a' differing by one data subject. A relaxed DP mechanism is the (ε, δ)-Differential Privacy, which adds a new parameter δ as follows:

$$P(M(a) \in S) \leq e^{\varepsilon} P\left(M\left(a'\right) \in S\right) + \delta \tag{2}$$

Statisticians have looked more carefully at the definition of (ε, δ)-Differential Privacy since the parameter δ can be thought of as a utility measure and allows a small degree of 'slippage' in the bounded ratio of the ε – Differential Privacy definition in (1). For example, in Rinott et al. (2018), δ is the probability of not perturbing beyond a certain cap in a table of frequency counts thus narrowing the range of possible perturbations for a given count. Any perturbation beyond the cap receives a probability of 0 and hence the ratio in (1): $P(M(a) \in S)/P\left(M\left(a'\right) \in S\right)$, is unbounded. In their example and under a discretized Laplace perturbation mechanism M, the chance of selecting a perturbation beyond the cap is defined by δ and is very small.

Differential privacy for histograms and frequency tables have been investigated in many research papers in the computer science literature (see for example Barak et al.

2007; Gaboardi et al. 2016 and references therein) and more recently in Rinott et al. (2018) with an application to a web-based table builder. However, research on DP for synthetic data is still ongoing and recently was advanced by the competition on the 'Herox' platform sponsored by the National Institute of Standards and Technology (NIST), see: https://www.herox.com/Differential-Privacy-Synthetic-Data-Challenge. In this paper, we investigate how we can include and adapt DP algorithms and mechanisms into the standard SDC tool-kit for the production of synthetic data.

Section 2 describes the method of generating synthetic data based on multivariate sequential chained-equation regression models. Section 3 proposes methods to include DP into the synthetic data production process. Section 4 presents a small simulation study to test the feasibility of the proposed methods. We conclude in Sect. 5 with a discussion and conclusions.

2 Generating Synthetic Data

Generating synthetic data under the statistical disclosure control (SDC) framework include parametric and non-parametric methods. Rubin (1993) first introduced synthetic data as an alternative to traditional SDC methods. Rubin's proposal is to treat all data as if they are missing values and impute the data conditional on the observed data. Since then, there have been many alternatives for generating synthetic data in the SDC tool-kit. Partially synthetic data where some of the variables remain unperturbed have higher utility but there is a need to assess disclosure risk with a focus on attribute disclosure, using for example measures of l-diversity and t-closeness The specification of a joint distribution of all variables for generating synthetic data is difficult in real data applications. Raghunathan et al. (2001) and Van Buuren (2007) introduced an approximation using a series of sequential chained-equation regression models for imputing missing data. Raghunathan et al. (2003) and Reiter (2005) show that the approach is useful for generating synthetic data in the SDC tool-kit and it is one of the methods used in the Synthpop R package (Nowok et al. 2016).

The synthesis is performed on each variable by fitting a sequence of regression models and drawing synthetic values from the corresponding predictive distributions. The models for drawing the synthetic values are based on conditional Bayesian posterior distributions and can handle continuous, binary and categorical variables as well as logical and other constraints. For a particular variable, the fitted models are conditioned on the original variables and the synthetic values that come earlier in the synthesis sequence. Synthetic values are drawn for the variable and it is then used as a predictor for the next variable, and so on. At the end of the first iteration, the last variable is conditioned on all other variables. These steps are repeated for a number of iterations. Multiple copies of the data are generated and inference carried out on each of the data sets and results combined for point and variance estimates under well-established combination rules.

Using the notation of Raghunathan et al. (2001) and under the simple case of a continuous variable Y in the data (possibly transformed for normality), we fit a linear regression model $Y = U\beta + e$ $e \sim N(0, \sigma^2 I)$ where U is the most recent predictor matrix including all predictors and previously generated variables. We assume that $\theta = (\beta, \log \sigma)$ has a uniform prior distribution. The coefficient β is estimated by solving the score function $Sc(\beta; \sigma) = \sum_i U_i'(Y_i - U_i\beta) = 0$ and obtaining $\beta = \left(U'U\right)^{-1}U'Y$. The residual sum of squares is $SSE = (Y - U\beta)'(Y - U\beta)$ having df = rows(Y)-cols(U). Let T be the Cholesky decomposition such that $T'T = (U'U)^{-1}$. To draw from the posterior predictive distributions we generate a chi-square random variable u with degrees of freedom df and define $\sigma_*^2 = \frac{SSE}{u}$. We then generate a vector $z = (z_1, \ldots, z_p)$ of standard normal random variables where $p = \text{rows}(\beta)$ and define $\beta_* = \beta + \sigma_* Tz$. The synthetic values for Y are $Y_* = U\beta_* + \sigma_* v$ where v is an independent vector of standard normal random variables with dimension rows(U). More details are in Raghunathan et al. (2001) as well as descriptions for other types of models for binary and categorical variables.

As mentioned, there are other ways to generate synthetic data in the SDC tool-kit (see Drechsler 2011 for an overview), but using the sequential regression modelling approach is conducive to our proposal for adding a layer of protection based on DP in Sect. 3.

Prior to carrying out the synthetic data generation, common SDC methods should first be applied to the microdata. In consultation with the users of the data, an initial SDC step is carried out based on defining which variables need to be in the data and how they should be defined. Direct identifiers are removed from the data. Age and other quasi-identifiers are typically coarsened into groups. The level of geographical information is to be determined according to the requirements of the users. At this stage, we can also apply k-anonymity and related approaches to ensure that coarsened quasi-identifiers have some a priori privacy protection. Under the DP definition, all variables are considered identifiable, but it may not be plausible in practice. For example, there may be some variables that do not need to be perturbed due to legislation, such as some demographic variables in the US Census. Therefore, the dataset may need to be split into two sets of variables: $Y = (y_1, \ldots, y_L)$ variables that will need to be masked and $X = (x_1, \ldots, x_R)$ variables that do not need to be masked.

3 Adding a Layer of Differential Privacy

In the differential privacy framework, we propose to use a functional mechanism of adding random noise to the estimating equations in the regression models under the multivariate sequential chained-equation regression modelling approach described in Sect. 2. Adding differentially private noise to estimating equations in linear and logistic regressions is presented in Zhang et al. 2012 where it is shown that the functional mechanism ensures a differentially private output that is more efficient than adding random noise directly to the results. Therefore, adding random noise to the estimating equations for the generation of synthetic data, albeit under a Bayesian framework, will result in differentially private synthetic data. In addition, any post-analyses on a differentially private dataset accessed by users will remain differentially private (see Proposition 2.1 in Dwork and Roth 2014).

The Bayesian framework used in the sequential chained-equation regression modelling approach automatically induces additional random noise (see the algorithm for the OLS linear regression in Sect. 2). Therefore the exact quantification of the privacy budget accounting for all sources of randomness in the synthetic data generation is still to be determined and will be a topic for future research. As an example, Machanavajjhala et al. (2008) describe the perturbation used in the Bayesian framework for generating the 'On the Map' application of the US Census Bureau where they added smoothing parameters to the priors and then measured privacy post-hoc using the DP likelihood function described in (1).

Following the approach in Chipperfield and O'Keefe (2014), we define a random perturbation vector $s = (s_1, \ldots, s_p)$ and solve the score function $Sc(\beta; \sigma) = \sum_i U_i'(Y_i - U_i\beta) = s$. The solution is $\beta^{DP} = \beta + (U'U)^{-1}s$. Under the DP definition, the random noise is drawn from a Laplace distribution with a shape parameter defined by the sensitivity divided by the privacy budget ε. Here, we define the sensitivity to be equal to the maximum prediction error calculated as $e_i = (Y_i - U_i\beta)$ which represents the contribution of a single individual into the regression equation. We compare two privacy budgets $\varepsilon = 1.0$ and $\varepsilon = 0.5$. Further investigation is needed through simulation studies to assess the level of protection and the utility under the proposed approach under different levels of sensitivity and ε. Since $E(s) = 0$ we obtain a value of β^{DP} that is an unbiased estimate of β. We then follow the same approach for generating synthetic values replacing the coefficient β with the coefficient β^{DP} in Sect. 2.

We can use the same intuition for adding Laplace noise in estimating equations for other types of models in the sequential chained-equation regression models for binary and categorical variables. However, more simulation studies are needed to assess the stability of the score functions for obtaining an estimate β^{DP} under different types of models and for the case of skewed data.

Adding random noise to estimating equations according to the Laplace Distribution may cause unbounded noisy estimation equations with no optimal solution (Zhang et al. 2012). In addition, ordinary least squares (OLS) linear regression in particular may have unbounded global sensitivity when calculating regression coefficients (Li et al. 2017). To circumvent this problem, ridge regression can be used which bounds the OLS estimates (see Zhang et al. 2012; Dandekar et al. 2018). In addition, Sheffet (2015) shows that ridge regression in itself can be used as a differentially private mechanism under certain conditions.

Ridge regression is a technique that add an additional penalty term in the estimating equations and controls for coefficients with extreme values. Ridge regression is generally used to control for multicollinearity and reduced rank in $U'U$. It is a 'shrinkage' approach since it converges the regression coefficients toward 0. The solution for ridge regression is $\beta^{Ridge} = (\mathrm{U'U} + kI)^{-1}\mathrm{U'Y}$ where U is now centered and scaled and k is a parameter that determines by how much the ridge parameters differ from the parameters of OLS. Figure 1 demonstrates how ridge regression bounds the global sensitivity for an OLS linear regression coefficient. In the simulation study in Sect. 4 we explore whether the modifications of ridge regression and additive random noise to the estimating equations impact on the standard approach for producing synthetic data using the multivariate sequential chained-equation regression models approach.

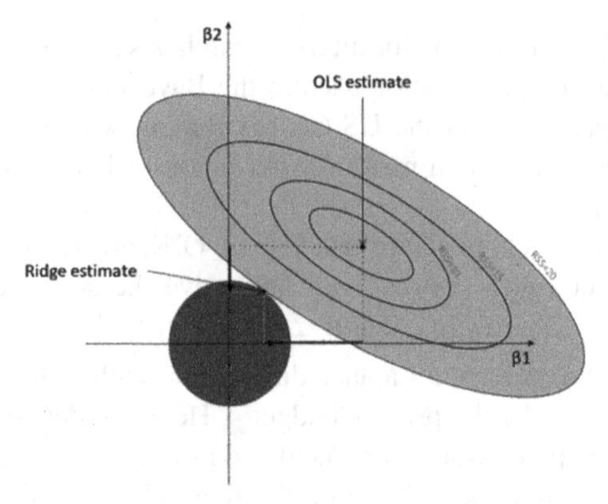

Fig. 1. Ridge Regression (figure taken from https://towardsdatascience.com/ridge-regression-for-better-usage-2f19b3a202db)

4 Small Simulation Study

For this small simulation study, we run one iteration of the standard sequential chained-equation regression modelling approach for generating synthetic data and modify it in three ways: (1) replace the OLS linear regressions with a ridge regression; (2) add Laplace noise to the estimating equations; (3) combination of both modifications. We assume that the variables of interest are multivariate normal to avoid skewed distributions and focus on OLS regression models.

We generate a population of N = 50,000 under the following settings: $(e_1, e_2, e_3) \sim MVN(0, \Sigma)$ where $\Sigma = \begin{pmatrix} 10 & 5 & 3 \\ 5 & 10 & 4 \\ 3 & 4 & 10 \end{pmatrix}$. Define $X_1 \sim U(0, 10)$ and $X_2 \sim U(0, 20)$. We generate three regression equations (Y_1, Y_2, Y_3) as follows: $Y_i = \beta_{0i} + \beta_{1i}X_1 + \beta_{2i}X_2 + e_i$ $i = 1, 2, 3$ and $\beta = \begin{pmatrix} 20 & 0.5 & 0.5 \\ 30 & 2 & 2 \\ 40 & 1 & 1 \end{pmatrix}$. We synthesize values for (Y_1, Y_2, Y_3) keeping X_1 and X_2 unperturbed. The true population values are in Table 1.

Table 1. True values of (Y_1, Y_2, Y_3) in the population N = 50,000.

	Mean	Std Dev	Minimum	Maximum
Y_1	27.5	4.5	11.3	43.5
Y_2	60.0	13.3	23.1	97.5
Y_3	55.0	7.2	32.7	77.3

We draw 200 simple random samples of size 5000 from the population and on each sample we generate synthetic data using the standard approach; replace the linear regression with ridge regression (k = 0.05); add Laplace noise to the score function with the shape parameter proportional to the maximal prediction value, which in this case is equal to 10, divided by ε; combination of the two modification approaches.

Table 2 presents the average relative absolute bias of the synthetic values $\hat{Y}$ compared to the true values Y using the formula: $\frac{1}{5000}\sum\left|Y-\hat{Y}\right|/Y$ averaged over 200 samples. In this simple simulation study, the biases are similar across the three modification approaches compared to the standard approach for generating synthetic data even when ε is reduced. This demonstrates that there is less impact in statistical analyses when random noise is added to the estimating equations compared to the results.

Table 2. Average relative absolute bias over 200 samples

Variables	Standard	Ridge regression	Noise addition		Ridge regression and noise addition	
			$\varepsilon = 1.0$	$\varepsilon = 0.5$	$\varepsilon = 1.0$	$\varepsilon = 0.5$
Y_1	0.131	0.136	0.131	0.137	0.136	0.141
Y_2	0.053	0.060	0.053	0.054	0.060	0.060
Y_3	0.061	0.059	0.061	0.062	0.059	0.060

Table 3 presents the regression parameters for the model $Y_1 \sim Y_2, Y_3, X_1, X_2$ for the standard and three modified synthetic data approaches and includes the true regression parameters from the population in the second column. The method of additive noise to the estimating equations are similar to the truth and the standard approach albeit with noted differences in the intercept. Under the ridge regression, there are larger deviations from the original regression parameters as expected due to the shrinkage with a larger intercept to compensate.

Table 3. Regression parameters for the model $Y_1 \sim Y_2, Y_3, X_1, X_2$ average over 200 samples

Variables	True	Standard	Ridge regression	Noise addition		Ridge regression and noise addition	
				$\varepsilon = 1.0$	$\varepsilon = 0.5$	$\varepsilon = 1.0$	$\varepsilon = 0.5$
Intercept	2.074	2.226	5.867	2.251	2.135	5.885	5.786
Y_2	0.447	0.443	0.252	0.443	0.448	0.252	0.256
Y_3	0.113	0.110	0.154	0.109	0.111	0.153	0.155
X_1	−0.505	−0.501	−0.139	−0.501	−0.505	−0.138	−0.142
X_2	−0.507	−0.502	−0.141	−0.502	−0.506	−0.141	−0.144

Table 4 presents correlations between the target variables to assess whether the correlation structure is preserved in the standard and modified approaches for generating synthetic data compared to the true correlations in the population. The reduced privacy budget has slightly increased the bias compared to the standard version. For example, under the ridge regression with additive noise in the estimating equations, we obtain a bias of 2.6% for the correlation between Y_1 and Y_2 and 4.1% for the correlation between Y_1 and Y_3.

Table 4. Correlations average over 200 samples

Variable pairs	True	Standard	Ridge regression	Noise addition		Ridge regression and noise addition	
				$\varepsilon = 1.0$	$\varepsilon = 0.5$	$\varepsilon = 1.0$	$\varepsilon = 0.5$
Y_1, Y_2	0.78	0.77	0.77	0.77	0.79	0.77	0.80
Y_1, Y_3	0.73	0.73	0.75	0.73	0.75	0.75	0.76
Y_2, Y_3	0.91	0.92	0.92	0.92	0.91	0.92	0.91

5 Discussion and Conclusions

NSIs are investigating the feasibility of including differential privacy algorithms and mechanisms into the SDC tool-kit, particularly when disseminating statistical data via web-based platforms and producing open synthetic data. Up till now, there have been various approaches proposed for generating differentially private (DP) synthetic data. For example, at the US Census Bureau, they are proposing to generate synthetic census data by perturbing count data under a DP mechanism in many census hyper-cubes and then reproducing the microdata from the perturbed hyper-cubes. This approach is problematic since differentially private counts can be negative and not all logical constraints, correlations and sub-group totals can be preserved without adding additional optimization solvers. Other approaches in the computer science and machine learning literature use Generative Adversarial Networks (GANs) (see Torkzadehmahani et al. 2020 and references therein).

In this paper, we have demonstrated the potential of including aspects of the DP framework into standard synthetic data generation under the multivariate sequential chained-equation regression modelling approach. We looked at two features and their combination: adding random noise to estimating equations according to a differentially private Laplace distribution and regularization through ridge regression.

For the shape of the Laplace distribution for generating random noise to add to the estimating equations, we defined the sensitivity as the maximum prediction error and set the privacy budget at $\varepsilon = 1.0$ and $\varepsilon = 0.5$. There were little differences in the outputs from the versions of synthetic data under the varying levels of the privacy budget although there was slightly larger bias in the correlation structure. In addition, there are many other mechanisms of randomization under the Bayesian framework for generating synthetic data based on non-informative priors and therefore, future research is needed to formulate the exact mathematics of the sensitivity and privacy budget taking into account the Bayesian elements which a priori smooths out the parameter space.

For the ridge regression, we tested one level of the ridge parameter at k = 0.5. Ridge regression aims to bound the global sensitivity in OLS and also stabilize the estimating equations given that we set them to a noise variable s instead of zero. In fact, the way we add the random noise to the estimating equation by adding $\left(U'U\right)^{-1}s$ to the original calculation of β (see Sect. 3) could mitigate the problem of unstable estimating equations, but this needs further investigation. In addition, the ridge regression also contributes to the privacy budget and needs to be accounted for when formulating the mathematics of the sensitivity and privacy budget in future research.

The aim of this paper was to show that we can use differentially private algorithms and mechanisms to modify the production of synthetic data according to a common approach in the SDC tool-kit without causing additional bias to the underlying data structure. The small simulation study verified this finding for the simple case of OLS linear regressions. More large scale simulation studies and real data applications are needed with other types of variables and regression models to assess feasibility under more practical settings.

References

Barak, B., Chaudhuri, K., Dwork, C., Kale, S., McSherry, F., Talwar., K.: Privacy, accuracy, and consistency too: a holistic solution to contingency table release. In: Proceedings of the 26th ACM SIGMOD-SIGACT-SIGART Symposium on Principles of Database Systems (PODS), pp. 273–282 (2007)

Chipperfield, J.O., O'Keefe, C.M.: Disclosure-protected inference using generalised linear models. Int. Stat. Rev. **82**(3), 371–391 (2014)

Dandekar, A., Basu, D., Bressan, S.: Differential privacy for regularised linear regression. In: Hartmann, S., Ma, H., Hameurlain, A., Pernul, G., Wagner, Roland R. (eds.) DEXA 2018. LNCS, vol. 11030, pp. 483–491. Springer, Cham (2018). https://doi.org/10.1007/978-3-319-98812-2_44

Drechsler, J.: Synthetic Datasets for Statistical Disclosure Control, Theory and Implementation. Lecture Notes in Statistics. Springer, New York (2011). https://doi.org/10.1007/978-1-4614-0326-5

Dwork, C., McSherry, F., Nissim, K., Smith, A.: Calibrating noise to sensitivity in private data analysis. In: Halevi, S., Rabin, T. (eds.) TCC 2006. LNCS, vol. 3876, pp. 265–284. Springer, Heidelberg (2006). https://doi.org/10.1007/11681878_14

Dwork, C., Roth, A.: The algorithmic foundations of differential privacy. Found. Trends Theoret. Comput. Sci. **9**, 211–407 (2014)

Gaboardi, M., Arias, E.J.G., Hsu, J., Roth, A., Wu, Z.S.: Dual query: practical private query release for high dimensional data. J. Priv. Confid. **7**, 53–77 (2016)

Li, N., Lyu, M., Su, D., Yang, W.: Differential Privacy: From Theory to Practice. Synthesis Lectures on Information Security, Privacy and Trust. Morgan and Claypool (2017)

Machanavajjhala, A., Kifer, D., Abowd, J., Gehrke, J., Vilhuber, L.: Privacy: theory meets practice on the map. In: Proceedings of the IEEE 24th International Conference on Data Engineering. ICDE, pp. 277–286 (2008)

Nowok, B., Raab, G.M., Dibben, C.: Synthpop: bespoke creation of synthetic data in r. J. Stat. Softw. **74**(11), 1–26 (2016)

Raghunathan, T.E., Lepkowksi, J.M., van Hoewyk, J., Solenbeger, P.: A multivariate technique for multiply imputing missing values using a sequence of regression models. Surv. Methodol. **27**, 85–95 (2001)

Raghunathan, T.E., Reiter, J.P., Rubin, D.B.: Multiple imputation for statistical disclosure limitation. J. Off. Stat. **19**, 1–16 (2003)

Reiter, J.: Releasing multiply-imputed, synthetic public use microdata: An illustration and empirical study. J. Roy. Stat. Soc. A **168**(1), 185–205 (2005)

Rinott, Y., O'Keefe, C., Shlomo, N., Skinner, C.: Confidentiality and differential privacy in the dissemination of frequency tables. Stat. Sci. **33**(3), 358–385 (2018)

Rubin, D.B.: Satisfying confidentiality constraints through the use of synthetic multiply-imputed microdata. J. Off. Stat. **91**, 461–468 (1993)

Sheffet, O.: Private approximations of the 2nd-moment matrix using existing techniques in linear regression (2015). https://arxiv.org/abs/1507.00056. Accessed 29 June 2020

Torkzadehmahani, R., Kairouz, P., Paten, B.: DP-CGAN: Differentially Private Synthetic Data and Label Generation (2020). https://arxiv.org/abs/2001.09700. Accessed 31 May 2020

Van Buuren, S.: Multiple imputation of discrete and continuous data by fully conditional specification. Stat. Methods Med. Res. **16**(3), 219–242 (2007)

Zhang, J., Zhang, Z., Xiao, X., Yang, Y., Winslett, M.: Functional mechanism: regression analysis under differential privacy (2012). https://arxiv.org/abs/1208.0219. Accessed 31 May 2020

Advantages of Imputation vs. Data Swapping for Statistical Disclosure Control

Satkartar K. Kinney(✉), Charlotte B. Looby, and Feng Yu

RTI International, Research Triangle Park, NC 27709, USA
skinney@rti.org

Abstract. Data swapping is an approach long-used by public agencies to protect respondent confidentiality in which values of some variables are swapped with similar records for a small portion of respondents. Synthetic data is a newer method in which many if not all values are replaced with multiple imputations. Synthetic data can be difficult to implement for complex data; however, when the portion of data replaced is similar to data swapping, it becomes simple to implement using publicly available software. This paper describes how this simplification of synthetic data can be used to provide a better balance of data quality and disclosure protection compared to data swapping. This is illustrated via an empirical comparison using data from the Survey of Earned Doctorates.

Keywords: Synthetic data · Applications · Methodology

1 Introduction

Data swapping has been used for many years by statistical agencies around the world to produce public-use files (PUFs) and restricted-use files (RUFs) [10], though some agencies, such as the U.S. Census Bureau are phasing it out [7]. Swapping is typically used as a supplementary disclosure protection measure when reductive methods such as suppression and coarsening, which reduce the detail of the information provided but do not alter it, are used as the primary disclosure avoidance mechanism. The method involves randomly selecting a small fraction of records and then swapping certain values for each selected record with a similar record [1]. This adds uncertainty and plausible deniability in the event of any ostensible re-identification since analysts do not know which records have been perturbed. The rate of data swapping, swapping variables, and other details are typically kept confidential. Critics of data swapping point out that this lack of transparency makes data swapping *non-ignorable*; i.e., analyses of swapped data are unable to take into account the uncertainty introduced by swapping. Agencies implementing swapping indicate it is a small source of error compared to other sources of error such as nonresponse [1]. Since both the number of records and number of variables that can be perturbed with swapping without

J. Domingo-Ferrer and K. Muralidhar (Eds.): PSD 2020, LNCS 12276, pp. 281–296, 2020.
https://doi.org/10.1007/978-3-030-57521-2_20

compromising the usefulness of the data are limited [10,11,15], it is generally known that the proportion of values that are swapped is small, but how small is not public information and thus this paper does not define 'small'.

The use of imputation as a statistical disclosure method, i.e., synthetic data, was first proposed by Rubin [22], as an extension of multiple imputation for missing data [23]. As it was originally proposed, entire datasets would be replaced with multiple implicates, each implicate a new sample from the frame. This approach is now called 'fully synthetic data', whereas partially synthetic data [14] involves replacing part of a dataset with multiple imputations. As with multiple imputation for missing data, analysts apply standard data methods to each implicate and then apply combining rules that incorporate the uncertainty due to imputation [19]. Recent applications of synthetic data include U.S. Internal Revenue Service tax data [5] and the U.S. Census Bureau Survey of Program Participation [2], which was also one of the first surveys to start using synthetic data in 2004 [3]. Synthetic data can provide a high level of protection against re-identification, since the data are simulated; however, in the case of partially synthetic data, the portion of data replaced greatly influences the level of protection. Criticism of synthetic data includes the reliance of inferences using synthetic data on the models used to generate the data, the time involved to produce good quality synthetic data, and reluctance of data users to use multiply-imputed or 'fake' data.

This paper describes a simplification of synthetic data that is similar to data swapping in application and intent. That is, a small portion of records have their values replaced with single imputations for selected variables in order to add a degree of uncertainty and deniability to any ostensible record linkage. Although we use synthetic data methods and software to generate the imputations, we refer to this approach as 'disclosure imputation' to distinguish from synthetic data applications that use multiple imputation and/or impute much larger portions of the data, as well as to aid in public acceptance. Users of public data are accustomed to analyzing datasets in which a substantial portion of values have been imputed (due to nonresponse), whereas 'synthetic' to many connotes fake. Disclosure imputation is quite simple to implement with automated routines such as those provided by the R package *synthpop* [16]. Data stewards who use or are considering the use of data swapping may find imputation to be more intuitive and simpler to implement, particularly when it comes to preserving relationships between perturbed and unperturbed variables. Section 2 describes the method of implementation and Sect. 3 describes an empirical experiment to illustrate that the method can provide better utility at comparable levels of disclosure protection compared to data swapping. Section 4 concludes with additional discussion.

2 Method

This section describes the steps to implement disclosure imputation using R synthpop: 1) select variables to be imputed; 2) select records to be imputed; 3) generate imputations; and 4) evaluate imputations.

2.1 Selecting Variables for Imputation

Ideally the variables selected for imputation should include identifying variables such as race, age, and sex in order to provide meaningful uncertainty to any attempted record linkage. When all selected variables are categorical, it may be possible to impute several variables such that each variable has a low perturbation rate but a large portion of records have at least one imputed value. When categorical variables are imputed, many records will have imputed values that are the same as the original value, so the effective perturbation rate will be lower than the proportion of records selected for imputation. If only one categorical variable is imputed, then the overall perturbation rate, which we define as the proportion of records with at least one value changed by imputation, will be the same as the variable-level perturbation rate. If several categorical variables are imputed, then the variable-level perturbation rates will be lower than the overall perturbation rate, so it could be advantageous from both a risk and utility perspective to select several variables for imputation. It may also be sensible to impute portions of sensitive variables, such as income or health status, which might also be coarsened or top-coded, as an added protection measure.

2.2 Selecting Records for Imputation

The simplest approach for selecting records is to use simple random sampling. An advantage of this is that the imputation models can be constructed using the full sample since the selected records are representative of the full dataset. If the records selected are not representative of the full dataset then better data utility will be obtained by building the model using only the records that are going to be imputed. In this case the imputation rate should be high enough to ensure enough records are available for modeling.

If there are subgroups considered to be at a high risk of re-identification, they may be protected in a simple random sample selection if variables defining the high risk group are selected for perturbation. Alternatively, records can be selected for imputation based on their risk. Records at high risk for identity disclosure via record linkage are those that are most unique on identifying variables. This can be evaluated using the principle of k-anonymity privacy protection [24]. That is, protection against record linkage is provided when each individual is identical to at least $k-1$ other respondents in the dataset on specified key identifiers. Individuals that are unique on a set of key variables are at a greater risk of identification than those that are identical to many individuals on key variables. This depends heavily on the set of identifying variables used. For example, a dataset may have k-anonymity privacy protection for key identifiers race, ethnicity, and sex with $k = 5$, but not when other potentially identifying variables are used, like marital status and age. The R package sdcMicro [25], which contains functions for assessing and addressing disclosure risk in microdata, provides an easy way to assess k-anonymity.

Other methods of selecting records may be used. For example, Drechsler and Reiter [9] propose an iterative algorithm for selecting values to replace with imputations. The algorithm seeks to minimize the number of values replaced under constraints defined by risk and utility measures.

2.3 Generating Imputations

The R package synthpop provides tools for generating fully or partially synthetic data. There are a variety of modeling options available in synthpop, including classification and regression trees (CART) which are a good default option for generating synthetic data. CART models and related methods have become increasingly popular for both missing data and synthetic data [4,8,20]. They provide a flexible modeling approach, are simple to specify, and do not require distributional assumptions. The CART algorithm automatically detects important variables, interaction terms, and nonlinear relationships, even with a large list of potential predictors, increasing the likelihood that important relationships will be preserved in the imputed data. The adaptation of CART for synthetic data proposed in Reiter [20] includes a Bayesian bootstrap step which yields 'proper' imputations. Subsequently, Reiter and Kinney [21] showed that this step is unnecessary and hence the default option in synthpop is 'improper' imputation. Proper imputation will yield greater variability between imputations.

Variables are imputed sequentially, conditional on previously imputed variables. This is the case generally for synthetic data: the joint posterior distribution for the variables being imputed can be written as $f(y_1|X) \cdot f(y_2|y_1, X) \cdot f(y_3|y_2, y_1, X) \ldots$ where X represents all the variables not being imputed and $y_1, y_2, \ldots$ are the variables being imputed. See Drechsler [8] for details and methods for synthetic data generation. This sequential approach is equivalent to modeling the joint distribution regardless of the order of the variables; however, the order of imputation can impact data quality in datasets with complex logical relationships. Additional variables that are not being imputed may be included as predictor variables. In surveys with sampling weights, weights may also be used as predictor variables. If design variables are imputed, another option is to recompute weights after imputation [15].

For each imputed variable, imputed values are drawn from leaves of a CART model fit on a set of predictor variables, including previously imputed variables and potentially additional variables used only as predictors. As is typical when using CART models for prediction, tuning parameters may be used to improve prediction, though synthpop provides useful defaults. Using larger trees increases the number of interactions that can be accounted for in the model. Note for variables with missing values, synthpop imputes missingness first and then imputes the non-missing values.

As of this writing, synthpop allows users to specify which variables should be imputed but not which records; however, this can be easily worked around. Sample code is provided in the appendix. Other synthetic data tools besides synthpop are available; however, we have not evaluated their use. The most user friendly and closest in nature to synthpop is IVEWare which does not at present support CART models but can be used for both missing data imputation and disclosure imputation and is available for SAS, Stata, and SPSS in addition to R and a standalone version [17].

2.4 Evaluating Imputations

When imputing data for disclosure purposes, the goal is to model the dataset itself, so evaluating data quality primarily entails comparisons of the data before and after imputation and assessing how well the imputed data preserves the statistical properties of the original data. That can be cumbersome when there are a large number of analyses, so risk and utility measures have been developed to assist in evaluating risk-utility trade-offs of candidate datasets [6]. If certain subgroups were targeted for imputation (Sect. 2.2) or are considered higher risk, then estimates for those subgroups should be evaluated separately. Since the protection mechanism of data swapping and disclosure imputation comes from the portion of data being imputed, a key step in risk evaluation is to verify the perturbation rates.

3 Experiments

This section describes a set of experiments conducted to compare applications of data swapping and imputation using restricted-use data from the National Center for Science and Engineering Statistics Survey of Earned Doctorates (SED). SED is an annual census of new research doctorate recipients in the United States. Data swapping and disclosure imputation strategies were made to be as similar as possible to facilitate comparison. The experiments described do not correspond to any actual data product.

The variables selected for perturbation (imputation or swapping) are shown in Table 1 below. Given the analytic importance of Race/Ethnicity, we also conducted the experiments without perturbing Race/Ethnicity.

Table 1. Variables selected for perturbation

Variable label	Data swapping	Imputation
Race/Ethnicity	RACE (categorical) was swapped and binary indicators were link swapped	Binary indicators were imputed and RACE re-derived after imputation
Father's Education	EDFATHER	EDFATHER
Mother's Education	EDMOTHER	EDMOTHER
Marital Status and number of dependents by age category	MARITAL swapped and DEPEND5, DEPEND18, and DEPEND19 link swapped	MARITAL, DEPEND5, DEPEND18, DEPEND19 all imputed
Graduate debt level (categorical)	GDEBTLVL	GDEBTLVL

Data swapping was performed using *DataSwap* software developed at the National Center for Education Statistics [12,13] and used with their permission. Imputation was performed using the R package *synthpop* [16].

3.1 Data Swapping

Four data swapping experiments were performed. Data swapping was performed within strata defined by the variables institution HBCU/Carnegie classification, institution number of doctorates (large/small), sex, and citizenship. The full cross-classification of these variables contained numerous small cells. These were collapsed following two strategies: one that collapsed institution information resulting in 20 strata, with cell sizes ranging from 63 to 13,516; and a second that had more institution information resulting in 88 strata, with cell sizes ranging from 39 to 9,707. Thus the four experiments were:

1. Experiment 1: Swap all variables in Table 1 using 20 swapping strata
2. Experiment 2: Swap all variables in Table 1 using 88 swapping strata
3. Experiment 3: Exclude race from swapping and use 20 strata
4. Experiment 4: Exclude race from swapping and use 88 strata

Each experiment was run with five different swap rates: 1.25%, 2.5%, 5%, 7.5%, and 10%. These correspond to target perturbation rates of 2.5%, 5%, 10%, 15%, and 20%, where the perturbation rate is defined as the percent of records with at least one value perturbed. For each experiment, records were randomly selected for data swapping within each swapping stratum at the given swap rate. Partners for each selected record were chosen based on partitions of the stratum created by cross-classifying the swap variables. These partitions are called swapping cells. For target records selected for swapping, candidate swapping partners were identified in the adjacent swapping cell within the same stratum such that partners had similar swap variable values numerically close to, but not equal to, the target record values. For each target record, the candidate with the smallest distance between swapping variables was selected as the swapping partner. The swapping variable listed last in the variable list will tend to have the highest swap rate, so the order of the swap variables is arranged in each partition such that the swap rates are more balanced across swapping variables. Each swapping pair exchanges values of the swapping variable and any link swapping variables.

For each experiment and swap rate, five random swapping runs were completed and the best run of the five was selected using data quality measures built into DataSwap. These quality measures include measures for pairwise associations based on Pearson's product correlation, Pearson's contingency coefficient, Cramer's V, as well as measures for multivariate association via linear regression models. Note Hellinger's Distance was not used in quality assessment because the SED is a census with no weights involved. Swapping does not impact Hellinger's Distance when the weight is one at the univariate level.

3.2 Imputation

Two experiments were conducted to impute the variables listed in Table 1: one imputing all variables in Table 1 and one imputing the same variables excluding race. Each experiment was run with five perturbation rates of 2.5%, 5%, 10%, 15%, and 20%. Records were selected at random for imputation from the whole dataset at a rate specified slightly higher than the target perturbation rate such that the target perturbation rate was met. For each experiment and perturbation rate, five random subsets of the data were replaced with imputed values and the best of the five selected using general and local utility measures available in synthpop.

Imputation was conducted as described in Sect. 2.3. Variables were imputed sequentially in the order HISPANIC, AMERIND, ASIAN, BLACK, HAWAIIAN, WHITE, EDMOTHER, EDFATHER, MARITAL, DEPEND5, DEPEND18, DEPEND19, GDEBTLVL. HISPANIC was imputed conditional on all of the predictor variables; AMERIND was imputed conditional on HISPANIC and all of the predictor variables, ASIAN was imputed conditional on HISPANIC, AMERIND, and all of the predictor variables, and so on. An additional 32 variables were used as predictor variables but not imputed.

Correlations between imputed and non-imputed variables are preserved by including related variables as model predictors even if they are not being imputed for disclosure. For example, when race is imputed, the relationship between race and institution HBCU and Carnegie Classification is preserved in the imputed data by including these variables as model predictors. In data swapping logically constrained variables are link swapped, i.e., swapped together between the same swapping partners. Marital status and number of dependents are not logically constrained; however, they were link swapped due to having a high correlation. For imputation, it would have been sufficient for data quality purposes to include the dependent variables (DEPEND5, DEPEND18, and DEPEND19) as predictors and not to impute them; however, they were included as imputation variables for consistency with the swapping experiment. Since the categorical race variable is a deterministic function of the race indicators, the race and ethnicity indicators were imputed and then the combined race/ethnicity variables were re-computed after imputation from the imputed race indicators. For variables with deterministic logical relationships that cannot be derived post-imputation, it may be necessary to impute the related variables together. For example, if the dataset included, say, spouse age, it would have been necessary to impute this along with marital status to maintain logical consistency.

Utility was evaluated by comparing marginal and conditional proportions of key analytical variables of interest. Most were found to be identical to two decimal places and all within ± 0.01. In addition, some three-way tables were constructed and found to be quite similar (not shown). Additional utility evaluation is described in Sect. 3.4. In order to assess post-imputation disclosure risk, we merged the pre-imputation data and post-imputation data in order to verify risk reduction and the rates of perturbation. These are described below in Sect. 3.3 and compared with the swapping experiments.

3.3 Perturbation Rates

All swapping and imputation experiments were successful in achieving target overall perturbation rates, that is, the target proportion of records with at least one value perturbed. As expected, the variable-level perturbation rates tend to be much lower than the overall perturbation rate. As seen in Table 2, the variable-level perturbation rates for most variables are higher for the imputation experiments. Results displayed correspond to an overall perturbation rate of 10%. Similar results were seen for other perturbation rates. The number of variables perturbed per record was also higher for imputation. As seen in Table 3, the swapping experiments have higher percents of records with one variable perturbed while the imputation experiments have higher percents of records with three or more variables perturbed.

Table 2. Variable-level perturbation rates by experiment

Variable	Swapping 1	Swapping 2	Swapping 3	Swapping 4	Imputation 1	Imputation2
GDEBTLVL	2.8	4.7	2.2	4.3	3.7	3.8
EDMOTHER	4.5	4.8	3.9	4.6	6.4	6.6
EDFATHER	3.8	3.1	4.0	3.1	7.2	7.6
MARITAL	2.0	1.8	2.7	2.3	3.9	4.0
DEPEND5	0.8	0.8	1.2	1.0	2.3	2.2
DEPEND18	0.5	0.5	1.0	0.8	1.3	1.3
DEPEND19	0.3	0.4	0.7	0.6	0.7	0.6
CITRACE	1.0	1.4	n/a	n/a	1.8	n/a

Table 3. Percent of records with different numbers of variables perturbed

Experiment	Number of variables perturbed		
	1	2	3+
Imputation 1	1.3	2.2	5.9
Imputation 2	1.7	3.1	5.1
Swapping 1	4.6	2.6	2.3
Swapping 2	4.0	2.5	2.8
Swapping 3	6.5	2.1	1.2
Swapping 4	5.6	2.5	1.6

3.4 Utility Comparison

In order to assess the impact of perturbation on data quality, six two-way tables of key analytic interest were reproduced for each experiment. While this provides concrete evidence of the impact of perturbation on actual data products, it is a lot of output to evaluate. Utility measures were computed to provide one-number summaries of the differences between the perturbed and unperturbed tables. Most of these tables involve Race/Ethnicity combined with Citizenship into one variable, CITRACE, in which temporary visa holders are in one category and U.S. citizens and permanent residents are in categories by race and ethnicity. This is how race and ethnicity are typically reported for SED.

This section provides a summary of the differences between the perturbed and unperturbed tables and utility measures for each key table across experiments. Tables for which the only perturbed variable is Race/Ethnicity are only compared for experiments that perturbed Race/Ethnicity. The utility measure used is the Freeman-Tukey measure [18]. This measure is related to a chi-square statistic and is calculated as $4\sum_i(\sqrt{obs_i} - \sqrt{syn_i})^2$, where summation is over table cells. This is a distance measure that provides a metric for comparing tables generated from different candidate perturbed datasets to the same table generated from the confidential data. Note that for distance measures, *larger values correspond to lower utility and lower values correspond to higher utility.*

In order to compare utility scores at comparable levels of perturbation, the utility scores for each data table were plotted against variable-level perturbation rates for the variables in the table. For tables with both variables perturbed, the perturbation rate is the average of the two variable-level perturbation rates. As expected, relationships between swapped and unswapped variables are not preserved well in the perturbed data, particularly as the perturbation level increases. This is seen in the relationships between CITRACE and SRFOS (field of study) and between CITRACE and PRIMSRCE (primary funding source). On the other hand, as seen in Fig. 1, much better utility is observed for the imputation experiments. That said, the tables produced for all experiments appeared to do reasonably well in estimating counts and percents for larger cells, but CITRACE*SRFOS in particular has many small counts. By chance, some of these avoid perturbation while others have meaningful differences.

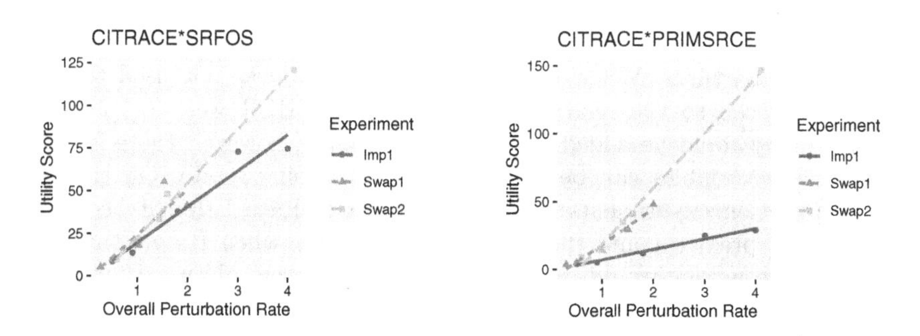

Fig. 1. Utility scores by experiment and perturbation level for CITRACE*SRFOS and CITRACE*PRIMSRCE

The distinction between swapping and imputation is less clear for the table CITRACE*PHDINST which contains counts by institutions and Race/Ethnicity. These are reasonably close to the unperturbed counts for all experiments, even for experiments that did not account for institution, likely because much of the correlation between race and institution is accounted for by the institution characteristics (HBCU, Carnegie classification) which are included in all experiments. A challenge for both approaches is that there are some 400 institutions in the dataset, many of them with few doctorates. For the imputation experiments small institutions were grouped together to facilitate inclusion of PHDINST in the imputation models. For swapping experiments, fewer and larger swapping strata provide a larger selection of swapping partners yielding better utility than smaller swapping strata. As seen in Fig. 2, the experiments all perform comparably in reproducing this table except at the highest perturbation level.

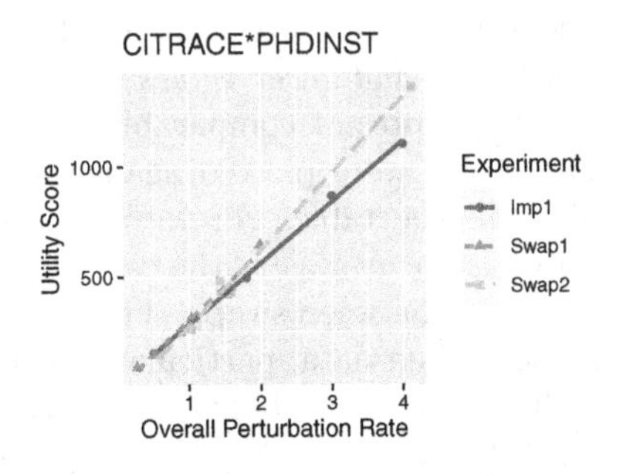

Fig. 2. Utility scores by experiment and perturbation level for CITRACE*PHDINST

Greater differences between perturbed and unperturbed tables were observed when the tables contained two perturbed variables. For EDMOTHER* CITRACE, the experiments with the poorest utility are swapping experiments in which Race/Ethnicity was *not* swapped, as seen in Fig. 3. When Race/Ethnicity was swapped, the swapping algorithm considered both variables when locating swapping partners. When it was not swapped, it was not used to define the swapping cell so the relationship with EDMOTHER was not preserved. Race/Ethnicity could be added as a stratum variable; however, there is a limit on how many variables can be used to define swapping strata. On the other hand, imputation experiments that did not impute Race/Ethnicity could use it as a model predictor and thus the utility is better when Race/Ethnicity is not perturbed, as one would expect. Similar results were observed for EDFATHER*CITRACE.

For GDEBTLVL*CITRACE (Fig. 4), another table in which both variables were perturbed, the effect of swapping Race/Ethnicity is more pronounced. When Race/Ethnicity was perturbed, the swapping experiment with the larger swapping strata actually appears to perform better than imputation; however, when Race/Ethnicity was not perturbed, the swapping experiment with large swapping strata appears much worse, while the imputation experiment results are as expected, namely that utility is better when Race/Ethnicity is not perturbed. A less expected result is that data utility for the table MARITAL*SRFOS (Fig. 5) depends on whether Race/Ethnicity was perturbed even though it does not involve Race/Ethnicity. This is due to the higher variable-level perturbation rate for MARITAL when Race/Ethnicity was excluded from perturbation. The difference is more pronounced for swapping due to the involvement of all swapping variables in the selection of swapping partners.

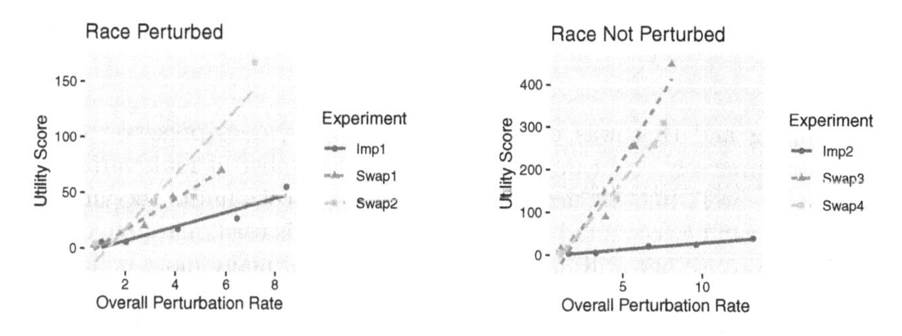

Fig. 3. Utility scores by experiment and perturbation level for EDMOTHER* CITRACE

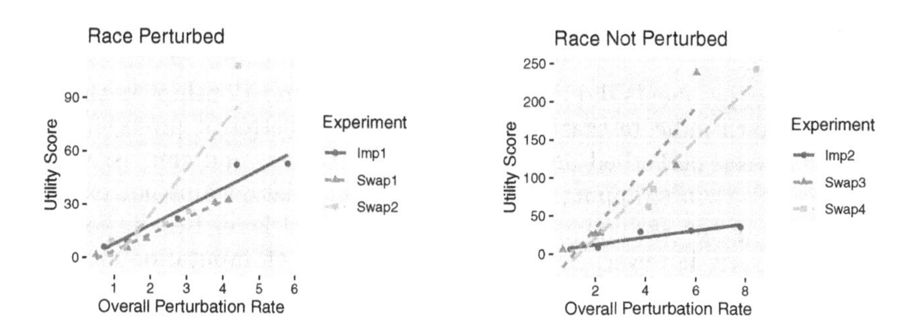

Fig. 4. Utility scores by experiment and perturbation level for GDEBTLVL*CITRACE

3.5 Experiment Summary

While data swapping and imputation yielded roughly comparable utility for the tables investigated when the perturbation level was low, imputation provided better utility at higher perturbation levels and for a wider range of analyses. At high perturbation levels, imputation virtually always provides better utility. Correlations between swapped and unswapped variables are a known problem

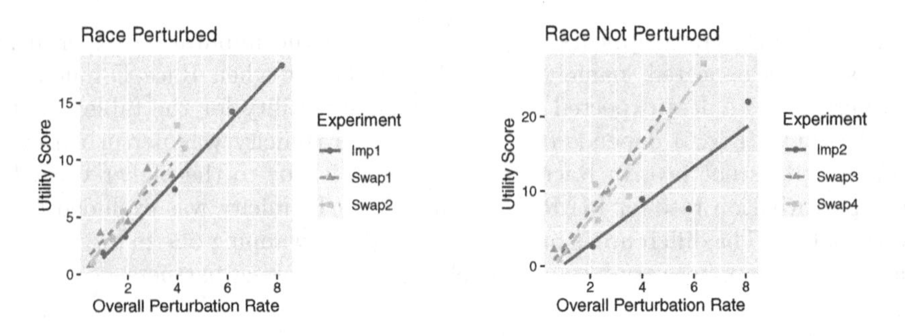

Fig. 5. Utility scores by experiment and perturbation level for MARITAL*SRFOS

with swapping and that was evident here. Swapping does preserve marginal distributions precisely and can preserve a limited number of multivariate relationships quite well, but the need for large swapping cells limits the number of variables that can be accounted for. When imputation is used, marginal distributions are not precise but still preserved quite well, and many more multivariate relationships can be preserved.

Within the swapping experiments, experiment 1 proved to be the best. The use of institution as a stratification variable proved to be unnecessary for preserving the distribution of race within institution, and the smaller swapping strata reduced quality overall. Including race as a swapping variable preserved some of the relationships between race and other swapping variables, whereas excluding race actually reduced data quality. The two imputation experiments were fairly comparable in utility aside from the impact on Race/Ethnicity when it was imputed.

While the experiments did not explore all possible relationships in the data, the results shown indicate that imputation performs better at preserving relationships between perturbed and unperturbed variables, and thus is likely to do a better job than swapping at preserving other relationships not explicitly accounted for in the swapping algorithm or evaluated here. Results for other datasets may vary; however, given that the properties of imputation and swapping with respect to preserving relationships between perturbed and unperturbed data are well known, it is expected that similar results will be seen in other datasets with many related variables.

4 Discussion

We have described a simplification of synthetic data that is efficient to implement and illustrated that it can be used as a practical alternative to data swapping. This implementation, termed disclosure imputation, can be used to provide a higher level of disclosure protection in the form of higher perturbation rates compared to data swapping at comparable utility levels, or better data quality at comparable disclosure protection levels. In particular, imputation can preserve a wide range of multivariate relationships between perturbed and unperturbed variables. For the purposes of comparison we have applied synthetic data in the manner of data swapping, but it is of course not limited by this. There are a range of possibilities between what we call 'disclosure imputation' and fully synthetic data. Disclosure imputation might be considered a starting point; data stewards have a great deal more flexibility than they would with data swapping to determine the appropriate portion of data to perturb.

The modeling burden of synthetic data and the dependence of inferences on imputation models is reduced when the proportion of values perturbed is reduced. Synthetic data tools make imputation simpler to specify than data swapping since one does not need to select which unperturbed variables should have their relationships with perturbed variables preserved. Another practical advantage of disclosure imputation is that it is similar in principle to imputation for missing data; efforts undertaken for missing data imputation, such as maintaining logical relationships and evaluating imputations can be re-used to facilitate disclosure imputation or more typical synthetic data products.

Multiple imputation would allow for analysts to properly account for the variance due to imputation; however, this is impractical for disclosure imputation since providing multiple implicates would make it simple to identify which records and variables have had their values altered. The strategy of only imputing a small portion of records relies on this information being protected. However, the variance due to imputation is quite small in our case due to the lower imputation rate as well as the use of improper imputation. This method has very little between-imputation variability for aggregate estimates, which data stewards can verify empirically as the generation of additional implicates is trivial.

An appeal of swapping is that it is simple to understand. In practice, however, the methods for selecting swapping pairs are rarely identified, and proper implementation of data swapping in a manner that preserves data quality can be complicated even with software like *DataSwap*. We are unaware of any similar software that is available to the public. By contrast, imputation can be performed using open-source software and code can be widely shared with minimal redaction (see appendix). Data users should also be familiar with imputation for nonresponse which should aid in user acceptance. Although the methodology offers improved transparency over data swapping, imputation still shares one of the major criticisms of data swapping which is the reliance of disclosure protection on secrecy about the portion of data that has been perturbed. As long as only a portion of records considered high risk for disclosure are imputed, it will be necessary to conceal which records those are. If enough variables are imputed it may be safe to provide users with information about which variables were imputed. We leave the effect of this on disclosure risk to future work.

Acknowledgments. This work was supported in part by the National Science Foundation's National Center for Science and Engineering Statistics. The authors are grateful for assistance and feedback from RTI colleagues David Wilson, Alan Karr, Neeraja Sathe, and Lanting Dai, and Wan-Ying Chen at the National Science Foundation.

Appendix

This code can be used to implement the method described in this paper. In lieu of confidential data, the dataset SD2011 included in R synthpop is used. This code has been tested using R 3.6.1 and synthpop 1.5.1

```
### Step 1: Impute  all records for selected variables
library(magrittr);
library(synthpop);
library(dplyr);
set.seed(138)

orig <- SD2011[1:10] %>% mutate_if(is.character, as.factor)
imp_vars <- c("sex", "age", "edu")
predmat <- matrix(1, 10, 10)
method1 <- rep("",10)
method1[colnames(orig) %in% imp_vars] <- "cart"

partsyn <- syn(data = orig, method = method1, m = 1, visit.sequence =
imp_vars, predictor.matrix = predmat)

### Step 2: Replace a random subset of observed records with imputed data
target_rate = .10 # desired overall perturbation rate
imp_rate = target_rate/mean(apply(orig != partsyn$syn, 1,
    function(x) ifelse(sum(x, na.rm=T) >=1, 1, 0)))
imp_rate # replacement rate
repinds <- rbinom(nrow(orig), 1, imp_rate)

partsyn2 <- orig
partsyn2[repinds==1, ] <- partsyn$syn[repinds==1,]

### Step 3: Check perturbation rates
overall_p_rate = mean(apply(orig != partsyn2, 1,
    function(x) ifelse(sum(x, na.rm=T) >=1, 1, 0)))
overall_p_rate %>% round(3)  #overall rate

v_pert = apply(orig[imp_vars] != partsyn2[imp_vars], 2, mean, na.rm=T)
v_pert %>% round(3)  #variable-level rates

### Step 4: Check utility
synth_eval = partsyn
synth_eval$syn = partsyn2
compare(synth_eval, orig, vars = imp_vars)
multi.compare(synth_eval, orig, var = "income", by = "edu")
```

References

1. Abowd, J.M., Schmutte, I.M.: Economic analysis and statistical disclosure limitation. Technical report, Brookings Papers on Economic Activity, pp. 271–293 (2015)
2. Benedetto, G., Stanley, J.C., Totty, E.: The creation and use of the SIPP synthetic beta v7.0. Technical report 18-03. U.S. Census Bureau Center for Economic Studies (2018)
3. Benedetto, G., Stinson, M.H., Abowd, J.M.: The creation and use of the SIPP synthetic beta. Technical report U.S, Census Bureau (2013)
4. Burgette, L., Reiter, J.P.: Multiple imputation for missing data via sequential regression trees. Am. J. Epidemiol. **172**, 1070–1076 (2010)
5. Burman, L.E., et al.: Safely expanding research access to administrative tax data: creating a synthetic public use file and a validation server. Technical report U.S, Internal Revenue Service (2019)
6. Cox, L.H., Karr, A.F., Kinney, S.K.: Risk-utility paradigms for statistical disclosure limitation: how to think, but not how to act (with discussion). Int. Stat. Rev. **79**(2), 160–199 (2011)
7. Dajani, A.F., et al.: The modernization of statistical disclosure limitation at the U.S. Census Bureau. Technical report U.S. Census Bureau (2017)
8. Drechsler, J.: Synthetic Datasets for Statistical Disclosure Control. Springer, New York (2011). https://doi.org/10.1007/978-1-4614-0326-5
9. Drechsler, J., Reiter, J.P.: Sampling with synthesis: a new approach for releasing public use census microdata. J. Am. Stat. Assoc. **105**(492), 1347–1357 (2010)
10. Fienberg, S.E., McIntyre, J.: Data swapping: variations on a theme by Dalenius and Reiss. J. Official Stat. **21**(2), 309–323 (2005)
11. Gomatam, S., Karr, A.F., Sanil, A.P.: Data swapping as a decision problem. J. Official Stat. **21**, 635–655 (2005)
12. Kaufman, S., Seastrom, M., Roey, S.: Do disclosure controls to protect confidentiality degrade the quality of the data? In: Proceedings of the Joint Statistical Meetings, Section of Government Statistics (2005)
13. Krenzke, T.R., et al.: Tactics for reducing the risk of disclosure using the NCES DataSwap software (2006)
14. Little, R.J.A.: Statistical analysis of masked data. J. Official Stat. **9**, 407–426 (1993)
15. Mitra, R., Reiter, J.P.: Adjusting survey weights when altering identifying design variables via synthetic data. In: Domingo-Ferrer, J., Franconi, L. (eds.) PSD 2006. LNCS, vol. 4302, pp. 177–188. Springer, Heidelberg (2006). https://doi.org/10.1007/11930242_16
16. Nowok, B., Raab, G.M., Dibben, C.: synthpop: Bespoke creation of synthetic data in R. J. Stat. Softw. **74**(11), 1–26 (2016)
17. Raghunathan, T., et al.: IVEware: Imputation and Variance Estimation Software (Version 0.3) (2016). https://www.src.isr.umich.edu/wp-content/uploads/iveware-manual-Version-0.3.pdf
18. Read, T.R.C., Cressie, N.A.C.: Goodness-of-Fit Statistics for Discrete Multivariate Data. Springer, New York (1988). https://doi.org/10.1007/978-1-4612-4578-0
19. Reiter, J.P.: Inference for partially synthetic, public use microdata sets. Surv. Methodol. **29**, 181–189 (2003)
20. Reiter, J.P.: Using CART to generate partially synthetic, public use microdata. J. Official Stat. **21**, 441–462 (2005)

21. Reiter, J.P., Kinney, S.K.: Inferentially valid, partially synthetic data: generating from posterior predictive distributions not necessary. J. Official Stat. **4**(28), 1–9 (2012)
22. Rubin, D.B.: Discussion: statistical disclosure limitation. J. Official Stat. **9**, 462–468 (1993)
23. Rubin, D.B.: Multiple Imputation for Nonresponse in Surveys, p. 258. Wiley (1987)
24. Sweeney, L.: K-anonymity: a model for protecting privacy. Int. J. Uncertain. Fuzziness Knowl.-Based Syst. **10**(5), 557–570 (2002)
25. Templ, M., Kowarik, A., Meindl, B.: Statistical disclosure control for micro-data using the R package sdcMicro. J. Stat. Softw. **67**(4), 1–36 (2015)

Data Quality

Evaluating Quality of Statistical Disclosure Control Methods – VIOLAS Framework

Olga Dzięgielewska(✉)

Military University of Technology, gen. Sylwestra Kaliskiego 2, 00-908 Warsaw, Poland
olga.dziegielewska@wat.edu.pl

Abstract. VIOLAS Framework (Volatile Index of Liable Accuracy for Statistical Disclosure Controls) defines an assessment methodology for evaluating quality of statistical disclosure control methods by measuring the key factors determining the effectiveness of such methods. This paper describes the base model and explains how the framework may be used and extended by additional measures that may be desirable in particular IT environments or for specific data sets.

Keywords: SDC quality framework · VIOLAS framework · Risk assessment

1 Introduction

When it comes to the modern IT systems security, emerging academic assumptions and ideas cannot be only evaluated under theoretical models. Currently, the commonly used approach to measure the overall security in the IT systems or its components are through the risk analyses or more offensive activities such as security testing. The systematic approach to measure the risk is through the application of risk assessment frameworks such as CVSS [4] or OWASP RRM [5]. Although it is possible to measure the risk related with particular statistical disclosure threats in a similar fashion, it must be noted that because the statistical disclosure threat is somehow beyond the scope of standardized approach of security testing, the results of such risk analyses may be overgeneralized rendering the results inconclusive and inaccurate.

Although designing a comprehensive risk analysis framework for statistical disclosure threats may be a task on its own, with the VIOLAS Framework we take a slightly different approach to measure the quality of the statistical disclosure controls methods and mechanisms. VIOLAS Framework (or Volatile Index of Liable Accuracy for Statistical Disclosure Controls) defines an assessment methodology for evaluating quality of statistical disclosure control methods by measuring the key factors determining the effectiveness of such methods. This paper describes the base model and explains how the framework may be used and extended by additional measures that may be desirable in particular IT environments or for specific data sets.

The motivation behind crating an assessment framework was a need to establish a standardized set of metrics to evaluate the overall security of statistical disclosure controls and objectively compare them between each other and against characteristics of different IT environments. The need to provide somewhat graphic comparison between

J. Domingo-Ferrer and K. Muralidhar (Eds.): PSD 2020, LNCS 12276, pp. 299–308, 2020.
https://doi.org/10.1007/978-3-030-57521-2_21

the methods originated from the business requirement to deliver a comprehensive way to argument the correct technological and scientific choices to the executive boards. The assumptions made for the definitions, the values of the evaluation characteristics within the framework and finally the design of the framework are an idea based on the observations and practice made during the IT systems security assessments.

2 VIOLAS Framework Definition

VIOLAS Framework (or **V**olatile **I**ndex **o**f **L**iable **A**ccuracy for **S**tatistical Disclosure Controls) defines an assessment methodology for evaluating effectiveness of statistical disclosure control methods.

The statistical liability index, denoted as *SLI*, is a value that denotes the quality of the method under assessment. The *SLI* value is a real number in a range [0, 1], where 1 means a perfect value, and 0 means the lowest score.

$$SLI \in [0,\ 1] \tag{1}$$

The framework measures the quality of a Statistical Disclosure Control (SDC) method (m_{SDC}) by determining the compliance of a method with the statistical security criteria. The baseline criteria are:

- ***Statistical confidentiality*** [*sc*] - the statistical confidentiality is determined by evaluating the risk of the primary and the secondary data identification.
- ***Statistical integrity*** [*si*] - the statistical integrity is achieved when the retrieved results from different queries asking the same data are identical.
- ***Statistical accuracy*** [*sa*] - the statistical accuracy is a measure of the quality of the retrieved statistics.
- ***Statistical transparency*** [*st*] - the privacy transparency which can be defined as protecting the selected critical data sets by elimination from the results.

It must be noted that the method of calculating the *SLI* allows for an easy expansion of a baseline model by additional criteria.

2.1 Statistical Confidentiality

The statistical confidentiality is determined by evaluating the risk of the primary and the secondary data identification. The primary data identification must be understood as a possibility to identify or derive a sensitive characteristic of an individual or predefined group of individuals from the raw data set that persists in the database. The secondary data identification must be understood as a possibility to identify or derive a sensitive characteristic of an individual or predefined group of individuals from the statistical data set, i.e. the results of the statistical data queries.

Let:

- m_{SDC} be a statistical disclosure control method
- C be a sensitive characteristic
- F_Q be a statistical query retrieval function
- Q be query result set
- q_n be result of a query function
- n be the size of result of a query function.

Then statistical confidentiality sc is satisfied when:

$$m_{SDC}(F_Q) \vdash sc \leftrightarrow q_n \notin C \, \forall q_i \in Q, i \in \{0, .., n\} \tag{2}$$

2.2 Statistical Integrity

The statistical integrity is achieved when the retrieved results from different queries asking the same data are identical or at least fit within the statistical margin.

Let:

- m_{SDC} be a statistical disclosure control method
- C be a statistical characteristic to be retrieved
- D be a dataset
- F_Q be a statistical query retrieval function
- Q be query result set
- N be the number of queries executed to retrieve the same characteristic C

Then statistical integrity si is satisfied when:

$$m_{SDC}(F_Q) \vdash si \leftrightarrow Q_A \simeq Q_B \tag{3}$$

$$\forall Q_i \in Q, i \in \{0, .., N\}, F_{Q_A}(D) = C_A, F_{Q_B}(D) = C_B, C_A = C_B \tag{4}$$

2.3 Statistical Accuracy

The statistical accuracy is a measure of the quality of the retrieved statistics. It can be determined with the appropriate confidence interval that the results must comply with. The confidence interval must be predetermined at the particular database system's level.

Let:

- m_{SDC} be a statistical disclosure control method
- S be a statistical database system
- CI_S be the predetermined confidence level of a system S
- F_Q be a statistical query retrieval function
- Q be query result set

- q_n be result of a query function
- n be the size of result of a query function

Then statistical accuracy *sa* is satisfied when:

$$m_{SDC}(F_Q) \vdash sa \leftrightarrow q_n \in CI_S \, \forall q_i \in Q, i \in \{0, .., n\} \tag{5}$$

2.4 Statistical Transparency

The statistical transparency can be defined as protecting the selected critical data sets by elimination from the results. The elimination of the selected data sets must not reveal any metadata allowing to identify the effect of the retrieved statistics before and after the elimination from the data sets. This characteristic is strictly related with the differential privacy model.

Let:

- m_{SDC} be a statistical disclosure control method
- D_1 be a dataset that differs with D_2 only by one element
- D_2 be a dataset that differs with D_1 only by one element
- S all subsets of m_{SDC} image
- F_Q be a statistical query retrieval function

Then statistical transparency *st* is satisfied when:

$$m_{SDC}(F_Q) \vdash st \leftrightarrow \Pr[m_{SDC}(F_Q(D_1)) \in S] \leq \varepsilon \cdot \Pr[m_{SDC}(F_Q(D_2)) \in S \tag{6}$$

2.5 System Criticality Matrix

Apart from the statistical security criteria, the system criticality is used to calculate the *SLI* value. The system criticality is understood as the level of data sensitivity stored in the database with conjunction of the data access model characteristic and can be assigned the values ranging from *None* to *Critical.*

The Table 1 depicts how to determine the overall criticality based on the system characteristic. The data are divided into four categories:

- ***Public [PU]*** - public data, not protected by any kind of regulations, the integrity of the data may be easily verified.
- ***Private and fully anonymized [PR FA]*** - sensitive individually identifiable data, that had been fully anonymized, the integrity of the data can be verified, but requires additional effort.
- ***Private and partially anonymized [PR PA]*** - sensitive individually identifiable data, that had been partially anonymized, the integrity of the data can be verified, but requires additional effort.

- ***Private and not obfuscated [PR NO]*** - sensitive individually identifiable data, that had not undergone any kind of obfuscation, the integrity of the data can be verified, but requires additional effort.

It must be noted that the effectiveness of the anonymization is not measured here, in particular, it is the system's owner responsibility to ensure that the anonymization process was successful and was executed as planned. In case of any errors during the anonymization process, the assessment of the system criticality may be affected.

Table 1. System criticality matrix

		DATA			
		PU	PR		
			FA	PA	NO
ACCESS	N				
	R				
	RW				

Legend:

None
Low
Medium
High
Critical

As for the data access, we divide the access into three categories: ***no data access [N]***, ***read access [R]*** and ***read and write access [RW]***.

3 SLI Derivation

Each characteristic (sc, si, sa, st) of a given m_{SDC} can be assigned a predetermined value from the set $V = \{0, 2, 5, 10\}$. The values are referenced as $v_{sc}, v_{si}, v_{sa}, v_{st}$, where

Table 2. System criticality value assessment

Value	*Criticality level*
1	None/Neutral
2	Low
4	Medium
8	High
10	Critical

the lower index of v denotes the characteristic. The values are given according to the Table 3. The value of the system criticality (cr) is assigned once per system, as this metric is independent from the m_{SDC} under assessment. The values are assigned according to the Table 2.

Table 3. VIOLAS Framework: criteria values definition

Value	*Meaning*
0	The method does not satisfy the characteristic for any case of statistical data set
2	The method satisfies the characteristic for a special case, but fails for majority of cases of statistical data sets
5	The method satisfies the characteristic for major cases, but fails for a special case of statistical data sets
10	The method satisfies the characteristic for all the cases of statistical data sets

For the overall result, each of the characteristic by default is calculated with the same weight ($w_{sc}, w_{si}, w_{sa}, w_{st}$). However, thanks to the modularity of the formula and the framework itself, the framework allows to extend the measuring method to assign custom weights for particular characteristics, e.g. in cases where some of the characteristics are more desirable and other ones are not critical. The total given weights must sum up to 1, therefore, by default the weight are given the value of 0,25:

$$w_{sc} = w_{si} = w_{sa} = w_{st} = 0{,}25 \tag{7}$$

$$w_{sc} + w_{si} + w_{sa} + w_{st} = 1 \tag{8}$$

The final index (SLI) of the m_{SDC} is measured according to the mechanism:

$$\begin{aligned} &if w_{sc} \cdot v_{sc} + w_{si} \cdot v_{si} + w_{sa} \cdot v_{sa} + w_{st} \cdot v_{st} == 10 \\ &\quad SLI = 1 \end{aligned} \tag{9}$$

$$\begin{aligned} &else \\ &SLI = \frac{(w_{sc} \cdot v_{sc} + w_{si} \cdot v_{si} + w_{sa} \cdot v_{sa} + w_{st} \cdot v_{st})}{10 \cdot cr} \end{aligned} \tag{10}$$

The $SLI = 1$ is considered the perfect value and means that the m_{SDC} satisfies all the necessary characteristics for statistical security.

Due to the building-blocks architecture of the framework, it can be easily extended by additional characteristics. Each additional characteristic ($s_n, n \in \mathbb{N}$) must:

1. Comply with the scoring mechanism, i.e. it can be assigned values from the set $V = \{0{,}2, 5, 10\}$
2. Appropriately extend the Eqs. 7–10, e.g. for a single additional characteristic s_1, the equations would be as follows:

a. $w_{sc} = w_{si} = w_{sa} = w_{st} = w_{s_1} = 0{,}2$
b. $w_{sc} + w_{si} + w_{sa} + w_{st} + w_{s_1} = 1$
c. $if w_{sc} \cdot v_{sc} + w_{si} \cdot v_{si} + w_{sa} \cdot v_{sa} + w_{st} \cdot v_{st} + w_{s_1} \cdot v_{s_1} == 10$
d. $SLI = \frac{(w_{sc} \cdot v_{sc} + w_{si} \cdot v_{si} + w_{sa} \cdot v_{sa} + w_{st} \cdot v_{st} + w_{s_1} \cdot v_{s_1})}{10 \cdot cr}$

4 Application

The VIOLAS Framework can be applied to measure any specific statistical disclosure method. To assess the value of each characteristic (sc, si, sa, st) for a particular SDC method m_{SDC} it must be proven that the method falls into appropriate category. Proposed proofs are shown in the Table 4.

It must be stressed that the assignment of the criteria values as well as the final *SLI* derivation is considered valid at the moment of the assessment. Given that the security conditions are constantly changing along with the threat landscape expansion and changes made within the IT environment itself the assessments must be performed at a regular basis to verify the accuracy of the initially assigned index.

Table 4. VIOLAS Framework requirements for proofs

Value	*Proof*
0	Example of full disclosure of data under given characteristic
2	Example of working condition and proof of disclosure of data under given characteristic for other cases
5	Example of disclosure of data under given characteristic and proof of compliance for other cases
10	Proof of compliance for all data sets for a given characteristic

The following sections describe fundamental classes of SDC methods and provides examples on how to apply the framework. However, it must be noted that for non-specific methods, only an estimate can be given, as the framework itself is designed to measure the quality of actual implementations in certain IT environments.

4.1 Fundamental Classes of Statistical Disclosure Protection Methods

As with any IT security related area, there is no universal method for protecting the statistical processing. However, there are statistical disclosure controls which when implemented properly can significantly decrease the risk of releasing confidential data from the database. The inference attacks threat can be mitigated by adopting series of mechanism, starting from the fundamental methods described in the early works.

The standard protection methods, can be divided into two groups (Domingo-Ferrer, A Survey of Inference Control Methods for Privacy-Preserving Data Mining, 2008), perturbative and non-perturbative. The first group assumes that the query result sets are slightly altered before the release in such a way that the computed statistics do not significantly differ from the original result yet protect the privacy of potentially sensitive statistics. In the non-perturbative protection methods, the query result sets are not altered. The protection is achieved by properly restricting statistics. Each of the control mechanisms, when analyzed or implemented separately, might not be efficient and appropriate for each type of the database and prevent existing attack vectors. Therefore, in the real-life systems the protection methods are combined. However, for the purpose of the evaluation, each of the methods will be evaluated as a single entity.

Perturbative Protection Methods

The first perturbative protection method is rounding [*P1*]. This method replaces either original values or original results with rounded values [2]. The rounding function is the crucial decision factor, as the level of security of this control method depends on it. If the method returns predictive values, an attacker might deduct the sensitive statistic from sufficiently large sample [3].

Data swapping and rank swapping [*P2*] is a transformation which exchanges values between individual records in such a way that the individual sensitive statistic cannot be retrieved but the statistical accuracy of the returned result is maintained. A variant of the data swapping is rank swapping which sorts the values of the characteristic C in ascending order and later the values are interchanged within the scope of its rank. Ranks are defined as percentage ranges of all the C values [3].

Non-perturbative Protection Methods

One of the most basic non-perturbative controls is query-set-size control [*N1*] which restricts such results that have less than n or more than $N - n$ records for some positive integer n [1]. This way only the easiest attacks are prevented, e.g. *S1*, because this control can be easily bypassed [3].

Another non-perturbative control is maximum-order control [*N2*], which does not allow queries which includes too many parameters [1]. The number of the parameters allowed in a single query should be defined separately for every database in a process which finds the minimum number of attributes that allows to identify a sensitive statistic of a particular entity. The drawback of this method is that it restricts many non-sensitive statistics and with each change in a database the process of finding the maximum of allowed parameters must be repeated [3].

The suppression mechanism [*N3*] dismisses such values from the query set result which can be categorized as sensitive and those which are non-sensitive, but might allow to derive a sensitive statistic from the retrieved data [1]. The suppression criterion used in this method for the $count(C)$ query is a minimum query-set size and for the $sum(C)$ query is the $n - respondent\ k\% - dominance$ rule, i.e. a sensitive statistic is calculated with n or less values which make up more than $k\%$ of the total values [3].

Another control method is sampling [*N4*] in which different sample records are used to compute the queried statistic, i.e. the sample records of the original query result set are used in the result set released from the database [2].

Last analyzed protection method is generalization [*N5*] in which several data categories are combined into one more general to increase the number of entities used for calculating the statistics [2]. The special case of this control is top-bottom coding used for the categories which can be ranked. In this case, minor categories are grouped in major categories by the rank values, i.e. top values and bottom values are combined in separate groups [2].

4.2 Protection Mechanisms Assessment

Described protection mechanisms can be applied to mitigate but not eliminate the risk of succeeding in performing inference attacks. Intuitively, because of the non-perturbative methods which do not distort the data are preferred by statistical databases user, the starting point of the search for the new universal method which would mitigate the risk of the high-risk issues should be within the non-perturbative method family. Perturbative methods however address considerably bigger range of threats, when applied separately, but may miss the accuracy when the trade-off between the security and the accuracy is not well defined.

VIOLAS Framework however describes the necessary characteristic of the perfect SDC function; therefore, the protection mechanisms can be somewhat objectively assessed not only based on the preference of the particular database user. The Table 5 shows the results of the assessment conducted for fundamental classes of SDC methods. As the methods present high-level solutions appropriate estimations must had been done to address most likely scenarios for each method, therefore the final outcome shows the *SLI* value as a range, being a result of:

Table 5. VIOLAS Framework calculation for fundamental classes of SDC methods

	v_{sc}	v_{si}	v_{sa}	v_{st}	cr *(sample)*	*SLI*
P1	{5, 10}	{5, 10}	{5, 10}	{5, 10}	4	[0,125–1]
P2	{5, 10}	{5, 10}	{5, 10}	{5, 10}	4	[0,125–1]
N1	2	{5, 10}	{5, 10}	{2, 5}	4	[0,0875–0, 169]
N2	2	5	{5, 10}	{2, 5}	4	[0,0875–0, 138]
N3	2	5	{5, 10}	{2, 5}	4	[0,0875–0, 138]
N4	{5, 10}	{5, 10}	{5, 10}	{2, 5}	4	[0,106–0, 219]
N5	2	{2, 5}	{5, 10}	{2, 5}	4	[0,069–0, 138]

- v_{sc}, v_{si}, v_{sa}, v_{st} for the particular method has been shown in a range possible to achieve under such SDC,
- as the cr value is independent of the SDCs, to simplify the range of the SLI, a medium value was selected as a sample value.

The estimations show, that contradictory to the intuition, the perturbative methods state a valid contender to the non-perturbative ones provided that the accuracy of the applied method complies with the statistical requirements of the system. As it was mentioned before, there is no singular solution to achieve perfect statistical disclosure control method under every condition, but the closest functions to achieve the perfect score within the VIOLAS Framework metrics are perturbative methods.

5 VIOLAS Framework Limitations

The VIOLAS Framework should be solely treated as a quantitative method of describing a security level of an SDC method at the time of performed analysis. As the environmental conditions may change overtime, the scoring of a particular method may change as well, therefore it is advised to reevaluate the *SLI* at minimum at every:

- change of the database system,
- change of the SDC method.

Additionally, depending on the business purpose of the database and the sensitivity of the data that it holds additional reevaluations may be done periodically and their frequency should be decided at the procedural level.

Apart from that, as in every case of an evaluation framework, a risk related with human factor must be considered. The designed framework strongly depends on the input provided by its users. As much as most of the input that the framework takes seem to be user-independent, the quality of analysis made to provide proofs produced to accurately evaluate the criteria play a key part in the assessment and must be prepared with proper diligence. Without that, the received results may remain erroneous.

References

1. Denning, D.: Cryptography and Data Security. Addison-Wesley Publishing Company Inc, Boston (1982)
2. Domingo-Ferrer, J.: A survey of inference control methods for privacy-preserving data mining. In: Aggarwal, C.C., Yu, P.S. (eds.) Advances in Database Systems, vol. 34, pp. 53–80. (2008). https://doi.org/10.1007/978-0-387-70992-5_3
3. Dzięgielewska, O., Szafrański, B.: A brief overview of basic inference attacks and protection controls for statistical databases. Comput. Sci. Math. Model. **4**, 19–24 (2016)
4. Common Vulnerability Scoring System SIG. https://www.first.org/cvss/
5. OWASP Risk Assessment Framework. https://owasp.org/www-project-risk-assessment-framework/

Detecting Bad Answers in Survey Data Through Unsupervised Machine Learning

Najeeb Moharram Jebreel, Rami Haffar, Ashneet Khandpur Singh, David Sánchez(✉), Josep Domingo-Ferrer, and Alberto Blanco-Justicia

Department of Computer Engineering and Mathematics, CYBERCAT-Center for Cybersecurity Research of Catalonia, UNESCO Chair in Data Privacy, Universitat Rovira i Virgili, Av. Països Catalans 26, 43007 Tarragona, Catalonia
{najeebmoharramsalim.jebreel,rami.haffar,ashneet.singh, david.sanchez,josep.domingo,alberto.blanco}@urv.cat

Abstract. Surveys are one of the most common ways of collecting data on individuals. Such data are of great value for economic and social research. However, the quality of the decisions and research results based on survey data depends on the ability to detect and filter out bad answers. The most common source of bad data are the respondents, who might provide imprecise or fabricated answers due to several reasons. In this paper we present a method to sanitize survey data that relies on combining the classification outcomes of three unsupervised machine learning algorithms (DBSCAN, PCA and IForest) aimed at detecting bad answers. Empirical results on real data show that our approach is able to improve the detection of both completely and partially bad answers with respect to the results provided by each algorithm independently.

Keywords: Survey data · Data quality · Bad answers · Fabrication · Unsupervised machine learning

1 Introduction

Survey data are central to many decision-making processes taking place at public and private institutions. They are also of paramount importance for research studies in fields such as economics, politics, social sciences or medicine. The quality of the decisions and research results based on survey data is, however, heavily dependent on the quality of the data themselves [1].

Experience shows that not all survey data are as good as desired. Errors during the survey design, such as choosing non-representative population samples or introducing unclear questions may lead to bad results [2]. Nonetheless, the most common source of bad data are the respondents, who might provide imprecise responses or fabricated answers when they are faced with sensitive questions, when they feel too much private information is requested or when they are not interested in the survey [3].

J. Domingo-Ferrer and K. Muralidhar (Eds.): PSD 2020, LNCS 12276, pp. 309–320, 2020.
https://doi.org/10.1007/978-3-030-57521-2_22

A literature survey in [4] shows that the percentage of fabricated survey data is typically below 5% for large-scale surveys with strict supervision, reaching up to 50% in small surveys with limited supervision (*e.g.* with inaccessible respondents). Regarding the effects of data fabrication, even small levels of fabrication may cause big effects. For example, in the German Socio-Economic Panel (SOEP) survey, removing 2.5% of the fabricated answers changed the effect of the years of education on gross wages by 80% [5]. The authors in [6] conclude that if roughly 1% of the answers are fabricated via, *e.g.* duplication, the probability of obtaining unbiased statistical estimates is of 41.6%; and if 10% or more of the answers are fabricated, the probability falls to 11.4%.

Given the importance of good data, effective methods are needed to sanitize survey data before conducting any meaningful empirical research or making decisions. Several methods have been proposed to detect bad answers in surveys, which involve re-interviewing respondents, recording and analyzing interviews with respondents, or analyzing statistical features of the responses [7]. However, the application of such verification checks incurs high costs.

Machine learning (ML) classification algorithms are a promising alternative. In particular, unsupervised ML algorithms, which extract patterns from unlabeled data with little human intervention, are widely used in anomaly detection; this includes detecting credit card fraud, cyberattacks, changes in medical conditions and unusual images [8]. In this work, we apply unsupervised ML algorithms to detect bad answers in survey data.

Although ML algorithms have been used to solve many regression and classification problems, no individual algorithm performs best under every circumstance. A common way to overcome this problem is to combine different algorithms [9]. Following this line, we propose aggregating a chosen set of ML algorithms to detect bad answers.

Contribution and Organization of This Paper

We introduce a new approach for detecting bad responses in survey data. We distinguish two kinds of bad responses: completely bad responses, in which the answers to all survey items have been fabricated in some way (*e.g.* by providing systematic random answers); and partially bad responses, where only part of the answers are fabricated (*e.g.* to hide sensitive features). Our proposal aggregates the results from three different unsupervised ML algorithms, namely linear-based, density-based and ensemble-based methods. With this approach, we aim to improve the detection of both completely and partially bad answers.

The main contributions of our work are:

- We show that unsupervised machine learning algorithms used to detect anomalies and fraud in categorical data can be effectively used to detect bad answers in survey data.
- We use three straightforward unsupervised machine learning methods, which have low computational cost, are easy to implement and have good performance.

- We show that the accuracy of detecting bad answers can be improved by aggregating the results from those three algorithms.

The results show that our approach detects completely bad answers with an overall accuracy over 96%, and a true positive rate (TPR) over 95%. Also, we obtain an overall accuracy over 80% and a true positive rate (TPR) over 75% in detecting partially bad answers.

The remainder of the paper is organized as follows. Section 2 discusses related works aimed at detecting fabricated survey data. Section 3 presents our approach and details the three ML algorithms we use. Section 4 describes and reports the results of the empirical experiments. Finally, Sect. 5 depicts the conclusions and proposes several lines of future research.

2 Related Work

The authors of [10] used Census Bureau Studies data to show the importance of re-interviewing respondents as a means of fabrication detection. In their study, most faked responses were detected through re-interviews, although some were detected due to anomalies in the data. Additionally, most of the cheating involved consisted of complete fabrications rather than partial fabrications of responses. In [11], the authors used data anomalies from the National Survey on Drug Use and Health in the US to determine suspicious behaviors. Specifically, they used the interview duration as a potential sign of fabrication allowing fraud detection in survey data.

Benford's law has been widely used to detect bad answers in surveys [12,13]. Benford's law involves examining the distribution of the leading (or left-most) digits of all the numbers reported on a survey form. By comparing against the expected distribution of the leading digits, one can identify unusual data, which may be fraudulent or generated by an error-prone process: suspect interviews are those in which the distribution of leading digits does not follow the expected distribution. Furthermore, [14] introduces an approach for detecting falsification in public opinion data. The proposed measure is the maximum percentage match statistic, which is the maximum percentage of questions on which each respondent matches any other respondent in the survey data.

Machine learning has been also employed to detect bad answers. The authors of [15] use cluster analysis on a data set that contains labels identifying the interviewers who fabricated entire interviews. They combine Benford-related features with other attributes (like item non-response rate) for which they expect differences between honest interviewers and cheaters. In cluster analysis, the classes of the groups of data points are not known to the algorithms, which means that the actual classes are not used for the classification task. In [16] the authors employ supervised and unsupervised classification to automatically detect interviewer fabrication in three real data sets containing both completely and partially bad answers. The authors report a better performance for supervised algorithms.

Each of the above methods above has its advantages and disadvantages. Re-interviewing respondents is effective to detect cheaters, but it requires substantial time and personnel effort. Methods based on Benford's law are only suitable for numerical data, but survey data typically contain a mix of numerical, ordinal and categorical data. Supervised ML algorithms are undoubtedly effective if labeled data are available, but access to such data is a major challenge.

Unsupervised ML algorithms are a promising alternative, as they are easy to apply and do not require labeled data.

3 Methodology

In this work, we use three fast, effective unsupervised ML algorithms to detect and remove bad answers in surveys. Each of the considered algorithms is widely used to detect anomalies in multidimensional data points. Then we aggregate the decisions of the three algorithms to improve the accuracy of detecting bad answers.

We model survey data as microdata tables, where each row corresponds to a single respondent and each column to each of the questions in the survey. Before we apply any of the detection algorithms, the data are preprocessed. Numerical data are normalized and categorical data are one-hot encoded.

The methods used in this paper are chosen based on their performance in clustering and anomaly detection, and on their execution time. The first method is DBSCAN, a widely used density-based clustering algorithm. DBSCAN is described in Sect. 3.1. The second method is Principal Component Analysis (PCA), often used in exploratory data analysis and dimensionality reduction. PCA is described in Sect. 3.2. The third method is IForest, an ensemble-based model that works on the principle of isolating anomalies in datasets. The IForest model is described in Sect. 3.3.

The answer i from every respondent in the survey (rows in our microdata table) is tested with each of these methods. Whereas DBSCAN produces a discrete score according to whether the input data is considered anomalous or not, PCA and IForest produce continuous values quantifying the anomaly degree. For the latter, we use the parameter τ to convert continuous outcomes to binary scores, being 0 a normal answer and 1 a bad one. First, we compute the interquartile range $IQR = Q3 - Q1$, where $Q1$ and $Q3$ are the first and third quartile of the algorithms' outcomes. After that if the algorithm's outcome is greater than $Q3 + \tau \times IQR$ for a given answer, we consider the answer a bad one and assign it a score $= 1$. Otherwise, we consider the answer a good one and assign it a score $= 0$.

Finally, the overall score of answer i is 1 if at least one of the methods scores it 1, in which case the respondent is classified as a cheater, and 0 otherwise. Since each method has its own way of detecting anomalies, it may happen that only one method detects a true bad answer, while the other two misclassify it as a good answer. Even in such cases, our score aggregation strategy enables us to detect as many bad answers as possible.

This process is formalized in Algorithm 1. Note that DBSCAN employs two hyperparameters (ϵ and $MinPts$), which we discuss in Sect. 3.1.

Algorithm 1. Bad survey data detection algorithm.

```
1     Input: Survey_data, τ, ε, MinPts;
2     Output: Survey_data (without cheaters);
3  numerical_data ← one_hot_encoding(Survey_data);
4  normalized_data ← normalize(numerical_data);
5  for i in normalized_data do
6      if DBSCAN(i, ε, MinPts) == −1 then
7          score_dbscan(i) ← 1;
8      else
9          score_dbscan(i) ← 0 ;
10     if Iforest(i) > Q3 + τ × IQR then
11         score_Iforest(i) ← 1;
12     else
13         score_Iforest(i) ← 0 ;
14     if pca(i) > Q3 + τ × IQR then
15         score_pca(i) ← 1;
16     else
17         score_pca(i) ← 0 ;
18     if score_dbscan(i) + score_Iforest(i) + score_pca(i) ≥ 1 then
19         Survey_data ← Survey_data.drop(i);
```

3.1 Density-Based Spatial Clustering of Applications with Noise (DBSCAN)

DBSCAN [17] is a clustering algorithm that groups data points according to their density with respect to a distance parameter ϵ. DBSCAN identifies outlying points as those lying in low-density regions, *i.e.* points that are unreachable from other points when taking ϵ-distance jumps. It has a worst-case time complexity $\mathcal{O}(n^2)$.

Next, we summarize the working principles of the algorithm:

1. *Neighboring points* are those at a distance smaller than ϵ.
2. *Core points* are those points that have at least $MinPts$ neighbors.
3. A point is *directly reachable* if it is within distance ϵ from a core point.
4. A point is p *reachable* from q if there is a path of core points between p and q.
5. A core point p forms a cluster with all points reachable from it.
6. Points that are not reachable, and thus do not belong to any cluster, are *outliers*.

3.2 Principal Component Analysis (PCA)

Principal Component Analysis was first proposed in [18] and it is widely used for dimensionality reduction, data visualization and exploratory data analysis.

PCA is a linear transformation that takes the data points into an orthonormal coordinate system that has the principal components as its basis. The data in the new coordinate system have their components uncorrelated (*i.e.* the covariance or correlation matrix is diagonal). The record components are ordered with respect to their variance, with the first component having the highest variance and the sum of the variances contributed by each of the components being equal to that of the original data. The principal components can be obtained from an eigenvector decomposition of the covariance or correlation matrices of the data [19].

We use an outlier detection mechanism based on PCA, which was proposed in [20] to identify attacks or outliers in data sets. First, a robust estimator of the correlation matrix is obtained via multivariate trimming and the principal components are obtained. Two detection functions are used: the first one depends on *major* components of the transformed data points (those which explain about 50% of the variance) and is able to detect whether the values for single components of the original data points are outliers. The second one, based on *minor* components (components that contribute each less than 0.2 of the variance) is able to detect anomalies in the correlation structure of the original data points. If either of these functions are above a certain threshold, the original data point is considered an outlier.

3.3 Isolation Forest (IForest)

Isolation forests are unsupervised anomaly detection algorithms. Anomaly detection algorithms typically model the normal behavior of data points and look for points that do not follow this normal behavior. In contrast, isolation forests explicitly isolate outliers [21].

An isolation forest is a collection of isolation trees, which are data-induced random binary trees. Isolation trees contain external nodes, which are nodes with no children, and internal nodes, which are nodes with a test ($q < p$) and exactly two children. To build such trees, the data set is recursively split by choosing a random attribute q and a random split value p in the range of the chosen attribute until i) the tree reaches a predefined maximum height, ii) the data set contains a single element, or iii) all remaining data points are equal.

The working principle of isolation forests is that, given such random trees, outliers will be isolated (be assigned to an external node) with less splits on average than normal data points, that is, their height (distance from the root) will be smaller on average.

To identify anomalies, data points are searched in the collection of isolation trees and their average height is computed. This average height is normalized to obtain an anomaly score s as follows:

$$s(x,n) = 2^{-\frac{E(h(x))}{c(n)}},$$

where $E(h(x))$ is the average height of x in the collection of isolation trees and $c(n)$ is the average path length of unsuccessful searches in binary search trees, given in [22]. Data points with anomaly scores close to 1 are considered anomalous.

4 Experiments

4.1 Dataset

To test the performance of the three considered algorithms and their aggregated results, we used the Adult income data set [23]. This data set contains 48,842 records, each of them reporting 6 continuous and 9 categorical demographic attributes on a certain individual. We viewed Adult as if it resulted from a survey, that is, as if each record represented the answer from an individual respondent.

After cleaning up the survey data from missing values we still had 45,222 answers. Then, we simulated two types of bad answers: *completely bad answers* and *partially bad answers*. For completely bad answers, we randomly chose 20% of the records and replaced values in them with random answers generated from the respective domains of the 15 attributes. For the partially bad answers, we randomly selected 20% of the records; after that, we randomly chose 6 out of the 15 attributes in the selected records and replaced their values with random values from the corresponding attribute domains. We then attached a label to each record. The label value was set to 1 if the answer was bad or 0 if the answer was good. Later, we used these labels to evaluate the performance of the methods.

For both completely and partially bad answers, we performed the same data preprocessing steps. First, we converted the categorical attributes into numerical using label encoding. After that, we applied min-max normalization for each attribute value, defined as

$$z_i = \frac{x_i - min(x)}{max(x) - min(x)},$$

where $x = (x_1, ..., x_n)$ and z_i is the i-th normalized value.

4.2 Performance Metrics

Since we were dealing with a classification problem (to classify an answer as good or bad), we used five well-known evaluation metrics that have been used to assess the performance of classification methods [24]:

- **Accuracy** (ACC) represents the percentage of correctly classified answers compared with the total number of answers. We used the accuracy to assess the overall performance of each method. Accuracy is given by:

$$ACC = \frac{TP + TN}{TP + TN + FP + FN} \times 100,$$

where TP, TN, FP and FN stand, respectively, for number of true positives, number of true negatives, number of false positives and number of false negatives.

- **True positive rate** (TPR), a.k.a. detection rate or sensitivity, represents the percentage of correctly classified bad answers compared with the total number of bad answers. TPR is given by:

$$TPR = \frac{TP}{TP + FN} \times 100.$$

- **True negative rate** (TNR), a.k.a. recall, represents the percentage of correctly classified good answers compared with the total number of good answers. TNR is given by:

$$TNR = \frac{TN}{TN + FP} \times 100.$$

- **False positive rate** (FPR), a.k.a. fall out, represents the percentage of good answers incorrectly classified as bad answers compared with the total number of good answers. FPR is given by:

$$FPR = \frac{FP}{FP + TN} \times 100.$$

- **False negative rate** (FNR) represents the percentage of bad answers incorrectly classified as good answers compared with the total number of bad answers. FNR is given by:

$$FNR = \frac{FN}{FN + TP} \times 100.$$

4.3 Hyper-parameters

We used different hyper-parameter values with each method.

For DBSCAN applied to completely bad answer detection, we used $\epsilon = 0.515$ and a minimum number of answers required to form a cluster $MinPts = 300$. For DBSCAN applied to partially bad answer detection, we used stricter values $\epsilon = 0.42$ and $MinPts = 500$.

For IForest, we used the default parameters as described in [25]. Then, we filtered the scores obtained from the decision function of IForest using $\tau = 0.2$ for completely bad answers and $\tau = 0.1$ for partially bad answers.

For PCA, we used the default parameters specified in PyOD [26] with a number of components equal to 2. PyOD implements [20] to use PCA as an outlier detection tool.

4.4 Results

Figure 1 shows the performance at detecting completely bad answers. We can see that both DBSCAN and IForest have a similar performance with an overall accuracy 96%, and with a slight advantage for DBSCAN regarding TPR and FPR: DBSCAN achieved TPR 91.3% and FPR 2.3%. PCA had a somewhat poorer performance.

Anyway, the aggregation of the three methods as detailed in Algorithm 1 results in a noticeable improvement. On the one hand, the overall accuracy is maintained at 96% like for DBSCAN and IForest. On the other hand, the TPR 95.3% outperformed the best score achieved by DBSCAN, which was 91.3%.

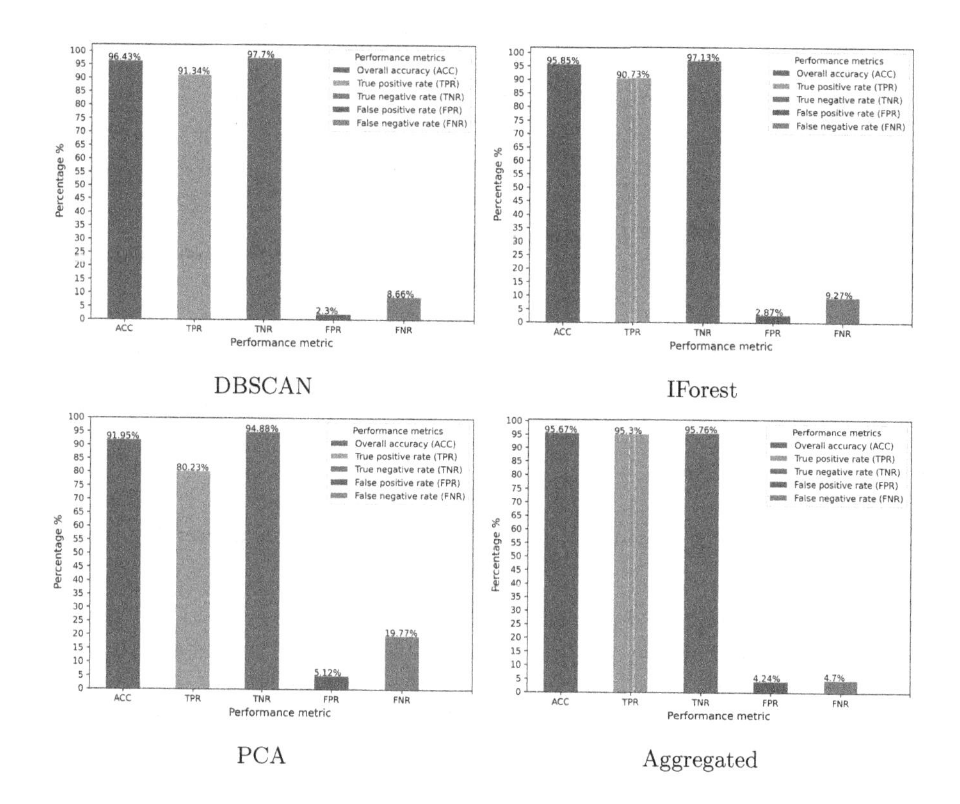

Fig. 1. Performance at detecting completely bad answers

Figure 2 shows the performance at detecting partially bad answers. We can notice the good performance of both DBSCAN and IForest: DBSCAN achieved an overall accuracy 85.1% while IForest achieved 84.3%. However, IForest achieved the best TPR performance at 64.8%. We also see a poor performance of PCA at detecting partially bad answers: TPR was just 34.4%; however, PCA achieved a very good TNR 96%.

Again, we see a noticeable improvement when aggregating the three methods: the aggregation achieved TPR 72.98%, while IForest was the second best TPR performer with only 64.78%.

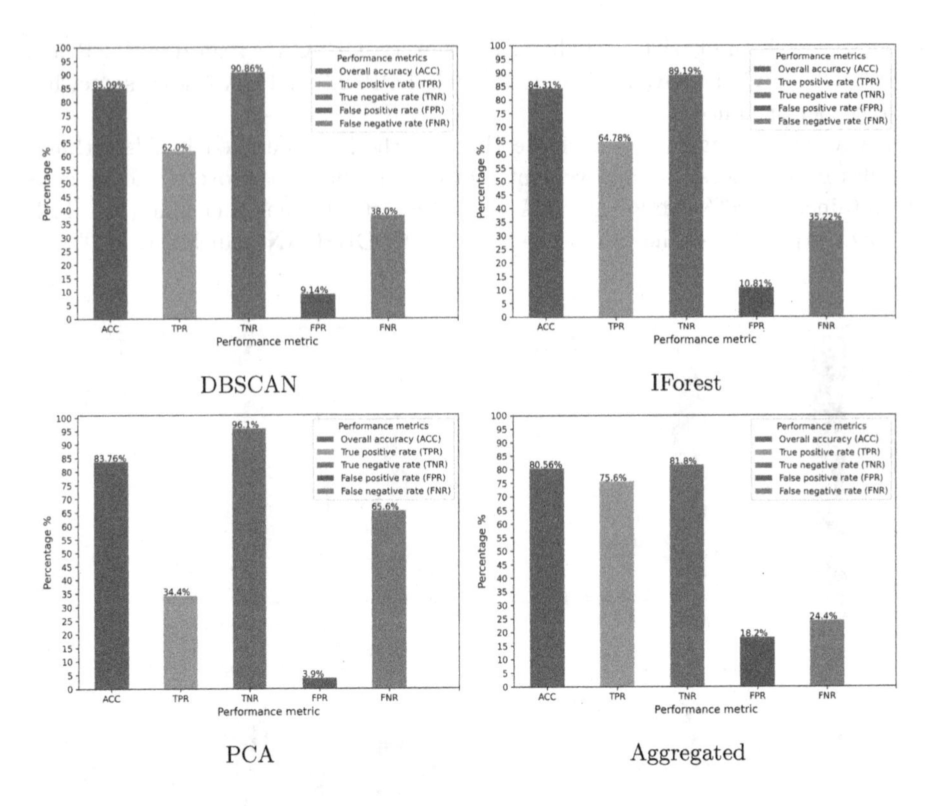

Fig. 2. Performance at detecting partially fabricated answers

From the results, it is clear that aggregating the results of the three methods improves the detection rate of bad answers. The aggregation helps to complement the different strategies implemented by the three methods at detecting bad answers and overcome some of their limitations. It is also worth noting that the performance of detecting partially bad answers is noticeably lower than detecting completely bad answers. This is not surprising because in this latter case only 40% of the values of each answer were randomly generated. Consequently, some partially random answers might still be close to good answers and, therefore, would be considered good.

5 Conclusions and Future Research

We have introduced a method to detect fabricated answers in surveys that aggregates the results from different unsupervised machine learning and anomaly detection algorithms. Our experiments show that the aggregated results improve the detection rate of bad answers with respect to the outcomes of each individual algorithm.

Despite the promising results, the choice of the hyperparameters remains a major challenge. The value of these parameters depends on several factors like the domain of the survey and the size of the answers. More experiments will be needed to analyze the influence of the parameters on the results and, particularly, on their aggregation.

As future work, we plan to investigate other aggregation strategies and unsupervised ML methods. We will also consider specific cases of data fabrication, such as answer duplication, for which specific methods have been proposed.

Acknowledgments and Disclaimer. We acknowledge support from: European Commission (project H2020-871042 "SoBigData++"), Government of Catalonia (ICREA Acadèmia Prize to Josep Domingo-Ferrer and grant 2017 SGR 705) and Spanish Government (projects RTI2018-095094-B-C21 and TIN2016-80250-R). The authors are with the UNESCO Chair in Data Privacy, but their views here are not necessarily shared by UNESCO.

References

1. Biemer, P.P., Lyberg, L.E.: Introduction to Survey Quality. Wiley, Hoboken (2003)
2. Winker, P.: Assuring the quality of survey data: incentives, detection and documentation of deviant behavior. Stat. J. IAOS **32**(3), 295–303 (2016)
3. Rogers, F., Richarme, M.: The honesty of online survey respondents: lessons learned and prescriptive remedies. Decision Analyst, Inc. White Papers (2009). https://www.decisionanalyst.com/whitepapers/onlinerespondents/
4. Bredl, S., Storfinger, N., Menold, N.: A literature review of methods to detect fabricated survey data, Discussion Paper, Technical report (2011). https://www.econstor.eu/handle/10419/74449
5. Schraepler, J.-P., Wagner, G.G.: Identification, characteristics and impact of faked interviews in surveys: an analysis by means of genuine fakes in the raw data of SOEP, IZA Discussion Paper no. 969 (2003). http://hdl.handle.net/10419/20205
6. Sarracino, F., Mikucka, M.: Estimation bias due to duplicated observations: a Monte Carlo simulation, Munich Personal RePEc Archive-MPRA (2016)
7. Schwemmer, C.: Identification of fabricated interviews in surveys - an analysis using the Socio-Economic Panel (SOEP), Working Paper, September 2015. https://www.researchgate.net/publication/318888495_Identification_of_fabricated_interviews_in_surveys_-_An_analysis_using_the_Socio-Economic_Panel_SOEP
8. Chandola, V., Banerjee, A., Kumar, V.: Anomaly detection: a survey. ACM Comput. Surv. **41**(3), 1–58 (2009)
9. Dietterich, T.G., et al.: Ensemble learning. Handb. Brain Theor. Neural Netw. **2**, 110–125 (2002)
10. Schreiner, I.D., Newbrough, J., Pennie, K.: Interviewer Falsification in Census Bureau Surveys. US Census Bureau, Suitland (1988). (custodian)
11. Murphy, J., et al.: A system for detecting interviewer falsification. In: American Association for Public Opinion Research 59th Annual Conference, pp. 4968–4975 (2004)
12. Bredl, S., Winker, P., Kötschau, K.: A statistical approach to detect interviewer falsification of survey data. Surv. Method. **38**(1), 1–10 (2012)
13. Schäfer, C., et al.: Automatic identification of faked and fraudulent interviews in surveys by two different methods, DIW Discussion Papers, Technical report (2004). http://hdl.handle.net/10419/18293

14. Kuriakose, N., Robbins, M.: Falsification in survey research: Detecting near-duplicate observations. In: Annual Meeting of the American Political Science Association, pp. 3–6. CA, San Francisco, September (2015)
15. Bredl, S., Winker, P., Kötschau, K.: A statistical approach to detect cheating interviewers, Discussion Paper, Technical report (2008). https://www.econstor.eu/handle/10419/39808
16. Birnbaum, B.: Algorithmic approaches to detecting interviewer fabrication in surveys, Ph.D. dissertation, University of Washington (2012). https://digital.lib.washington.edu/researchworks/handle/1773/22011
17. Ester, M., Kriegel, H.-P., Sander, J., Xu, X.: Density-based spatial clustering of applications with noise. In: International Conference Knowledge Discovery and Data Mining-KDD 1996, pp. 226–231 (1996)
18. Pearson, K.: LIII. On lines and planes of closest fit to systems of points in space. Lond. Edinb. Dublin Philos. Mag. J. Sci. **2**(11), 559–572 (1901)
19. Wold, S., Esbensen, K., Geladi, P.: Principal component analysis. Chemometr. Intell. Lab. Syst. **2**(1–3), 37–52 (1987)
20. Shyu, M.-L., et al.: A novel anomaly detection scheme based on principal component classifier, University of Miami Coral Gables, Department. of Electrical and Computer Engineering, Technical rep (2003). https://apps.dtic.mil/dtic/tr/fulltext/u2/a465712.pdf
21. Liu, F.T., Ting, K.M., Zhou, Z.-H.: Isolation forest. In: 2008 Eighth IEEE International Conference on Data Mining, pp. 413–422. IEEE (2008)
22. Preis, B.: Data Structures and Algorithms with Object-Oriented Design Patterns in Java. Wiley, Hoboken (1999)
23. Dua, D., Graff, C.: UCI machine learning repository. University of California, School of Information and Computer Science, Irvine, CA (2019)
24. Metz, C.E.: Basic principles of ROC analysis. In: Seminars in Nuclear Medicine, vol. VIII, no. 4, pp. 283–298 (1978)
25. Liu, F.T., Ting, K.M., Zhou, Z.-H.: Isolation-based anomaly detection. ACM Trans. Knowl. Discov. Data **6**(1), 1–39 (2012)
26. Zhao, Y., Nasrullah, Z., Li, Z.: Pyod: a python toolbox for scalable outlier detection. J. Mach. Learn. Res. **20**(96), 1–7 (2019)

Case Studies

Private Posterior Inference Consistent with Public Information: A Case Study in Small Area Estimation from Synthetic Census Data

Jeremy Seeman(✉), Aleksandra Slavkovic, and Matthew Reimherr

Department of Statistics, Pennsylvania State University, University Park, USA
jhs5496@psu.edu

Abstract. Methods for generating differentially-private (DP) synthetic data have received recent attention as large government agencies such as the U.S. Census have decided to release DP synthetic data for public usage. In the synthetic data generation process, it is common to post-process the privatized results so that the final synthetic data agrees with what the data curator considers public information. Our contributions are three fold: 1) we show empirically that using post-processing to incorporate public information in contingency tables can lead to sub-optimal inference, 2) we propose an alternative Bayesian sampling framework that directly incorporates both noise due to DP and public information constraints, leading to improved inference, and 3) we demonstrate the proposed methodology on a study of the relationship between mortality rate and race in small areas given priviatized data from the CDC and U.S. Census.

1 Introduction

Differential privacy [10] is a mathematical framework for producing statistical results with provable privacy guarantees. Since its inception, numerous methods have been proposed for different private inferential procedures, e.g., [4,6,15,32, 33], to name a few. These methods typically require a trusted data curator to have access to the non-private data, and the query responses can only be received once while consuming a fixed proportion of a predetermined privacy budget. Since data curators cannot anticipate the needs of all data users, nor could they feasibly allocate privacy budgets to all possible requested analyses, there has been a growing interest in methods for producing differentially-private synthetic data [3,14,23]. This allows data curators to release data sets with the same privacy guarantees as individual procedures, implying that any statistic derived from this synthetic data will automatically be private.

DP synthetic data is particularly useful for multipurpose data sets, where the number of possible queries of interest is large; for example, the U.S. Census plans to release its 2020 Census of Population and Housing as DP synthetic data [2]. However, synthetic data is not usually released as-is once the initial privacy

J. Domingo-Ferrer and K. Muralidhar (Eds.): PSD 2020, LNCS 12276, pp. 323–336, 2020.
https://doi.org/10.1007/978-3-030-57521-2_23

mechanism has been applied; instead, post-processing is performed so that the final privatized data agree with a set of constraints often capturing known public information. In the case of the U.S. Census, this could correspond to structural inequality relationships between cells (such as the ages of respondents who can live in certain housing types) or marginal invariant values (such as the populations of certain geographic areas being exact). As many practitioners have noted [19,20,25–27,31,34], the Census' proposed post-processing method [1] preserves reasonable utility aggregated across large geographic areas, but largely fails for supporting inference on small areas.

One of the core problems is the difficulty of characterizing the distribution of post-processed data given the non-privatized data. Post-processing makes it significantly harder to calculate or even numerically estimate $P(\boldsymbol{Z} \mid \boldsymbol{X})$, even when the privatization mechanism itself is relatively easy to characterize, as is the case when DP noise is an independent, additive perturbation; e.g., see [18]. This is analogous to classical problems in measurement error modeling when considering whether or not the measurement error distribution is known or unknown [7]. Unlike most measurement error models, however, the privatization noise distribution is known exactly, which suggests that better inferential methods are possible. On the other hand, we will demonstrate that while post-processing can improve utility from a loss optimization perspective, it can degrade utility from a statistical inference perspective.

Our contributions in this paper focus on re-framing the aforementioned problem. The underlying goal of post-processing is to enforce constraints that make the final synthetic data agree with public information. Instead of using post-processing to retroactively transform the data to meet these constraints, we propose incorporating the constraints directly into the data generating process via Bayesian analysis using the private synthetic data without post-processing. Note that both approaches are still DP. Our approach, which builds on ideas from [16], is flexible since it allows us to distinguish between constraints that depend on the underlying model of interest and constraints that depend on the data generating process. Furthermore, our approach integrates nicely into standard tools for analysis of generative models, such as simulation-based inference for empirical test statistics as we will see in a case study.

In Sect. 2.1, we apply a rejection algorithm of [16] to sample from the posterior distribution of a parameter given DP synthetic data. In Sect. 2.2 we compare and contrast two different kinds of public information, one that affects the prior parameter distribution (which we call "model-dependent"), and one that affects the DP error distribution (which we call "data-dependent"). We also highlight the ways in which post-processing may not support valid inference. In Sect. 2.3, we incorporate this public information into a new sampling algorithm, and demonstrate its effectiveness in performing inference about mortality rates in Sect. 3. We conclude a paper with a brief discussion in Sect. 4.

2 Posterior Sampling Methods

Throughout this section, we let $[n] \triangleq \{1, 2, \ldots, n\}$. Suppose we observe T different multinomial tabular queries $\{\boldsymbol{X}_t\}_{t=1}^T$, each of which has the form

$\boldsymbol{X}_t \mid \boldsymbol{\theta}_t \sim \text{Multinom}(n_t, \boldsymbol{\theta}_t)$. Finally, Suppose each table has its own privacy budget $\epsilon_t > 0$, and define $\eta_t \triangleq \exp(-\epsilon_t/2)$ for $t \in [T]$. Let $\boldsymbol{Y}_t = \boldsymbol{X}_t + \boldsymbol{\varepsilon}_t$, where $\varepsilon_{tj} \overset{\text{iid}}{\sim} \text{DL}(\eta_t)$ and $\text{DL}(\eta_t)$ is the discrete Laplace distribution with PMF [17]:

$$P(\varepsilon_{ij} = k) = \frac{1 - \eta_t}{1 + \eta_t} \eta_t^{|k|} \qquad k \in \mathbb{Z} \tag{1}$$

Then by the properties of the geometric mechanism and composition [9], $\{\boldsymbol{Y}_t\}_{t=1}^T$ is an ϵ-DP synthetic data set, where $\epsilon = \sum_{t=1}^T \epsilon_t$. For convenience, let $\boldsymbol{X}$ and $\boldsymbol{Y}$ be the concatenated 1-dimensional count vectors from all tables, i.e. $\boldsymbol{X}, \boldsymbol{Y} \in \mathbb{Z}^K$ for some total length K.

2.1 Posterior Sampling Methods for DP Synthetic Data

Differentially private analysis should ideally treat each $\boldsymbol{Y}_t$ as a noisy measurement of $\boldsymbol{X}_t$ and propagate the measurement error through any subsequent transformations. Our approach relies on techniques borrowed from approximate Bayesian inference [12,35], with the end goal of sampling from the distribution of a parametric statistic of interest conditional on $\{Y_t\}_{t=1}^T$. In this section, we demonstrate a rejection algorithm (Algorithm 1), proposed by Gong [16], for sampling from a distribution of a given statistic with a known mean zero additive perturbation as shown below. This approach, while computationally simple, allows for easy generation of point estimates for individual small area sub-populations and linear combinations of small area sub-populations (when there is not further public information). In particular, since the perturbations to the cell counts are additive and independent, sum queries on the contingency table also have additive errors following a generalized discrete Laplace distribution, $\text{GDL}(k, \eta)$, where k is the number of cells in the summation query [21].

Data: Observed DP data $\boldsymbol{Y}$, prior π, likelihood $f_{\boldsymbol{X}|\boldsymbol{\theta}}$, error density $g_{\boldsymbol{Y}|\boldsymbol{X}}$
Result: N samples from $\boldsymbol{\theta} \mid \boldsymbol{Y}$
while $i \leq N$ **do**
  Sample $\boldsymbol{\theta}^{(i)} \sim \pi$, $\boldsymbol{X}^{(i)} \mid \boldsymbol{\theta}^{(i)} \sim f(\cdot \mid \boldsymbol{\theta}^{(i)})$, $U \sim \text{Unif}(0,1)$;
  if

$$U \leq \frac{g(\boldsymbol{Y} \mid \boldsymbol{X}^{(i)})}{\sup_{\boldsymbol{Y}} g(\boldsymbol{Y} \mid X^{(i)})} \tag{2}$$

  then
    Accept sample $\boldsymbol{\theta}^{(i)}$, $i \mapsto i + 1$;
  else
    Reject sample $\boldsymbol{\theta}^{(i)}$;
  end
end

Algorithm 1: DP posterior sampling given additive perturbation [16]

Figure 1 gives simulated 95% credible intervals for a single-cell count and a 5-cell summation query using three different methods: the non-private true credible interval, the naive DP approach which treats the privatized data as observed

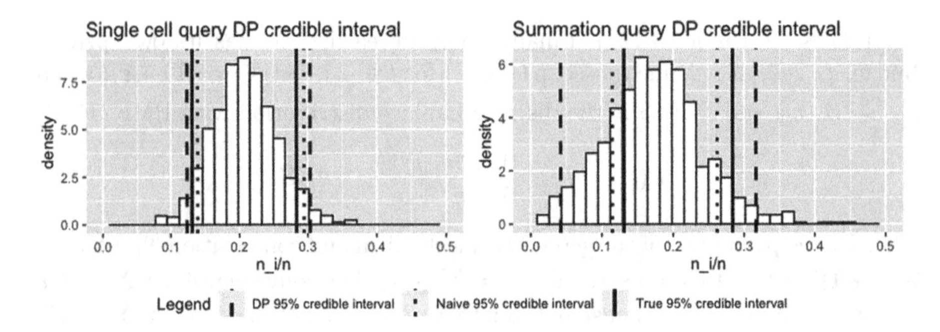

Fig. 1. Example empirical credible intervals for a single-cell count and a 5-cell summation query with true proportion $\theta_{\text{true}} = .25$, prior $\theta \sim \text{Beta}(1, 3)$, and likelihood $X \mid \theta_{\text{true}} \sim \text{Binom}(100, \theta_{\text{true}})$. With $\epsilon = 1$, the single-cell DP perturbation has distribution $\epsilon \sim \text{DL}(\exp(-1/2))$, and the summation query DP perturbation has distribution $\epsilon \sim \text{GDL}(5, \exp(-1/2))$. Empirical CIs derived from $N = 1000$ simulated samples.

without noise, and the empirical DP CI using Algorithm 1. It is important to note that directly substituting the synthetic data into a statistical analysis (i.e. the "naive approach" where we ignore the privacy mechanism) underestimates the uncertainty, leading to deficient coverage probabilities of the CI. By comparison, the posterior CI using Algorithm 1 produces a wider interval that properly captures the uncertainty introduced by the privacy mechanism.

2.2 Post-processing and Constraints in Data Generating Processes

In Sect. 2.1 we assumed that the perturbed cell counts, $\{\boldsymbol{Y}_t\}_{t=1}^T$, are directly available in the absence of additional public information. Next, we consider incorporating public information. For the purposes of this paper, we consider only public information that can be represented as linear constraints. Following the notation from [1], these take the form of indicator functions evaluated on specific random vectors: $C_j(\boldsymbol{Y}) \triangleq \mathbb{1}\{Q_j^T \boldsymbol{Y} \text{ op}_j\ c_j\}$ where $Q_j \in \mathbb{R}^{K \times d_j}$, op_j represents an element-wise comparison operator (i.e., one of $=, \leq, \geq$) and $c_j \in \mathbb{R}^{d_j}$. Under this definition, a constraint is satisfied if $Q_j^T \boldsymbol{Y} \text{ op}_j\ c_j$ is true, in which case $C_j(\boldsymbol{Y}) = 1$ (0 otherwise). This definition is commonly used and flexible enough to accommodate most public information about contingency tables, such as invariant summation queries, hierarchical geographic structures, structural inequalities, and more. Note that the constraints C_j as formulated can be evaluated against any arbitrary random variable of the same dimension; this will be used later.

Constraint enforcement is commonly achieved through post-processing, usually in the form of an optimization problem that depends only on $\boldsymbol{Y}$ and the constraints:

$$\boldsymbol{Z} = \arg\min_{\boldsymbol{Z}^*} ||\boldsymbol{Z}^* - \boldsymbol{Y}||_2^2 \quad \text{s.t.} \quad C_j(\boldsymbol{Z}^*) = 1 \quad \forall j \in [J]. \tag{3}$$

In particular, the Census TopDown algorithm's post-processing, applied to the first release of 2010 Census DP synthetic data [5], performs the above operations twice for each pair of sub-geographic levels: once to find a fractional histogram, and once to round the fractional histogram; see [1] for full details. This kind of optimization procedure has many attractive properties from a computer science perspective. First, the linear constraints can be directly analyzed to determine properties of the optimization solution, such as NP-completeness and computational complexity. Furthermore, $\boldsymbol{Z}$ can often be calculated using well known optimization techniques, such as quadratic or mixed-integer programming, and the constraints can often be simplified using techniques like network flow analysis [24]. However, the distribution of $\boldsymbol{Z} \mid \boldsymbol{X}$ can often be neither analytically nor computationally tractable, which makes it prohibitively difficult to directly perform inference given only the post-processed observations. Instead of relying on post-processing, we first look at two kinds of constraints in detail and see how they affect the sampling process in the posterior algorithm.

Example 1: Suppose we have the constraint $C(\boldsymbol{Z}^*) = \mathbb{1}\{\boldsymbol{a}'\boldsymbol{Z}^* \leq \boldsymbol{b}'\boldsymbol{Z}^*\}$ for two fixed vectors $\boldsymbol{a}, \boldsymbol{b} \in [0,1]^K$. Such a constraint arises when cell counts have to maintain structural inequalities; in the US Census, for example, this could encode the constraint that all eligible voters must be above the age of 18. This can be equivalently written as $P(\boldsymbol{a}'\boldsymbol{\theta} \leq \boldsymbol{b}'\boldsymbol{\theta}) = 1$, which contains the same information in an expression that depends only on $\boldsymbol{\theta}$.

Example 2: Suppose we have the constraint $C(\boldsymbol{Z}^*) = \mathbb{1}\{\boldsymbol{a}'\boldsymbol{Z}^* = n_a\}$ for some fixed vector $\boldsymbol{a} \in [0,1]^K$ and $n_a \in \mathbb{Z}^+$. Such a constraint arises when certain marginals need to be released exactly as public information; in the US Census, for example, this could encode the constraint that state populations must be released exactly. Let $\delta \triangleq \boldsymbol{a}'\boldsymbol{Y} - n_a$; then δ is a function of the observed DP data $\boldsymbol{Y}$, and provides information about the set of possible $\boldsymbol{\varepsilon}$ values that could have generated such a $\boldsymbol{Y}$. Therefore the constraint can be equivalently written as $P(\boldsymbol{a}'\boldsymbol{\varepsilon} = \delta) = 1$, which contains the same information in an expression that depends only on $\boldsymbol{\varepsilon}$.

Comparing and contrasting the two examples above suggests that the constraints need only provide information about certain parts of the data generating process, as the first example did for $\boldsymbol{\theta}$, and the second example did for $\boldsymbol{\epsilon}$. If our end goal is to generate posterior samples from $\boldsymbol{\theta}$ given the DP synthetic data, then we could incorporate these constraints into the sampling distributions in algorithm 1 instead of relying on post-processing. Next, we propose how to do that.

2.3 Posterior Sampling Under Public Information Constraints

As suggested by the examples above, we define two collections of constraints, the model-dependent constraints $\mathcal{M}$ and the data-dependent constraints $\mathcal{E}$ to be applied to $\boldsymbol{\theta}$ and $\boldsymbol{\varepsilon}$, respectively, where we indicate their satisfaction by:

$$C_{\mathrm{M}}(\boldsymbol{\theta}) \triangleq \prod_{m \in \mathcal{M}} m(\boldsymbol{\theta}), \qquad C_{\mathrm{E}}(\boldsymbol{\varepsilon}) \triangleq \prod_{e \in \mathcal{E}} e(\boldsymbol{\varepsilon}) \tag{4}$$

For the purposes of this paper, we assume that any constraint is either one of model-dependent or data-dependent (future work is needed to relax this assumption). Here, we propose Algorithm 2 which we refer to as a *constrained posterior method*, a modification of Algorithm 1, to incorporate these constraints.

Data: Observed DP data $\boldsymbol{Y}$, conditional prior $\pi_{\boldsymbol{\theta}|C_{\mathrm{M}}(\boldsymbol{\theta})=1}$, model-dependent constraints $\mathcal{M}$, data-dependent constraints $\mathcal{E}$, likelihood $f_{\boldsymbol{X}|\boldsymbol{\theta}}$, conditional error density $g_{\boldsymbol{\varepsilon}|C_{\mathrm{E}}(\boldsymbol{\varepsilon})=1}$,
Result: N samples from $\boldsymbol{\theta} \mid \boldsymbol{Y}, C_{\mathrm{M}}(\boldsymbol{\theta}) = 1, C_{\mathrm{E}}(\boldsymbol{\theta}) = 1$
while $i \leq N$ **do**
  Sample $\boldsymbol{\theta}^{(i)} \sim \pi(\cdot \mid C_{\mathrm{M}}(\boldsymbol{\theta}) = 1)$, $\boldsymbol{X}^{(i)} \mid \boldsymbol{\theta}^{(i)} \sim f(\cdot \mid \boldsymbol{\theta}^{(i)})$, $U \sim \mathrm{Unif}(0, 1)$;
  if

$$U \leq \frac{g(\boldsymbol{\varepsilon} = \boldsymbol{Y} - \boldsymbol{X}^{(i)} \mid C_{\mathrm{E}}(\boldsymbol{\varepsilon}) = 1)}{\sup_{\boldsymbol{Y}} g(\boldsymbol{\varepsilon} = \boldsymbol{Y} - \boldsymbol{X}^{(i)} \mid C_{\mathrm{E}}(\boldsymbol{\varepsilon}) = 1)} \tag{5}$$

  then
    Accept sample $\boldsymbol{\theta}^{(i)}$, $i \mapsto i + 1$;
  else
    Reject sample $\boldsymbol{\theta}^{(i)}$;
  end
end

Algorithm 2: Constrained posterior method – DP posterior sampling given additive perturbation and public knowledge

Our proposed algorithm is flexible enough to support known public information in a more statistically principled manner. Additionally, posterior simulation techniques allow for simulation-based inference, giving end users the ability to construct arbitrary inferential procedures, hypothesis tests, etc. This added flexibility does not come for free, though, since the new conditional prior and conditional error densities are more complex than their original counterparts. However, we will demonstrate the improved utility of this technique by presenting an end-to-end example using this algorithm.

3 Data Analysis

For a real data example, we use demographic survey data from the U.S. Census and the Center for Disease Control (CDC) to perform some inferential tasks about mortality rates given DP synthetic data (all data and code are available here). The CDC publishes non-private three year averages of the number of deaths per U.S. county by race, of which we use 2009–2011 [8]. Additionally, the U.S. 2010 Census publishes demographics data in Summary File 1, which contains the population of each U.S. county (indexed by FIPS code) by race (for this study, restricted to three categories: Hispanic, non-Hispanic White, and non-Hispanic black) [30]. We use this data as-is, ignoring the effects of any traditional disclosure limitation methods such as truncation or swapping since their implementation details are typically non-public.

For both analyses we focus on data for Alaska residents, yielding two 14×3 tables (14 counties $\times$ 3 race groups), which can be aggregated to a single 2×3 table of mortality status by race for the entire state. We demonstrate the advantage of our methods for such small, less densely populated areas. For both case studies, we assume DP synthetic data is created from the geometric mechanism outlined in Sect. 2. Sub-population data for both tables is available in Table 4, and the simulation code is available in a GitHub repository – see the URL in Appendix A. For both case studies, the inferential question of interest, public information constraints, and sampling hyperparameters are available in Table 3.

3.1 Case 1: Joint Estimation of Cell Counts for χ^2 Test

In this first case, the 14×3 Census and CDC tables are privatized with $\epsilon = .5$ (yielding a total budget of $\epsilon = 1$), and we assume the national mortality rates by race, state total populations, and state total deaths are all public information. While we list all constraints in the Appendix A Table 3, an example would be that the state total deaths from the DP generated synthetic tables match the publicly known state total deaths.

The goal is to test the null hypothesis H_0 that the national mortality rates by race agree with the Alaska-specific mortality rates by race (with H_1 otherwise). This will be achieved by estimating the observed Pearson Chi-Square test statistic, χ^2_{obs} via three different methods:

1. *Naive method:* directly use DP synthetic data without post-processing to produce the test statistic.
2. *Post-processing method:* post-process the naive DP synthetic data using Eq. 3 to enforce the constraints, then use the post-processed data to produce the test statistic.
3. *Constrained posterior method:* generate samples from the posterior distribution of the joint cell counts given DP synthetic data and public information using Algorithm 2, then sample from this empirical distribution to generate an empirical distribution of test statistics.

Notice immediately that our approach has the structural difference of producing samples from the possible values of the χ^2 test instead of point estimates only. For the purpose of this analysis, we compare the constrained posterior samples to the non-private test statistic using the posterior mode, estimated numerically from the samples (assuming unimodality) using the mean-shift algorithm [13].

We empirically generate 100 different synthetic DP data sets, apply the methods listed above, and present the observed distribution of estimated test statistics in Fig. 2 and Table 1. As expected, both the post-processing method and the posterior mode method offer better performance than naive substitution of DP synthetic data into the test statistic. This is a scenario in which we expect post-processed data to perform better than the naive perturbed counts, since in general post-processing methods do a good job of preserving aggregate information about higher-level summary queries as shown empirically [28].

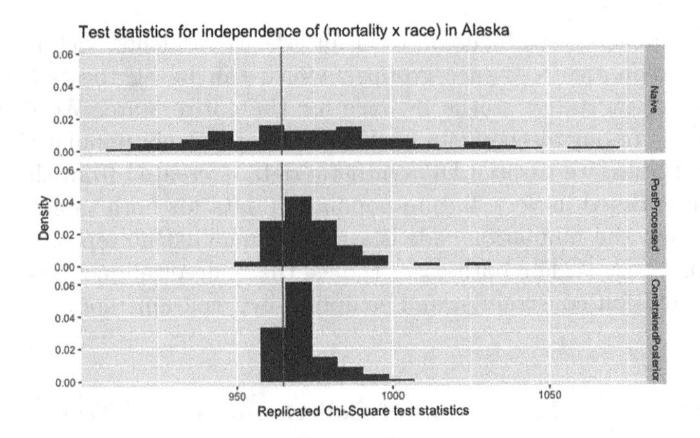

Fig. 2. Distributions of DP estimates of $\hat{\chi}^2_{\text{obs}}$ test statistic from 100 synthetic DP data sets (true χ^2_{obs} test statistic at vertical line)

Table 1. Comparison of DP estimates of $\hat{\chi}^2_{\text{obs}}$ test statistic from 100 synthetic DP data sets (true value $\chi^2_{\text{obs}} \approx 964$ on $(2-1) \times (3-1)$ degrees of freedom (two mortality statuses by 3 race groups)

ind	SampleMSE	SampleVar	SampleBias2
Naive	1053	986	77
PostProcessed	201	130	72
ConstrainedPosterior	102	62	41

Additionally, the constrained posterior mode better estimates the non-private test statistic compared to the other two methods. Although our method offers favorable performance as a point estimator, note that there are plenty of other choices for a centrality estimate. Analysis of different measures of centrality requires further investigation; however, the posterior sampling method still has the structural advantage of estimating a distribution rather than only estimating the point statistic, allowing for further capturing and characterizations of uncertainty such as the construction of empirical confidence intervals.

3.2 Case 2: Small Area Estimation of Cell Counts for Binomial Testing

In this second case, both tables are privatized with $\epsilon = .1$ (yielding a total budget of $\epsilon = .2$), but we now assume that county level non-Hispanic white populations and the national non-Hispanic white mortality rate are public information (again, the full specification is in Table 3). The goal is to test each county under the null hypothesis H_0 that the observed non-Hispanic white mortality rate is less than or equal to the national average non-Hispanic white mortality rate (with H_1 otherwise).

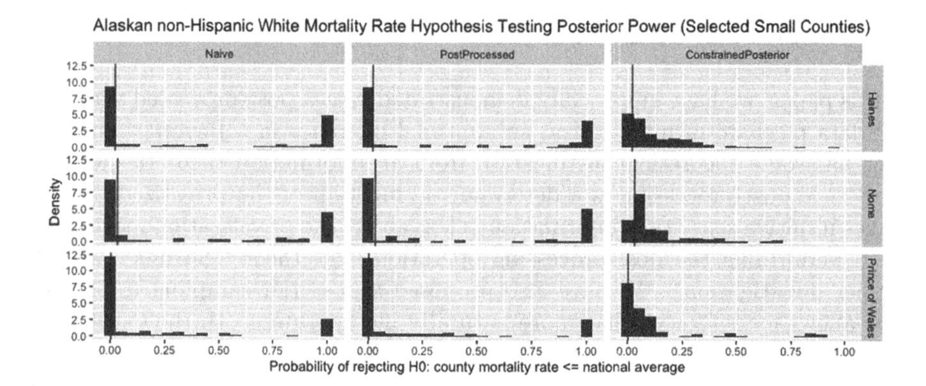

Fig. 3. Posterior distributions of $\hat{P}(H_1 \mid \boldsymbol{Y})$ using the three different methods outlined above. Vertical lines represent the true probabilities for each county.

Table 2. Comparison of DP estimates of $\hat{P}(H_1 \mid \{\boldsymbol{Y}_t\}_{t=1}^T)$ from 100 synthetic DP data sets for small counties in Alaska

County	Method	MSE	Variance	Bias
Haines	Naive	0.33	0.20	0.13
Haines	PostProcessed	0.32	0.19	0.13
Haines	**ConstrainedPosterior**	**0.04**	0.03	0.02
Nome	Naive	0.31	0.19	0.11
Nome	PostProcessed	0.31	0.20	0.11
Nome	**ConstrainedPosterior**	**0.03**	0.02	0.01
Prince of Wales	Naive	0.17	0.13	0.04
Prince of Wales	PostProcessed	0.17	0.13	0.04
Prince of Wales	**ConstrainedPosterior**	**0.06**	0.05	0.01

Again, we simulate 100 DP synthetic data sets, calculate the posterior probability of rejecting H_0 under the same three methods as in Case 1, and analyze the observed posterior distributions (complete county distributions are in the appendix in Fig. 4). Figure 3 and Table 2 show results for the subset of the three least populous counties. Here, in particular, we see the advantage of the posterior sampling method for small area estimation. In the previous case, post-processing ensured aggregated summary statistics were a reasonable proxy for the true summary statistics, but now this is no longer true. In comparing how the three different methods estimate the non-private posterior probability of rejecting H_1, we see that the constrained posterior sampling method clearly outperforms the other two; e.g., note very low MSE, variance and bias values.

4 Conclusion and Discussion

In this paper, we have empirically demonstrated that a post-processed synthetic DP data mechanism may lead to sub-optimal inference in a Bayesian analysis. This has some theoretical justification from an information theoretic perspective; since post-processing is a deterministic function of the privatized data, it makes intuitive sense that information is lost about the privatization mechanism when performing the post-processing operation (this theoretical link warrants further investigation). We have also proposed a posterior sampling framework that allows for inference that more naturally incorporates both the exact noise generating mechanism and public information simultaneously, and demonstrated its use on two small area estimation examples. This preliminary work, along with extensions of it, could help users of DP synthetic data perform inference in cases where the error distributions might be obscured by the post-processing operations.

The major limiting factors of this work are tied to the ease of computational implementation. In particular, the case studies we have analyzed are limited in scope for two practical reasons. First, sampling from the distributions given the constraints is a much more complex problem even under relatively simple constraints. For contingency table data, it is often easiest to sample from the marginal distribution of individual cells, whereas sampling from the joint distribution of cells under constraints is computationally difficult. This suggests investigating copula methods, such as those in [11], that could make it easier to generate samples from the constrained distribution, given that the copula function necessarily encodes constraints about the dependence of the cell values. Second, the rejection rates in Algorithm 1 can be prohibitively large, particularly in higher-dimensional problems. From a computational perspective, further investigation is needed to extend Algorithm 1, and thus Algorithm 2, to accommodate more efficient sampling schemes, such as MCMC-based methods [22] or hybrid importance-rejection sampling [12].

Aside from the computational aspects, this work also raises questions for data curators and administrators who hope to release DP synthetic data. In particular, the results above suggest that data curators *should release DP synthetic data both with and without post-processing when possible.* Given the results above, it is unclear as to whether or not there are scenarios in which working directly with post-processed data can guarantee similar inference to the posterior method above; this work would be necessary to inform practitioners of how best to use DP synthetic data.

Our results are also informative from a privacy mechanism perspective. Since certain public information constraints induce specific noise distributions, one could consider analyzing privatization mechanisms that explicitly add noise under these constraints. Additionally, previous studies have looked at empirical privacy guarantees in the presence of public information, such as public marginals [29]; such approaches could be useful for evaluating the effectiveness of different privatization mechanisms that enforce public information.

In conclusion, we have demonstrated that the posterior sampling approach to inference on DP synthetic data can be extended to include public information in the form of sampling constraints. Not only does this allow for better point estimation than post-processing based methods, but it also provides natural tools for performing more complex inferential tasks, such as constructing confidence intervals from DP synthetic data. Further research can yield more computationally efficient extensions of this algorithm, as well as tools to compare the efficiency with which different DP mechanisms incorporate public information.

Acknowledgements. Thanks to Roberto Molinari at Penn State for helpful discussions, John Abowd and Philip Leclerc at the U.S. Census for discussions about their DP methodology, and Alexis Santos at Penn State for providing data. This work was supported in part by NSF Award No. SES-1853209 to The Pennsylvania State University.

A Additional tables and figures

All code and data available on GitHub, here.

Table 3. Case study specifications

	Case 1	Case 2
H_0	mortality rate by race at national level equals mortality rate by race at state level	white mortality rate at national level equals white mortality rate by race at county level
Target dimensions	2×3	2×14
Statistic	χ^2_{obs}	$P(H_1 \mid \{Y_t\}_{t=1}^2)$
Budget	$\epsilon = 1$	$\epsilon = .2$
Public information	– State deaths – State total population	– County level white population
Constraints	– County level deaths between 0 and state total – County level population between 0 and state total – County level deaths smaller than county level population – State level marginal deaths agree with public data – State level marginal population agrees with public data	– County level white deaths between 0 and county level white population – County level marginal populations agree with public data
Prior	Dirichlet with hyperparameter corresponding to 50% of empirical privatized virtual observations	Beta with hyperparameter corresponding to 10 virtual observations from national white mortality rate
Posterior samples	10000	500
DP replicates	100	100

Table 4. Raw joined data from CDC (first four non-identifier columns) and US Census (last four columns)

County	FIPS	White-Deaths	Hisp-Deaths	Black-Deaths	Total-Deaths	WhitePop	HispPop	BlackPop	TotalPop
Anchorage	2020	1004	43	75	1122	192710	22061	19515	234286
Fairbanks NS	2090	327	0	14	341	75432	0	5091	80523
Matanuska	2170	433	0	0	433	76654	0	0	76654
Kenai	2122	339	0	0	339	47556	0	0	47556
Juneau	2110	112	0	0	112	22504	0	0	22504
Ketchikan	2130	81	0	0	81	9517	0	0	9517
Kodiak	2150	27	0	0	27	7599	0	0	7599
VC	2261	45	0	0	45	7283	0	0	7283
Sitka	2220	46	0	0	46	6048	0	0	6048
SE Fairbanks	2240	33	0	0	33	5723	0	0	5723
WP	2280	39	0	0	39	4669	0	0	4669
Prince of Wales	2201	16	0	0	16	2960	0	0	2960
Haines	2100	13	0	0	13	2167	0	0	2167
Nome	2180	10	0	0	10	1685	0	0	1685

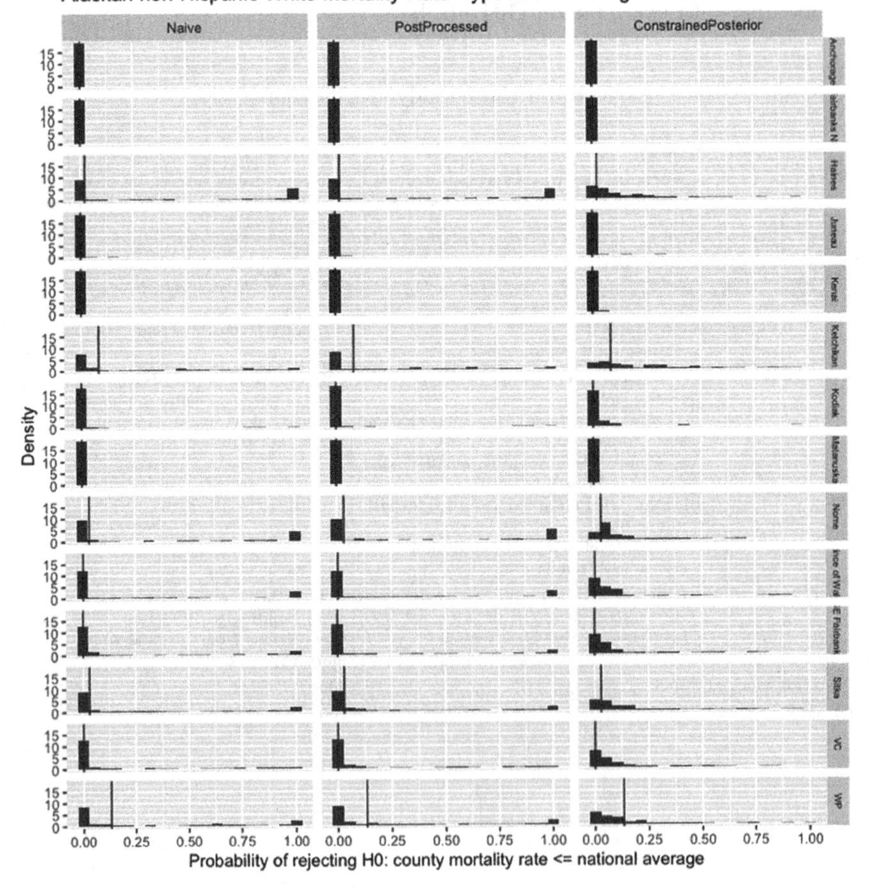

Fig. 4. Posterior distributions of $\hat{P}(H_1 \mid \boldsymbol{Y})$ using the three different methods outlined above. Vertical lines represent the true probabilities for each county.

References

1. Abowd, J., Kifer, D., Moran, B., Ashmead, R., Sexton, W.: Census TopDown Algorithm: Differentially Private Data, Incremental Schemas, and Consistency with Public Knowledge (2019)
2. Abowd, J.M.: The U.S. census bureau adopts differential privacy. In: Proceedings of the 24th ACM SIGKDD International Conference on Knowledge Discovery & Data Mining, pp. 2867–2867 (2018)
3. Abowd, J.M., Vilhuber, L.: How protective are synthetic data? In: Domingo-Ferrer, J., Saygın, Y. (eds.) PSD 2008. LNCS, vol. 5262, pp. 239–246. Springer, Heidelberg (2008). https://doi.org/10.1007/978-3-540-87471-3_20
4. Awan, J., Slavković, A.: Differentially private uniformly most powerful tests for binomial data. In: Bengio, S., Wallach, H., Larochelle, H., Grauman, K., Cesa-Bianchi, N., Garnett, R. (eds.) Advances in Neural Information Processing Systems, vol. 31, pp. 4208–4218. Curran Associates, Inc. (2018)
5. UC Bureau: 2010 demonstration data products. https://www.census.gov/programs-surveys/decennial-census/2020-census/planning-management/2020-census-data-products/2010-demonstration-data-products.html
6. Canonne, C.L., Kamath, G., McMillan, A., Smith, A., Ullman, J.: The structure of optimal private tests for simple hypotheses. In: Proceedings of the 51st Annual ACM SIGACT Symposium on Theory of Computing, pp. 310–321. ACM (2019)
7. Carroll, R.J., Ruppert, D., Stefanski, L.A., Crainiceanu, C.M.: Measurement Error in Nonlinear Models. Chapman and Hall, London (2006)
8. Center for Disease Control and Prevention: CDC WONDER Online Databases. https://wonder.cdc.gov/
9. Dwork, C., Lei, J.: Differential privacy and robust statistics. In: Proceedings of the Forty-First Annual ACM Symposium on Theory of Computing. STOC 2009, pp. 371–380. Association for Computing Machinery, New York (2009). https://doi.org/10.1145/1536414.1536466
10. Dwork, C., McSherry, F., Nissim, K., Smith, A.: Calibrating noise to sensitivity in private data analysis. In: Halevi, S., Rabin, T. (eds.) TCC 2006. LNCS, vol. 3876, pp. 265–284. Springer, Heidelberg (2006). https://doi.org/10.1007/11681878_14
11. Elfadaly, F.G., Garthwaite, P.H.: Eliciting Dirichlet and Gaussian copula prior distributions for multinomial models. Stat. Comput. **27**(2), 449–467 (2016). https://doi.org/10.1007/s11222-016-9632-7
12. Fearnhead, P., Prangle, D.: Constructing summary statistics for approximate Bayesian computation: semi-automatic approximate Bayesian computation. J. R. Stat. Soc. Ser. B (Stat. Methodol.) **74**(3), 419–474 (2012)
13. Fukunaga, K., Hostetler, L.D.: The estimation of the gradient of a density function, with applications in pattern recognition. IEEE Trans. Inf. Theor. **21**(1), 32–40 (1975)
14. Fung, B.C.M., Wang, K., Chen, R., Yu, P.S.: Privacy-preserving data publishing: a survey of recent developments. ACM Comput. Surv. **42**(4) (2010). https://doi.org/10.1145/1749603.1749605
15. Gaboardi, M., Lim, H., Rogers, R., Vadhan, S.: Differentially private chi-squared hypothesis testing: goodness of fit and independence testing. In: Balcan, M.F., Weinberger, K.Q. (eds.) Proceedings of The 33rd International Conference on Machine Learning. Proceedings of Machine Learning Research, vol. 48, pp. 2111–2120. PMLR, New York, 20–22 June 2016
16. Gong, R.: Exact Inference with Approximate Computation for Differentially Private Data via Perturbations (2019). http://arxiv.org/abs/1909.12237

17. Inusah, S., Kozubowski, T.J.: A discrete analogue of the Laplace distribution. J. Stat. Plan. Infer. **136**(3), 1090–1102 (2006)
18. Karwa, V., Slavković, A.: Inference using noisy degrees: differentially private β-model and synthetic graphs. Ann. Stat. **44**(1), 87–112 (2016). https://doi.org/10.1214/15-AOS1358
19. Krieger, N.: Census and differential privacy: public health and health equity questions. In: Workshop on 2020 Census Data Products: Data Needs and Privacy Considerations (2019)
20. Leclerc, P.: The 2020 decennial census TopDownDisclosure limitation algorithm: a report on the current state of the privacy loss-accuracy trade-off. In: Workshop on 2020 Census Data Products: Data Needs and Privacy Considerations (2019)
21. Lekshmi, S., Sebastian, V.S.: A skewed generalized discrete Laplace distribution. Int. J. Math. Stat. Invention **2**(3), 95–102 (2014)
22. Marjoram, P., Molitor, J., Plagnol, V., Tavaré, S.: Markov chain Monte Carlo without likelihoods. Proc. Natl. Acad. Sci. U. S. A. **100**(26), 15324–15328 (2003)
23. McClure, D., Reiter, J.P.: Differential privacy and statistical disclosure risk measures: An investigation with binary synthetic data. Trans. Data Priv. **5**(3), 535–552 (2012)
24. Papadimitriou, C., Steiglitz, K.: Combinatorial Optimization: Algorithms and Complexity, vol. 32, January 1982
25. Salvo, J.J.: Establishing priorities for the privacy budget: the case for good age data. In: Workshop on 2020 Census Data Products: Data Needs and Privacy Considerations (2019)
26. Santos-Lozada, A.R.: Differential privacy and mortality rates in the United States. In: Workshop on 2020 Census Data Products: Data Needs and Privacy Considerations (2019)
27. Sojourner, A., Van Riper, D.: Child poverty by local school district and allocation of federal Title1 funding. In: Workshop on 2020 Census Data Products: Data Needs and Privacy Considerations (2019)
28. Spence, M., Perry, M.: Demographic findings of the 2010 census demonstration product. In: Workshop on 2020 Census Data Products: Data Needs and Privacy Considerations (2019)
29. Uhler, C., Slavković, A., Fienberg, S.E.: Privacy-preserving data sharing for genome-wide Association Studies. J. Priv. Confidentiality **5**(1), 137–166 (2013)
30. US Census: US Census Summary File 1 Dataset. https://www.census.gov/data/datasets/2010/dec/summary-file-1.html
31. Van Riper, D., Spielman, S.: Geographic review of differentially private demonstration data. In: Workshop on 2020 Census Data Products: Data Needs and Privacy Considerations (2019)
32. Vu, D., Slavkovic, A.: Differential privacy for clinical trial data: preliminary evaluations. In: Proceedings of the 2009 IEEE International Conference on Data Mining Workshops, ICDMW 2009, pp. 138–143. IEEE Computer Society, Washington, DC (2009)
33. Wang, Y., Lee, J., Kifer, D.: Revisiting Differentially Private Hypothesis Tests for Categorical Data. ArXiv e-prints, November 2015
34. Wiley, K.: Decennial census, rural housing data and differential privacy. In: Workshop on 2020 Census Data Products: Data Needs and Privacy Considerations (2019)
35. Wilkinson, R.D.: Approximate Bayesian Computation (ABC) gives exact results under the assumption of model error. Stat. Appl. Genet. Mol. Biol. **12**(2), 129–141 (2013)

Differential Privacy and Its Applicability for Official Statistics in Japan – A Comparative Study Using Small Area Data from the Japanese Population Census

Shinsuke Ito[1(✉)], Takayuki Miura[2], Hiroto Akatsuka[3], and Masayuki Terada[3]

[1] Chuo University, Hachioji, Tokyo, Japan
ssitoh@tamacc.chuo-u.ac.jp
[2] NTT Secure Platform Laboratories, Musashino, Tokyo, Japan
takayuki.miura.br@hco.ntt.co.jp
[3] NTT DOCOMO, Inc., Yokosuka, Kanagawa, Japan
{hiroto.akatsuka.fb,teradam}@nttdocomo.com

Abstract. As part of its preparations for the 2020 U.S. Population Census, the U.S. Census Bureau uses the methodology of differential privacy to create privacy-preserved official microdata. It is expected that the use of differential privacy for official statistical data will become a topic in Japan in the future.

In this paper, we survey the current discussion on the use of differential privacy for creating official statistical data. We describe the differential privacy method used by the U.S. Census Bureau, develop a method to apply differential privacy to Japanese population grid data, and conduct a comparison between different differential privacy methods by applying them to Japanese small area data.

Results demonstrate that for population grid data, both the non-negative Wavelet method (and its public-n variant) and the US census top-down method preserve the non-negativity and sparsity of the original data, but each results in a different degree of error. This provides an important criterion for choosing the most appropriate method among the three options.

Keywords: Differential privacy · Population grid data · Laplace mechanism · Non-negative wavelet method · Top-down census method · 2D-Privelet

1 Official Statistics in Japan

In Japan, two types of statistical data are provided: statistical tables (open data) and microdata created based on questionnaire information (original microdata). Microdata are provided in a variety of formats, including anonymized microdata, original (non-anonymized) microdata, tailor-made statistical tables, and via on-site facilitation.

In order to increase the provision and use of anonymized official microdata in Japan, several empirical studies on the effectiveness of disclosure limitation methods for official microdata were conducted by [8, 9]. In preparation for the release of anonymized

J. Domingo-Ferrer and K. Muralidhar (Eds.): PSD 2020, LNCS 12276, pp. 337–352, 2020.
https://doi.org/10.1007/978-3-030-57521-2_24

microdata from the 2010 Population Census, empirical research was conducted by the Statistics Bureau of Japan and the National Statistics Center [10, 11], and empirical research on using top coding and recoding to create anonymized microdata that contain more detailed geographical information was conducted by [10]. Empirical research on the potential of perturbative methods such as data swapping and PRAM was conducted by [11].

Differential privacy is currently discussed as a disclosure limitation methodology for official statistics in the U.S. and Europe. Differential privacy was originally developed in computer science as a method to add controlled noise to data based on a standard of confidentiality. So far, differential privacy has not been discussed in the context of Japanese official statistics, with the exception of [12] who investigated the potential of differential privacy as a disclosure limitation method for official statistics in Japan.

This paper provides a mathematical explanation of differential privacy, and describes how differential privacy is used by the U.S. Census Bureau. On this basis, we develop a process to apply differential privacy to population grid data, and conduct a comparison between different differential privacy methods by applying them to Japanese small area data.

The results from this research aim to provide additional options for creating and publishing statistical tables specifically for small areas in a way that maximizes both confidentiality and data utility.

2 Definition of Concepts

In this section, we mathematically define concepts such as aggregate data and differential privacy.

2.1 Aggregate Data

Aggregate data refers to a set of values obtained by enumerating numbers of records corresponding to an attribute (or combination of attributes) in a database comprising a set of records having one or more attributes.

Let D be a dataset which has m records and they are tuples of d attributes. For the ith attribute, let A_i be a set of all possible attribute values and set $A := A_1 \times \cdots \times A_d$. Then we can consider D as a map $\mathrm{D} : [m] \to A$. Here note that $[m] := \{\mathrm{X} \in \mathrm{N} | 1 \leq \mathrm{x} \leq \mathrm{m}\}$. In the remainder of this section, **D** denotes the set of all possible databases that we want to protect.

Definition 1 (count query, aggregate data): Letting $\mathrm{C} \subset \mathrm{A}$, the "count query" for C a map $q_{\mathrm{C}} : D \to Z_{\geq 0}$ defined as $q_{\mathrm{C}}(D) = \#\{i \in [m] | D(i) \in \mathrm{C}\}$ for any $\mathrm{D} \in \mathbf{D}$ and $\mathrm{V}\left(\overrightarrow{C, \mathrm{D}}\right) := \left(q_{\mathrm{C}1}(\mathrm{D}), \ldots, q_{\mathrm{C}n}(\mathrm{D})\right) \in Z^n_{\geq 0}$ for $\vec{C} = (\mathrm{C}_1, \ldots, \mathrm{C}_n)$. We refer to this as the "aggregate data" for D, $\vec{C}$.

2.2 Differential Privacy

Differential privacy is a privacy protection index extended to include the concept of indiscernibility [5]. Adjacency between data is defined as follows:

Definition 2 (adjacent data): When two databases D, $D' \in \mathbf{D}$ differ by exactly one record, D and D' are referred to as adjacent data. The set of all adjacent data for $D \in \mathbf{D}$ is defined as N(D).

Differential privacy is defined as follows:

Definition 3 (ε-differential privacy): Let $\boldsymbol{K} : \boldsymbol{D} \rightarrow \boldsymbol{S}$ be a randomization function, where $\boldsymbol{S}$ is the set of all possible output values (in the context of this paper, $\boldsymbol{S} = \boldsymbol{Z}, \boldsymbol{R}$, etc.), and let ε be a positive real number.

$\boldsymbol{K}$ satisfies ε-differential privacy if for all $D \in D$, $\boldsymbol{D}' \in \mathrm{N}(\mathbf{D})$, and $\mathrm{S} \subset \boldsymbol{S}$, it holds that

$$\Pr[\boldsymbol{K}(\mathrm{D}) \in \mathrm{S}] \leq e^{\varepsilon} \Pr[\boldsymbol{K}(D') \in \mathrm{S}].$$

Intuitively speaking, this is based on the idea that a query from a database containing data for an individual is safe when the results are nearly indistinguishable from those when querying a database that does not contain data for that individual (individual privacy is maintained).

Smaller values for ε increase the degree of disturbance and thus more strongly protect the data. As $\varepsilon \rightarrow 0$ we get $e^0 = 1$, which means there is no differentiation between the two (they are information-theoretically indistinguishable); conversely, as ε increases the protection becomes weaker. The US Census Bureau calls this ε the "privacy loss budget".

2.3 The Laplace Mechanism

Here we introduce the Laplace mechanism as the most typical example satisfying the above-described differential privacy.

Definition 4 (Laplace distribution): Let $f : \boldsymbol{R} \rightarrow \boldsymbol{R}$ be a probability distribution defined as

$f(x) = \frac{1}{2\lambda} e^{-\frac{|x-\mu|}{\lambda}}$.

We call the distribution f the λ-scale Laplace distribution with mean μ or the double exponential distribution.

Definition 5 (Laplace mechanism): Let $\Delta_q := \max_{D \in \boldsymbol{D}} \max_{D' \in N(D)} \left\| q(D) - q(D') \right\|_1$ for a query $q : \boldsymbol{D} \rightarrow \boldsymbol{R}$ (this is called the L1 sensitivity of q). Then defining for $\varepsilon \in \boldsymbol{R}_{>0}$ a random function $\boldsymbol{K} : \boldsymbol{D} \rightarrow \boldsymbol{R}$ as $\boldsymbol{K}(\mathrm{D}) = \mathrm{q}(\mathrm{D}) + \mathrm{R}$, $\mathrm{R} \sim \mathrm{Lap}(\frac{\Delta_q}{\varepsilon})$, we obtain a privacy protection mechanism satisfying ε-differential privacy. This is called the Laplace mechanism.

3 Differential Privacy in Official Statistics from the U.S. Census Bureau

Abowd [1] showed that combining statistical tables which contain multiple geographic categories can result in an increased risk of identifying individuals even if each table on its own does not contain identifying information. The application of disclosure limitation methods such as added noise can reduce this risk, but often negatively impacts data accuracy.

As a specific example, Abowd [1] describes database reconstruction attacks on statistical tables, which can expose identifying information contained in the data by combining a small number of random queries [3]. From this, the basic idea of differential privacy – adding noise to the query based on a mathematically optimized privacy-loss budget ε – arose [4].

The following section describes the differential privacy method that the U.S. Census Bureau will apply to the 2020 U.S. Census (referred to as the top-down census method).

Processing in the top-down census method comprises the following six procedures. Inputs are a histogram with a hierarchical structure according to administrative divisions and a privacy loss budget ε, and the output is a histogram group satisfying ε-differential privacy. Note that the hierarchical structure by administrative division here has a tree structure with equal leaf depth k.

1. For a parent histogram H, denote child histograms as $H_1, \ldots, H_k$.
2. Apply noise satisfying $\frac{\varepsilon}{k}$-differential privacy to each component in the child histograms.
3. For each component, optimize based on the L2 norm so that sums of child histograms match their parent histograms.
4. Perform L1-norm optimization so that each component of all child histograms has an integer value.
5. Recursively perform procedures 1–4, treating each child as a parent.
6. Complete when all leaves have been processed.

Proposition 1: The above algorithm is a protection mechanism that satisfies ε-differential privacy under input privacy loss budget ε.

For details on the optimization based on the L1-norm and L2-norm, see Sect. 5 on Abowd et al. [2].

4 Application of Differential Privacy to Population Grid Data and Theoretical Study on the Non-negative Wavelet Method

When using geographic information to analyze relations between data and population, geographic areas are often divided into in a grid data format. This section develops a method to apply the U.S. Census Bureau's differential privacy method to population grid data, and compares this method with the non-negative wavelet method [14], which has similar characteristics to the top-down census method; i.e., both preserve privacy, alleviate the scale of noise injected to partial sums of grid cells, and prevent the output from involving negative values.

4.1 Basic Concepts for Application of Differential Privacy to Population Grid Data

Definition 7 (population grid data): Division of a geographic area into m × m vertical and horizontal areas, along with data including populations in each area, is called population grid data, which can be represented in matrix form as $A \in \mathbf{Z}_{\geq 0}^{m \times m}$. Here, we consider $m = 2^k$ in particular.

Since the grid data has a flat structure itself while the top-down census method assumes that the input data has a tree structure, some mapping from a two-dimensional grid data to a tree data has to be introduced for the top-down census method to process grid data. We use a locality preserving mapping scheme, called "Morton order mapping" for this purpose (see Appendix 1 for the definition).

Morton order mapping [13] is one of the most commonly used locality preserving mapping schemes to arrange multi-dimensional data in a one-dimensional sequence, and is utilized in several geo-location coding schemes including GeoHash.

4.2 Application of the Top-Down Census Method to Population Grid Data

In this section, we apply the top-down census method to population grid data ordered by Morton order mapping. Of particular note here is that the U.S. Census Bureau algorithm performs recursive processing of parent-child relations, so the algorithm needs be formulated for only a single parent-child relation. In other words, we need to formulate an algorithm for only procedures 1 through 4.

When considering application to population grid data, we assume a histogram with only one attribute (and therefore not requiring application of the special constraint). By this assumption, each histogram is simply a non-negative integer value. Algorithm 1 (C) shows procedures 1–4 for the top-down census method rewritten under these prerequisites. Outputs $\tilde{a}, \tilde{b}$ are respectively optimal solutions $\mathrm{h}_1, \mathrm{h}_2$ for the optimization problem.

Algorithm 1: The top-down census method (C)

Require: $\mathrm{a}, \mathrm{b}, \tilde{n} \in \boldsymbol{Z}_{\geq 0},\ \varepsilon \in \boldsymbol{R}_{>0}$

Ensure: $\tilde{a}, \tilde{b} \in Z \geq 0$ s.t. $\tilde{a} + \tilde{b} = \tilde{n}$, and output satisfies ε-differential privacy.

1: $r_1, r_2 \leftarrow Lap\left(\frac{2}{\varepsilon}\right)$ …generate Laplace noise
2: $a^* \leftarrow a + r_1, b^* \leftarrow b + r_2,$ …add noise
3: $a^\dagger \leftarrow \frac{1}{2}(a^* - b^* + \tilde{n}),\ b^\dagger \leftarrow \frac{1}{2}(-a^* + b^* + \tilde{n})$ …projection to line $\mathrm{x} + \mathrm{y} = \tilde{n}$
4: $\tilde{a}, \tilde{b} \leftarrow \boldsymbol{N}\left(a^\dagger, b^\dagger\right)$ …grid-point search for nearest non-negative integer
5: Output $\tilde{a}, \tilde{b}$

Algorithm $\boldsymbol{N}$ in Algorithm 1 performs a grid-point search for the nearest non-negative integer on the line. Input here is real numbers x, y satisfying $\mathrm{x} + \mathrm{y} \in \boldsymbol{Z}_{\geq 0}$, and output is $\tilde{x}, \tilde{y} \in Z_{\geq 0}$ nearest to the input and satisfying $\tilde{x} + \tilde{y} = \tilde{n}$.

4.3 Comparison with the Non-negative Wavelet Method

In this section, we introduce the non-negative wavelet method, which shares similar characteristics to the top-down census method. This method utilizes the Haar wavelet transform and noise is injected onto the transformed wavelet coefficients similar to Privelet [15], but non-negativity of the output is preserved. In addition, this method uses Morton mapping to reorder 2D-grid cells into 1D-histograms before the Haar decomposition to avoid sensitivity enlargement (and thus noise scales enlargement) led by

2D-wavelet transform, which makes Privelet difficult to apply to geographic data in a practical manner.

Roughly, the processing flow in the non-negative wavelet method is as follows:

1. Using Morton order mapping, sort population grid data in a histogram format.
2. Apply the Haar decomposition (see Appendix 2) to the histogram.
3. Apply scale-adjusted Laplace noise according to the Matrix mechanism [7], perform refinement for the non-negativity constant, and revert through inverse Haar decomposition.
4. Recursively repeat procedures 2 and 3 until return to the form prior to Haar wavelet transform.

The non-negative wavelet method also performs recursive processing of parent–child relations in a fully binary tree.

For comparison, we consider one stage of the non-negative wavelet method. One stage of operations 2–4 is equivalent to Algorithm 2 (**W**), which is also based on differential privacy.

Algorithm 2: The non-negative wavelet method (**W**)

Require: a, b, $\tilde{n} \in \mathbf{Z}_{\geq 0}$, s.t. a + b = $\tilde{n}$, ε ∈ R >0
Ensure: $\tilde{a}, \tilde{b} \in Z_{\geq 0}$ s.t. $\tilde{a} + \tilde{b} = \tilde{n}$ and the output satisfies ε-differential privacy.

1: $a' \leftarrow \frac{1}{2}(a+b), b' \leftarrow \frac{1}{2}(a-b)$ …Haar decomposition
2: $r \leftarrow Lap\left(\frac{1}{\varepsilon}\right)$ …generate Laplace noise
3: $a^* \leftarrow a', b^* \leftarrow b' + r$ …apply noise to b′
4: $b^{**} \leftarrow \min\{a^*, |b^*|\} \times \mathrm{sgn}(b^*)$ …preserve non-negativity of the output
5: $a^\dagger \leftarrow a^* + b^{**}, b^\dagger \leftarrow a^* - b^{**}$ …inverse Haar decomposition
6: $\tilde{a}, \tilde{b} \leftarrow N\left(a^\dagger, b^\dagger\right)$ …grid-point search for nearest integer
7: Output $\tilde{a}, \tilde{b}$

Proposition 2: The output of $\boldsymbol{W}(a, b, n, \varepsilon)$ satisfies ε-differential privacy for $\varepsilon \in \boldsymbol{R}_{>0}$.

4.4 Comparison Between the Top-Down Census Method and Non-negative Wavelet Method

When comparing the top-down census method and non-negative wavelet method, both clearly satisfy the characteristics of 'no negative populations' or other nonsensical output and no contradictions regarding values above and below in the hierarchy. This section compares the accuracy of the two methods with regards to the third characteristic, namely, maintained high accuracy of partial sums. Specifically, we compare variance values of the noise added to each output. As described above, each method recursively repeats operations on a hierarchical structure, so it is sufficient to compare the variance of noise added to the output in one step, in other words, in the two algorithms $\boldsymbol{C}$ and $\boldsymbol{W}$. Under same privacy-loss budgets, we can interpret smaller noise variance as indicating higher accuracy. Both algorithms use a grid-point search for the nearest non-negative N, and the behavior in both is the same, so we can compare variance at one prior stage.

Theorem 1: Fixing a privacy-loss budget $\varepsilon \in R_{>0}$, variance of noise applied to the output of top-down census method $\boldsymbol{C}$ is $\frac{8}{\varepsilon^2}$, and that for the non-negative wavelet method $\boldsymbol{W}$ is $\frac{4}{\varepsilon^2}$. Note that noise here refers to output prior to the grid-point search for nearest non-negative.

1. Top-down census method $\boldsymbol{C}$

Noise corresponding to the mean of two Laplace noises of scale $\frac{2}{\varepsilon}$ is added to the output ($R_1 - R_2 = R_1 + (-R_2)$, and the Laplace distribution is symmetric with respect to the positive and negative, so differences can be considered as sums).

2. Non-negative wavelet method $\boldsymbol{W}$

Laplace noise of scale $\frac{1}{\varepsilon}$ is added to the output.

The sum of two Laplace noises does not follow a Laplace distribution, so the top-down census method does not add noise to the output following the Laplace distribution. In contrast, the non-negative wavelet method adds Laplace noise as-is due to the Haar decomposition.

The Haar decomposition causes a 45° reduction rotation of point (a, b) about the origin of the x–y plane $\mathbf{R}^2$, and a counterrevolution operation for y = x. By that operation, the condition of a + b = n for total number of constraints can be transformed into a straight line parallel to the y-axis in the $\mathbf{R}^2$ plane. The y component is thus sufficient for the added noise, and this mechanism can be considered as simply equivalent to the Laplace mechanism for count queries. From this consideration, the results of Theorem 1 are consistent with the suboptimality of the Laplace mechanism for count queries [6].

5 Application of Differential Privacy to Japanese Population Grid Data

Based on the above discussion, we apply the non-negative wavelet method, the top-down census method, the (simple) Laplace mechanism and the 2D-Privelet to population grid data from the Japanese Population Census.

While the noise scales of the non-negative wavelet method are derived from the Matrix mechanism which presupposes a private-n regime (i.e. the 'adjacent' database is defined as addition or removal of one record), the noise scales of the top-down census method imply a public-n regime (i.e. the 'adjacent' database is defined as replacement of one record). In order to bridge the gap between these different definitions, we conduct an evaluation of the "public-n version" of the non-negative wavelet method, which is almost the same as the (original) non-negative wavelet method, but has its noise scales derived from the Matrix mechanism under the public-n regime. We hereafter refer to the (original) non-negative wavelet method, the public-n non-negative wavelet method, and the top-down census method as NN-Wavelet, NN-Wavelet+ , and USCB Topdown, respectively.

We apply these methods to data from the 2010 Japanese Population Census for a 1/2 standard area grid data (n = 512 × 512 = 262,144 grids, approximately 500 m per side) over a 256-km^2 area in the Tokyo region.

Figure 1 shows the original data, with lighter colors indicating higher populations. The data include 47,016,253 persons as the total population, and contain 95,317 nonzero values for a density ratio of approximately 36%. This is a remarkably high density for a census dataset, and is due to the fact that the region includes areas with the highest population densities in Japan (the Tokyo metropolitan area). However, it also includes uninhabitable areas (lakes and marine areas), which were processed as non-structural zeroes.

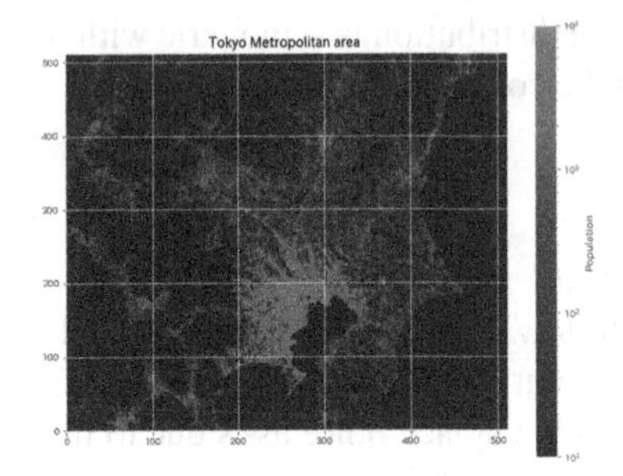

Fig. 1. Results for original data

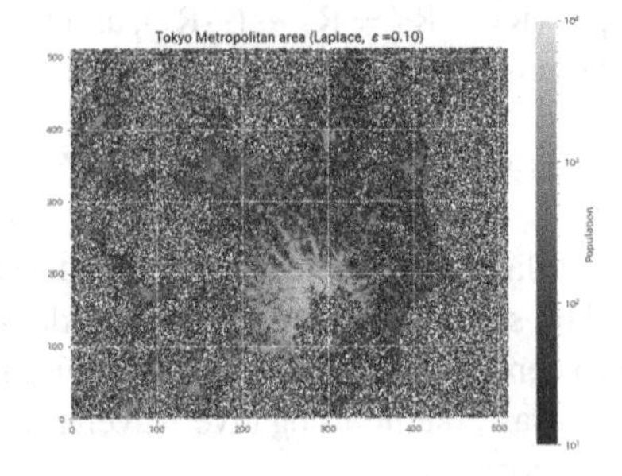

Fig. 2. Results for the laplace mechanism

Figure 2 shows the results for the application of the Laplace mechanism ($\varepsilon = 0.1$). Areas shown in white are those violating non-negativity constraints. These account for 91,744 (approximately 35%) of all cells. As the Laplace mechanism generates symmetrical noise for each cell, approximately 64% of the zero-valued cells in the original data have changed to negative (the others have changed to positive). Since all of the cell values have changed to nonzero, density is 100%.

Figure 3 shows the results for the application of the 2D-Privelet ($\varepsilon = 0.1$). These results include more white areas (with violation of non-negativity constraints) than the results from the Laplace mechanism; the number of the negative cells is higher at 118,805 (45%), which constitutes a worse result compared to the Laplace mechanism. At 100%, density is the same as for the Laplace mechanism.

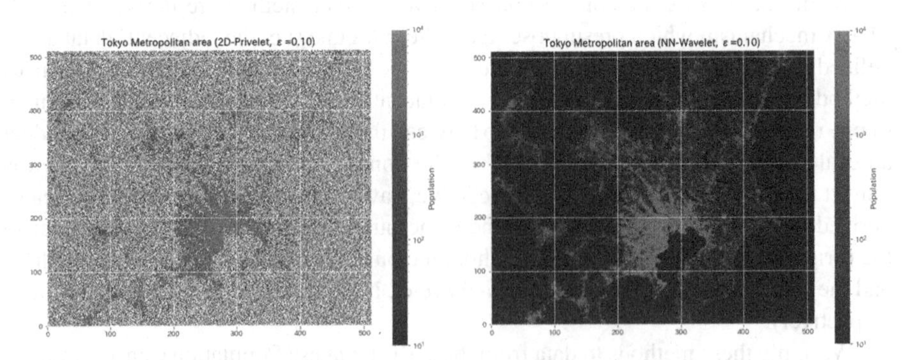

Fig. 3. Results for 2D-privelet

Fig. 4. Results for the NN-Wavelet method

Figures 4, 5 and 6 show the results for the application of the NN-Wavelet, NN-Wavelet+ , and USCB Topdown methods (ε = 0.1). Results appear similar and the absence of white areas indicates that no violations of non-negativity constraints occurred. Non-zero cells (and non-zero ratio) are 71,929 (27%), 60,832 (23%), and 52,789 (20%) respectively, which implies that the sparsity of the original data is largely maintained.

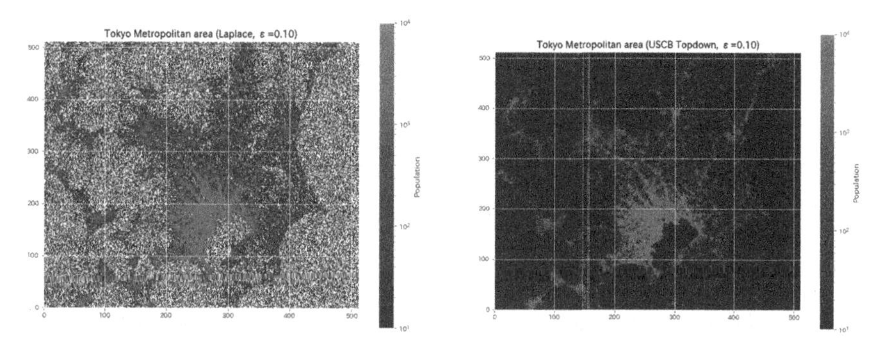

Fig. 5. Results for the NN-Wavelet+ method

Fig. 6. Results for the USCB Topdown method

These results indicate that the Laplace mechanism and 2D-Privelet result in negative values, and that this effect is more pronounced for population data that is sparse (i.e. includes many zero-values). Results also suggest that the Laplace mechanism and 2D-Privelet both fail to maintain the sparsity of the original data, while the NN-Wavelet, NN-Wavelet+ and USCB Topdown me- thods all preserve both non-negativity and sparsity of the original data.

To address errors for partial sums, for each method we conduct an empirical assessment of the mean absolute error (MAE) and the root mean squared error (RMSE) for privacy loss budgets ε of {0.1, 0.2, ln 2, ln 3, 3, 5, 10}. The results show similar trends for all values of ε and no significant difference between MAE and RMSE, so in the following section we discuss only results for the MAE of ε = 0.1. The other results are included in the Appendix Fig. 8, 9, 10, 11, 12 and 13 and Appendix Table 1.

Figure 7 shows the size of the error (mean absolute error over 100 trials) for partial sums for ε = 0.1. The horizontal axis x shows the area size of the partial sums (logarithmic base 2). For example, a value of x = 4 indicates an area which contains 16 (=2^4) grid cells.

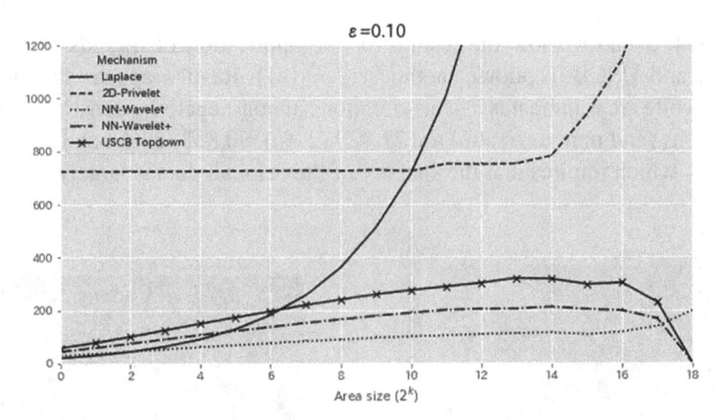

Fig. 7. Size of the Error (Mean Absolute Error Over 100 Trials) for partial sums for $\varepsilon = 0.1$

For the Laplace mechanism, the size of the error for partial sums is small for small areas, but greatly increases for large areas. This reflects the fact that the Laplace mechanism does not have any particular means to alleviate the noise of partial sums.

For 2D-privelet, the size of the error for partial sums is comparatively flat; even for large areas, the degree of increase in the error is small. At the same time, it is significantly larger for 2D-privelet than for the NN-Wavelet, NN-Wavelet+ , and USCB Topdown methods regardless of area size.

All three methods allow to control in the degree of increase in the error for partial sums similar to 2D-Privelet. Among the three methods, the size of the error is smallest for most areas in case of the NN-Wavelet method, and gradually increases for the NN-Wavelet+ and USCB Topdown methods. It is remarkable that for the NN-Wavelet+ and USCB Topdown methods, the size of the error significantly decreases for large area sizes close to the whole area sum ($x = 18$). This is due to the fact that the NN-Wavelet+ and USCB Topdown methods do not add noise to the total population according to a public-n regime (where the size of the 'adjacent' database does not change and thus the total population can be safely preserved), while the (original) NN-Wavelet method adds noise to the total population in order to follow the rule of private-n regime (where the disclosure of the total population can violate the privacy).

The above results also hold when we use RMSE as the indicator of errors, as shown in the Appendix Table 2.

6 Conclusion

This paper describes the methodology of differential privacy, and applies three differential privacy methods (non-negative Wavelet method, its public-n variant and top-down census method) to Japanese population grid data. Results show that the top-down census method - despite its complex structure - can be applied to population grid data, and demonstrate that the non-negative Wavelet method delivers higher accuracy at identical security levels than the top-down census method, while having only half the variance of added noise. This paper also demonstrates that for population grid data, both the non-negative Wavelet method (and its public-n variant) and the top-down census method preserve the non-negativity and sparsity of the original data, but each results in a different degree of error. This finding offers an important criterion that can make it easier to choose the most appropriate method among these three options.

The results of this research provide additional options for creating and publishing statistical tables specifically for small areas, while optimizing data confidentiality and data utility of the statistical data that is created.

Appendix 1. Definition of Morton Mapping

Definition A2 (Morton order mapping [13]): A mapping $F : \boldsymbol{R}^{m\times m} \to \boldsymbol{R}^{m^2}$ having the following characteristics. For given $A = (a_{ij}) \in \boldsymbol{R}^{m\times m}$, a characteristic of $F(A) = (v_k) \in \boldsymbol{R}^{m^2}$ is that for given $i = (i_k i_{k-1} \cdots i_1)_2$ and $j = (j_k j_{k-1} \cdots j_1)_2$, $a_{ij} = v_k$ holds for $k = (i_k j_k i_{k-1} j_{k-1} \cdots i_1 j_1)_2$, with $(\cdot)_2$ indicating a binary expansion.

Appendix 2. Definition of Haar Decomposition

Definition A1 (Haar decomposition): Let m : = 2^k for k $\in \boldsymbol{N}$. A mapping $H_m : R^m \to R^m$ defined as

$$H_m(x_1, \cdots x_m) = \left(a_1, \cdots a_{\frac{m}{2}}, d_1, \cdots, d_{\frac{m}{2}}\right),$$

where $a_i = \frac{x_{2i-1}+x_{2i}}{2}$, $d_i = \frac{x_{2i-1}-x_{2i}}{2}$ for each $1 \le i \le \frac{m}{2}$ is called "Haar decomposition." This is a invertible linear transformation, the inverse mapping for which is called "inverse Haar decomposition".

This paper particularly considers a Haar decomposition where k = 1, namely,

$$H_2(x_1, x_2) = \left(\frac{x_1 + x_2}{2}, \frac{x_1 - x_2}{2}\right).$$

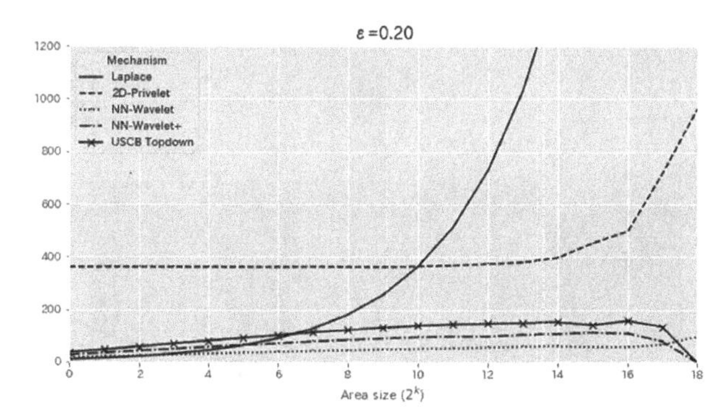

Fig. 8. Comparison for MAE between different differential privacy methods, ε = 0.20

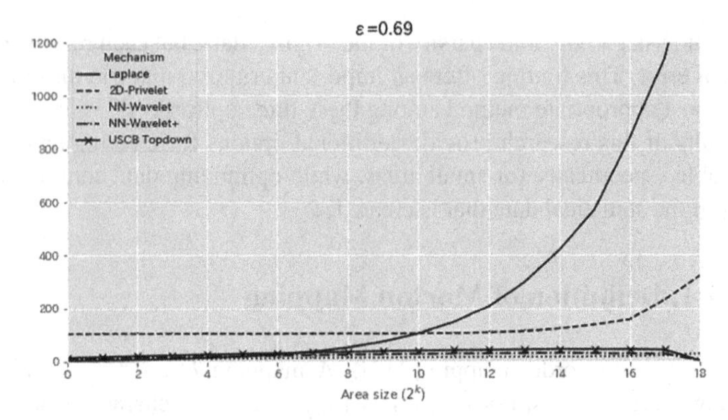

Fig. 9. Comparison for MAE between different differential privacy methods, $\varepsilon = 0.69$

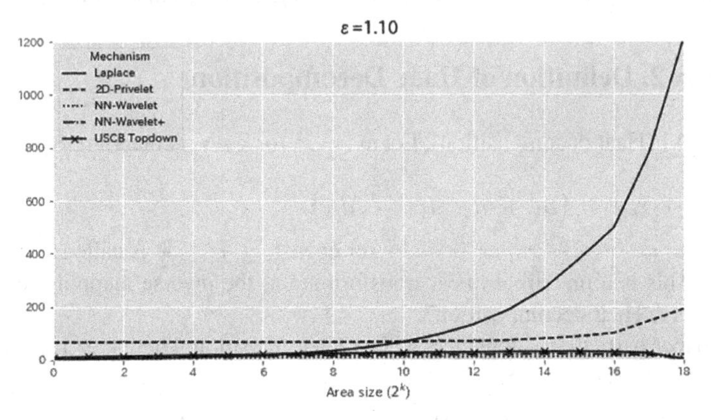

Fig. 10. Comparison for MAE between different differential privacy methods, $\varepsilon = 1.10$

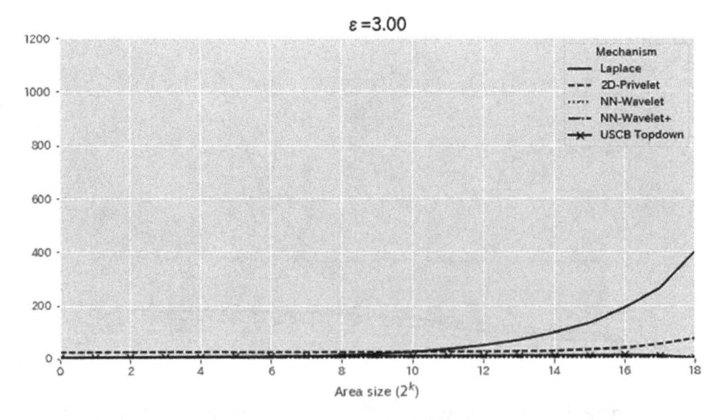

Fig. 11. Comparison for MAE between different differential privacy methods, $\varepsilon = 3.00$

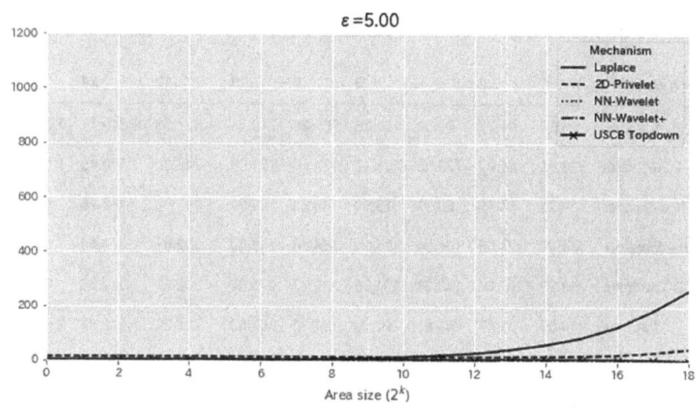

Fig. 12. Comparison for MAE between different differential privacy methods, ε = 5.00

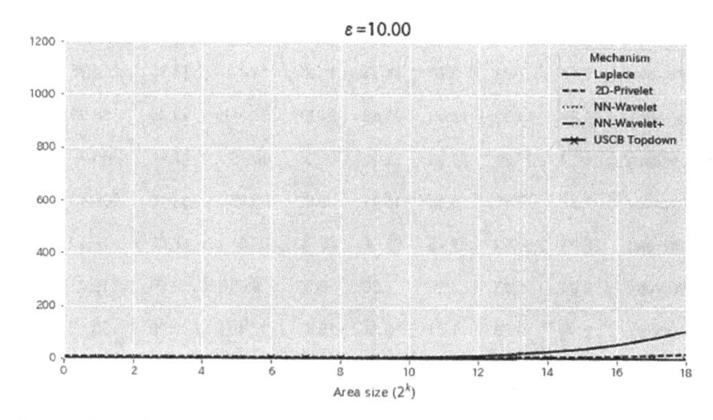

Fig. 13. Comparison for MAE between different differential privacy methods, ε = 10.00

Table 1. Size of the Error (MAE) for partial sums for ε = 0.1, 0.2, ln 2, ln 3, 3, 5, 10

	Area size (2^k)	0	2	4	6	8	10	12	14	16	18
	Laplace	20.00	43.77	89.59	180.26	361.26	713.78	1,427.53	2,849.18	5,728.36	11,826.14
	2D-Privelet	723.08	723.07	723.52	725.17	724.60	723.00	748.39	780.48	1,140.27	1,997.97
ε=0.10	NN-Wavelet	28.75	44.48	60.21	75.10	88.43	99.39	107.86	113.28	116.43	196.16
	NN-Wavelet+	44.89	74.06	106.04	135.99	164.64	188.49	202.81	210.44	196.72	0.00
	USCB Topdown	58.52	100.04	148.34	194.86	239.27	275.24	302.27	316.98	302.72	0.00
	Laplace	10.00	21.87	44.81	90.20	181.00	363.63	727.52	1,449.82	2,944.48	5,778.12
	2D-Privelet	361.43	361.68	361.41	361.71	361.97	364.36	374.27	396.61	497.61	962.07
ε=0.20	NN-Wavelet	17.01	24.62	31.83	38.78	44.97	50.64	55.96	60.01	57.69	97.11
	NN-Wavelet+	27.64	42.56	57.31	71.01	83.95	95.10	98.47	106.62	108.33	0.00
	USCB Topdown	36.79	58.57	81.05	102.41	122.15	139.81	148.13	153.22	158.07	0.00
	Laplace	2.89	6.31	12.92	25.95	52.22	104.01	206.95	419.09	840.32	1,639.46
	2D-Privelet	104.32	104.27	104.21	104.18	104.76	104.99	108.50	122.27	156.98	321.12
ε=0.69	NN-Wavelet	5.97	7.93	9.79	11.56	13.26	14.49	15.96	16.35	18.17	28.01
	NN-Wavelet+	10.39	14.32	18.06	21.68	24.92	27.51	29.49	30.98	28.28	0.00
	USCB Topdown	14.37	20.31	25.93	31.30	36.09	40.19	42.98	45.72	45.08	0.00
	Laplace	1.82	3.98	8.16	16.41	32.93	65.96	132.10	266.43	496.90	1,204.48
	2D-Privelet	65.79	65.82	65.95	65.92	65.70	66.61	68.63	78.19	98.59	188.72
ε=1.10	NN-Wavelet	3.94	5.13	6.26	7.36	8.33	9.05	9.70	10.41	11.57	18.47
	NN-Wavelet+	6.99	9.36	11.61	13.81	15.61	17.49	18.90	20.47	17.68	0.00
	USCB Topdown	9.79	13.36	16.70	19.94	22.80	25.37	27.43	28.79	27.78	0.00
	Laplace	0.67	1.46	2.98	6.01	12.04	24.19	47.92	95.72	189.22	396.20
	2D-Privelet	24.10	24.10	24.09	24.02	24.14	24.57	25.27	27.69	38.35	73.58
ε=3.00	NN-Wavelet	1.54	1.95	2.35	2.74	3.05	3.31	3.53	3.70	3.89	5.93
	NN-Wavelet+	2.81	3.62	4.40	5.14	5.78	6.36	6.88	7.15	6.53	0.00
	USCB Topdown	4.01	5.20	6.36	7.49	8.46	9.30	10.17	10.76	11.62	0.00
	Laplace	0.40	0.87	1.79	3.60	7.22	14.40	28.97	59.45	124.08	257.14
	2D-Privelet	14.46	14.46	14.47	14.42	14.43	14.52	15.16	16.09	21.93	44.28
ε=5.00	NN-Wavelet	0.94	1.18	1.42	1.65	1.85	2.01	2.15	2.39	2.31	3.69
	NN-Wavelet+	1.74	2.20	2.67	3.10	3.47	3.80	4.07	4.31	4.29	0.00
	USCB Topdown	2.49	3.19	3.86	4.51	5.04	5.50	5.99	6.41	6.02	0.00
	Laplace	0.20	0.44	0.90	1.80	3.59	7.16	14.32	28.19	55.20	108.77
	2D-Privelet	7.23	7.23	7.23	7.23	7.22	7.25	7.85	8.06	11.71	20.96
ε=10.00	NN-Wavelet	0.48	0.60	0.71	0.83	0.92	1.00	1.07	1.16	1.31	1.71
	NN-Wavelet+	0.89	1.12	1.34	1.56	1.74	1.91	2.03	2.18	2.28	0.00
	USCB Topdown	1.29	1.63	1.96	2.28	2.55	2.79	3.01	3.13	2.88	0.00

Table 2. Size of the Error (RMSE) for partial sums for $\varepsilon = 0.1$

Area size (2^k)	0	2	4	6	8	10	12	14	16	18
Laplace	28.29	56.59	113.14	226.28	452.22	894.18	1,784.63	3,535.44	6,763.48	11,826.14
2D-Privelet	942.77	942.83	943.05	944.16	945.22	937.39	959.50	965.34	1,287.08	1,997.97
NN-Wavelet	66.27	87.57	106.25	121.21	133.25	141.68	147.79	144.77	133.83	196.16
NN-Wavelet+	109.96	150.44	190.58	221.83	249.55	269.31	279.50	272.99	225.24	0.00
USCB Topdown	147.08	203.78	262.89	311.00	351.92	379.99	400.24	394.42	343.72	0.00

References

1. Abowd, J.M.: Staring-down the database reconstruction theorem. In: Joint Statistical Meetings, Vancouver, Canada (2018)
2. Abowd, J., et al.: Census TopDown: Differentially private data, incremental schemas, and consistency with public knowledge (2019). https://github.com/uscensusbureau/census2020-dase2e/blob/master/doc/20190711_0945_Consistency_for_Large_Scale_Differentially_Private_Histograms.pdf
3. Dinur, I., Nissim, K.: Revealing information while preserving privacy. In: Proceedings of the Twenty-Second ACM SIGMOD-SIGACT-SIGART Symposium on Principles of Database Systems, pp. 202–210. ACM (2003)
4. Dwork, C.: Differential privacy. In: ICALP (2006)
5. Dwork, C., McSherry, F., Nissim, K., Smith, A.: Calibrating noise to sensitivity in private data analysis. In: Halevi, S., Rabin, T. (eds.) TCC 2006. LNCS, vol. 3876, pp. 265–284. Springer, Heidelberg (2006). https://doi.org/10.1007/11681878_14
6. Geng, Q., Viswanath, P.: The optimal mechanism in differential privacy. In: 2014 IEEE International Symposium on Information Theory, pp. 2371–2375. IEEE (2014)
7. Hay, M., Rastogi, V., Miklau, G., Suciu, D.: Boosting the accuracy of differentially private histograms through consistency. Proc. VLDB Endowment **3**(1-2), 1021–1032 (2010)
8. Ito, S., Hoshino, N.: The potential of data swapping as a disclosure limitation method for official microdata in Japan: an empirical study to assess data utility and disclosure risk for census microdata. In: Paper presented at Privacy in Statistical Databases 2012, Palermo, Sicily, Italy, pp. 1–13 (2012)
9. Ito, S., Hoshino, N.: Data swapping as a more efficient tool to create anonymized census microdata in Japan. In: Paper Presented at Privacy in Statistical Databases 2014, Ibiza, Spain, pp. 1–14 (2014)
10. Ito, S., Hoshino, N., Akutsu, F.: Potential of disclosure limitation methods for census microdata in Japan. In: Paper presented at Privacy in Statistical Databases 2016, Dubrovnik, Croatia, pp. 1–14 (2016)
11. Ito, S., Yoshitake, T., Kikuchi, R., Akutsu, F.: Comparative study of the effectiveness of perturbative methods for creating official microdata in Japan. In: Domingo-Ferrer, J., Montes, F. (eds.) PSD 2018. LNCS, vol. 11126, pp. 200–214. Springer, Cham (2018). https://doi.org/10.1007/978-3-319-99771-1_14
12. Ito, S., Terada, M.: The potential of anonymization method for creating detailed geographical data in Japan. In: Joint UNECE/Eurostat Work Session on Statistical Data Confidentiality 2019, pp. 1–14 (2019)
13. Morton, G.M.: A Computer Oriented Geodetic Data Base; and a New Technique in File Sequencing, International Business Machines Co. Ltd., Canada, Technical Report (1966)

14. Terada, M., Suzuki, R., Yamaguchi, T., Hongo, S.: On publishing large tabular data with differential privacy. IPSJ J. **56**(9), 1801–1816 (2015). Information Processing Society of Japan (in Japanese)
15. Xiao, X., Wang, G., Gehrke, J., Jefferson, T.: Differential privacy via wavelet transforms. IEEE Trans. Knowl. Data Eng. **23**(8), 1200–1214 (2011)

Disclosure Avoidance in the Census Bureau's 2010 Demonstration Data Product

David Van Riper(✉), Tracy Kugler, and Steven Ruggles

Minnesota Population Center, University of Minnesota, Minneapolis, MN 55455, USA
{vanriper,takugler,ruggl001}@umn.edu

Abstract. Producing accurate, usable data while protecting respondent privacy are dual mandates of the US Census Bureau. In 2019, the Census Bureau announced it would use a new disclosure avoidance technique, based on differential privacy, for the 2020 Decennial Census of Population and Housing [19]. Instead of suppressing data or swapping sensitive records, differentially private methods inject noise into counts to protect privacy. Unfortunately, noise injection may also make the data less useful and accurate. This paper describes the differentially private Disclosure Avoidance System (DAS) used to prepare the 2010 Demonstration Data Product (DDP). It describes the policy decisions that underlie the DAS and how the DAS uses those policy decisions to produce differentially private data. Finally, it discusses usability and accuracy issues in the DDP, with a focus on occupied housing unit counts. Occupied housing unit counts in the DDP differed greatly from 2010 Summary File 1 differed greatly, and the paper explains possible sources of the differences.

Keywords: Differential privacy · 2020 US Decennial Census · Accuracy

1 Background

1.1 History of Census Disclosure Avoidance Techniques

Using a disclosure avoidance technique based on differential privacy represents a major break from methods used in prior decennial censuses. From 1970 through 2010, the Bureau used a variety of techniques, including whole table suppression (1970–1980), swapping (1990–2010), and blank and impute (1990–2010), to protect the confidentiality of respondents [24]. To implement these methods, the Bureau identified potentially disclosive variables and then found cells with small counts based on those variables. They would then suppress tables with these

Supported by the Minnesota Population Center (R24 HD041023), funded through grants from the Eunice Kennedy Shriver National Institute for Child Health and Human Development.

J. Domingo-Ferrer and K. Muralidhar (Eds.): PSD 2020, LNCS 12276, pp. 353–368, 2020.
https://doi.org/10.1007/978-3-030-57521-2_25

small counts or swap households matched on key demographic characteristics between geographic units.[1]

Traditional disclosure techniques introduced uncertainty into published data. Whole table suppression withheld information about certain aspects of the population. Swapping introduced error into some counts because households would not match on all demographic characteristics. It is impossible to precisely quantify the error introduced by these methods because the swap rates and key characteristics are confidential, but the Census Bureau concluded that "the impact in terms of introducing error into the estimates was much smaller than errors from sampling, non-response, editing, and imputation" [24, p. 11].

1.2 Reconstruction and Re-identification Attack

Concerned about increased computing power, increased availability of individual-level publicly available databases, and the massive volume of statistics published from decennial censuses, the Bureau executed a reconstruction and re-identification attack on the 2010 decennial census [2,17,23]. Working from the database reconstruction theorem [14],[2] the Bureau reconstructed 308,745,538 microdata records using their published census block and tract tabulations. Each reconstructed record had a census block identifier and values for sex, age, race, and Hispanic ethnicity.[3]

The Bureau then linked the reconstructed records to a commercial database by matching on age,[4] sex, and block ID. Forty-five percent (138,935,492) of the reconstructed records shared the same age, sex, and block ID as a record in the commercial database, which also included names. Finally, the Bureau attempted to link these 138 million records to the confidential microdata on all attributes–census block ID, age, sex, race, Hispanic ethnicity, and name. For 52 million persons, the Census Bureau was able to confirm that the two records referred to the same person. In other words, the Bureau reconstructed microdata that allowed it to match of 17% (52 million out of the 309 million) of the population enumerated in the 2010 decennial census to an external source.

The Census Bureau was able to verify the linkage of the reconstructed microdata and the commercial database because they have access to names via the confidential microdata. As Acting Director of the Census Bureau Ron Jarmin pointed out, an external attacker would not have access to such data; thus, they would not know for sure which 17% of the matches were true [19]. Of course, the

[1] Space constraints prevent us from a complete discussion of the Bureau's disclosure avoidance techniques. Interested readers are directed to McKenna [24,25].

[2] The database reconstruction theorem states that respondent privacy is compromised when too many accurate statistics are published from the confidential data. For the 2010 decennial census, more than 150 billion statistics were published [23].

[3] The Bureau reconstructed microdata from a set of 2010 decennial tables. Tables P1, P6, P7, P9, P11, P12, P12A-I, and P14 for census blocks and PCT12A-N for census tracts were used in the reconstruction [23].

[4] The Bureau linked the two datasets by exact age and by age plus or minus one year. [23].

attacker may have access to other datasets for verification purposes, but those datasets will differ from the confidential microdata.

1.3 Differential Privacy

For the 2020 Census, the Bureau has adopted disclosure control methods conforming to differential privacy. Differential privacy is a class of methods for introducing noise into published data [15].[5] Differential privacy is best understood as a description of the privacy-protecting properties of the algorithm used to generate published data, rather than a specific algorithm for implementing disclosure control. While differential privacy guarantees risk is below a certain level, it does not guarantee absolute protection from re-identification for all individuals.[6]

While differential privacy is not a specific algorithm, implementations of differential privacy generally follow a pattern of calculating cross-tabulations from "true" data and injecting noise drawn from a statistical distribution into the cells of the cross-tabulation. To illustrate with a simple example, let's say we asked 100 people about their sex and school attendance status and created the cross-tabulation of sex by school attendance (Table 1, confidential panel). For each of the six cells in the cross-tabulation, we draw a random value from a Laplace distribution with a pre-specified scale and add it to the value of the cell (Table 1, diff. private panel).

Table 1. Confidential and differentially private cross-tabulations from a simple survey.

Sex	Never attended	Attending	Attended in past	Total
Confidential				
Male	3	12	33	48
Female	4	17	31	52
Total	7	29	64	100
Diff. private				
Male	$3 - 1 =$ **2**	$12 + 0 =$ **12**	$33 + 1 =$ **34**	$48 + 0 =$ **48**
Female	$4 + 8 =$ **12**	$17 + 2 =$ **19**	$31 - 2 =$ **29**	$52 + 8 =$ **60**
Total	$7 + 7 =$ **14**	$29 + 2 =$ **31**	$64 - 1 =$ **63**	$100 + 8 =$ **108**

[5] Readers interested in learning more about differential privacy are directed to Wood et al. [32] and Reiter [27]. These papers provide a relatively non-technical introduction to the topic. A critique of differential privacy can be found in Bambauer et al. [5].

[6] The Census Bureau executed a reconstruction and re-identification attack on the 2010 Demonstration Data Product, which was generated from a differentially private algorithm. Approximately 5% of the reconstructed microdata records were successfully matched to confidential data. The 5% re-identification rate represents an improvement over the 17% re-identified from the 2010 decennial census data, but it still represents approximately 15 million census respondents [21].

Three key points from this simple example are worth noting. First, the noise introduced into each cell is independent of the original value of the cell. Therefore, it is possible to introduce relatively large noise values into relatively small cells. In the example, the number of females who had never attended school tripled from four to twelve in the noisy data. Second, introducing noise perturbs not only the values of the cells in the cross-tabulation but the size of the overall population. In the example, our original sample included 100 individuals, but the differentially private data described 108 synthetic records. Finally, though not illustrated in this example, it is possible to introduce noise such that cell values become negative. If a data producer wishes to maintain the total population count and avoid negative values in published data, they must use a post-processing algorithm that enforces total population as an invariant and non-negativity.

2 2010 Demonstration Data Product

The Census Bureau released the 2010 Demonstration Data Product (DDP) in October 2019 to help users examine impacts of the new disclosure avoidance algorithm on decennial census data [9].[7] The DDP consists of two datasets - the Public Law 94-171 Redistricting (PL 94-171) dataset and a partial version of the Demographic and Housing Characteristics (DHC) dataset.[8] Each dataset contains multiple tables for multiple geographic levels. Details about the tables and geographic levels are available in the technical documentation for the 2010 DDP [10].

The Bureau generated the DDP by running the 2010 Census Edited File (CEF)[9] through its Disclosure Avoidance System (DAS). The DAS takes in a data file and a set of parameters detailing the privacy loss budget, its allocation, and invariants and constraints. It then injects noise drawn from particular two-sided geometric distributions into the counts for geographic units. After the noise injection, the DAS uses a "Top-Down Algorithm" developed by Census to construct differentially private tabulations for specified geographic levels that are internally consistent and satisfy other invariants and constraints. Finally, it

[7] The 2010 Demonstration Data Product was the Census Bureau's third dataset produced by the Disclosure Avoidance System (DAS). The DAS consists of the Bureau's differentially private algorithm and the post-processing routines required to enforce constraints. The first dataset contained tabulations from the 2018 Census Test enumeration phase, carried out in Providence County, Rhode Island. The second dataset consists of multiple runs of the DAS over the 1940 complete-count census microdata from IPUMS. Details are available in [3,22].

[8] The Demographic and Housing Characteristics dataset is the replacement for Summary File 1.

[9] All decennial census products, except for Congressional apportionment counts, are derived from the Census Edited File (CEF). The CEF is produced through a series of imputations and allocations that fill in missing data from individual census returns and resolve inconsistencies. Readers interested in a more detailed discussion of the CEF production are directed to pages 10–11 of boyd [6].

generates the Microdata Detail File (MDF), which is differentially private. The Bureau's tabulation system reads in the MDF and constructs the PL94-171 and DHC tables for specified geographic levels. The DDP consists of these PL94-171 and DHC tables.

The 2010 Summary File 1 and PL94-171 datasets were also tabulated from the 2010 CEF and contained the same set of geographic levels and units as the DDP. By publishing the same set of tabulations for the same set of geographic units based on the same input data, the Bureau facilitated comparisons between a dataset produced using traditional disclosure avoidance techniques (Summary File 1 and PL94-171) and one produced using a differentially private algorithm (DDP). Data users could compare statistics derived from both products and determine whether the differentially private data would fit their needs.[10]

2.1 Policy Decisions

Disclosure control algorithms require parameters that control the amount of noise, suppression, or swapping applied to the input data. Values for these parameters have impacts on the quality and accuracy of the output data, and it is critical that data users understand both the significance of the parameters and their possible range of values. This section will discuss the Top-Down Algorithm's parameters and their values that were used to generate the DDP.

Global Privacy Loss Budget. The global privacy-loss budget (PLB), usually denoted by the Greek letter ϵ, establishes the trade-off between the privacy afforded to Census respondents and the accuracy of the published data. Values for ϵ range from essentially 0 to infinity, with 0 representing perfect privacy/no accuracy and infinity representing no privacy/perfect accuracy.[11] Once the global PLB is established, it can then be spent by allocating fractions to particular geographic levels and queries. Geographic levels or queries that receive larger fractions will be more accurate, and levels or queries that receive smaller fractions or no specific allocation will be less accurate.

For the DDP, the Census Bureau's Data Stewardship Executive Policy Committee set the global PLB to 6.0, allocating 4.0 to person tables and 2.0 to household tables. The person and household PLBs were then allocated to combinations of geographic levels and queries [12,16]. The geographic level-query allocations ultimately determine the magnitude of the noise injected into counts.

[10] The National Academies of Sciences, Engineering, and Medicine's Committee on National Statistics (CNStat) hosted a 2-day workshop on December 11–12, 2019. Census Bureau staff members presented details of the algorithm used to create the DDP. Census data users presented results from analyses that compared the 2010 DDP with 2010 Summary File 1 and PL94-171 data products. Privacy experts discussed issues surrounding the decennial census and potential harms of re-identification. Videos and slides from the workshop are available at https://sites.nationalacademies.org/DBASSE/CNSTAT/DBASSE_196518.

[11] Technically, ϵ must be greater than 0. If ϵ was zero, then no data would be published.

Geographic Levels. If we think of the cross-tabulations into which noise is injected as a set of rows and columns, the geographic levels define the rows. Each row in a cross-tabulation is a geographic unit within a geographic level (e.g., Minnesota is a geographic unit in the State geographic level). Seven geographic levels in the Census Bureau's hierarchy [8] received direct allocations of the PLB. The nation and state levels received 20% each, and the remaining five levels (county, census tract group,[12] census tract, block group, block) received 12% each. These allocations are the same for the person and household tables.

For geographic levels that receive no direct allocation of the PLB, they accumulate accuracy from the units that comprise them. Levels created from census blocks will inherit accuracy of census blocks. Within a particular level, units with larger populations will be more accurate than units with smaller populations.

Queries. If geographic levels define the rows of a cross-tabulation, then queries define the columns. Queries are essentially combinations of demographic or housing variables, and the PLB is allocated to these queries. Queries receiving a larger fraction of the PLB will be more accurate.

The DAS defines two types of queries. "Detailed" queries consist of all unique combinations of variables, and "DP queries" are specific combinations of variables. The "detailed" queries allow the Bureau to reconstruct the underlying microdata, and "DP queries" allow policy makers to select the statistics and relationships that will be more accurate in the published data.

Queries defined in the DAS do not have a one-to-one relationship with the tables published in the 2010 DDP. The queries are used in the noise injection and optimization processes, and the published tables are created from the synthetic microdata created by those processes. Categories in the published tables can and will differ from those used in the queries.

The Census Bureau designed seven queries to support the production of the person tables and six queries to support the production of the household tables in the 2010 DDP. Each query received a direct allocation of the PLB. The *voting age * Hispanic origin * race * citizenship*[13] query received the largest allocation (50%) of the person PLB. The *sex of householder * Hispanic origin of householder * race of householder * household type* and the *Hispanic origin*

[12] The census tract group is not a standard unit in the Census Bureau's geographic hierarchy. It was created specifically for the DAS to control the number of child units for each county. The census tract group consists of all census tracts with the same first four digits of their code (e.g., tract group 1001 consists of tracts 1001.01 and 1001.02). The DDP does not include data for tract groups.

[13] At the time the DAS for the 2010 Demonstration Data Product was designed, the Census Bureau assumed the citizenship question would be included on the 2020 Decennial Census questionnaire. Even though the US Supreme Court ruled in favor of the plaintiffs and removed the question, the Bureau did not have time to remove the citizenship variable from the DAS. No actual citizenship data was used to create the 2010 DDP; instead, the Bureau imputed citizenship status for records in the CEF [10].

*of householder * race of householder * household size * household type* queries received the largest allocations (25% each) of the household PLB. A list of all DDP queries and their allocations can be found in Appendix A.

Invariants and Constraints. Invariants and constraints play key roles in the DAS, particularly in post-processing routines applied to the noisy counts. Invariants are counts computed directly from the CEF into which no noise is injected. Constraints control the types and range of values in the final tabulations.

While the Bureau has not finalized the invariants for the 2020 Decennial Census, they did establish four invariants for the DDP. Total population is invariant at the state-level, and total housing units, total group quarters facilities, and total group quarters facilities by type are invariant at the census block-level.[14] These same four counts were invariant at the *census block-level* in the 2010 Decennial Census. Additionally, voting age population and occupied housing units (i.e., households) were invariant at the census block-level [1].

Constraints are the set of rules that the data produced by the DAS must follow. For the DDP, constraints included non-negativity, integer, and hierarchical consistency constraints. The non-negativity and integer constraints require all counts to be positive integer values. The hierarchical consistency constraint imposes consistency among geographic and attribute hierarchies. For geographic hierarchies, counts for child units must sum to the counts of the parent unit. For attribute hierarchies, counts for child attributes must sum to the counts of the parent attribute. If we sum the counts of occupied and vacant housing units for a given geographic unit, the sum must equal the published total housing unit count for the same unit.

2.2 2010 DDP Disclosure Avoidance System (DAS)

The DAS that generated the 2010 DDP consists of three steps: generating counts from the CEF, injecting noise, and post-processing to satisfy constraints.

Generating Counts. The first step in the DAS produces counts from the CEF. The DAS consumes the CEF, the queries, and the geographic levels and creates a set of histograms - one for each combination of query and geographic level. The cells each histogram contain the counts of a particular set of categories (e.g, 40 year-old females) for a given geographic unit. The number of cells in these histograms may be massive, particularly at the census block level, and the counts in the cells may be small or even zero.

Noise Injection. The second step in the DAS injects noise into the cell counts generated in the first step. These "noisy counts" are, by definition, differentially private, but they may not satisfy invariants or constraints set by policy. The noise injection step is accomplished through the following sub-steps.

[14] The geographic level associated with an invariant is the lowest level at which the invariant holds. All geographic levels composed of the lowest level will also be invariant.

*Compute ϵ for Each Geographic Level * Query Combination.* Policy decisions establish the global privacy loss budget and the fractional allocations of that PLB to specific geographic levels and queries. The DAS computes the ϵ value for each histogram as the product of the corresponding query and geographic allocations from the PLB.

Compute the Scale Parameter for the Statistical Distribution. Noise-injection values are generated by randomly drawing a value from a statistical distribution. The shape of the distribution is controlled by the scale parameter calculated in Eq. (1).

$$\frac{2}{\epsilon} = s \tag{1}$$

For this sub-step, ϵ is the histogram-specific value computed in the previous sub-step. The numerator is the sensitivity of the query, which is always 2 for histograms [13].[15] Scale parameters are shown in Table 2. Nation and state parameters are in column $Scale_{nation}$, and parameters for the other five geographic levels are in column $Scale_{county}$. Larger scale parameters represent distributions with higher variances, which yield potentially larger noise values.

Generate and Inject Random Noise into Each Histogram Cell. The final step in the noise injection process is to actually draw random values from the statistical distribution for a given *geographic level * query* combination and inject those values into the histogram cells computed from the CEF in Step 1. The scale parameter computed in the previous step determines the shape of a particular distribution.[16]

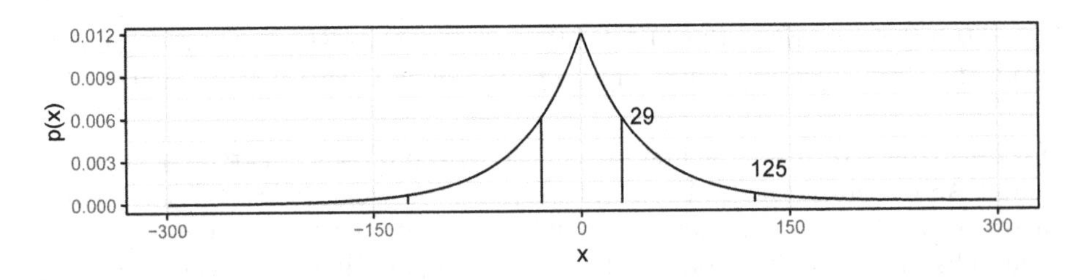

Fig. 1. Noise distributions used for the detailed histogram at the county, tract group, tract, block group, and block geographic level. The scale parameter for this distribution is 41.67.

[15] Sensitivity is the value by which a query changes if we make a single modification to the database. Histogram queries have a sensitivity of 2 - if we increase the count in a cell by 1, we must decrease the count in another cell by 1.

[16] Two types of distributions - the two-tailed geometric and the Laplace - are typically used to achieve differential privacy. The two-tailed geometric distribution is used when integers are required, and the Laplace distribution is used when real numbers are required. Source code for the 2010 DDP includes functions for both types of distributions [11].

Figure 1 depicts the Laplace distributions used for the detailed histogram at the county, tract group, tract, block group and block geographic levels. The labeled vertical lines illustrate the 75th and 97.5th percentiles. Fifty percent of all random draws will fall between 29 and −29, and 95% of all random draws will range from 125 to −125.

Post-processing. The output of step two is a series of noisy, differentially private, histograms, one for each *geographic level * query* combination. The raw noise-injected histograms are not integer-valued and may include negative values. Furthermore, because noise is injected independently into each histogram, they are inconsistent with each other, both within and across geographic levels. The histograms also do not satisfy invariants such as total population. Finally, the set of queries and geographic levels used for noise injection does not match the set of cross-tabulations and geographic levels desired for publication. In order to produce the final dataset, the Census conducts a series of optimization steps that ultimately generate synthetic microdata, which can then be tabulated to create tables for publication.

The Census refers to their post-processing algorithm as a "Top-Down Algorithm" because it starts at the national level and works down the geographic hierarchy, successively generating data for finer geographic levels. A diagram depicting the flow of data through noise injection and optimization can be found in Appendix B.

Generate National-Level Detailed Histogram. The first post-processing step produces an optimized version of the national-level detailed histogram. The detailed histogram is essentially a cross-tabulation of all possible categories of all the variables, which fully defines synthetic microdata.

The national-level detailed histogram is generated by solving an optimization problem to minimize the differences between the detailed histogram and the set of noisy histograms [22].[17] The optimization problem includes constraints to enforce invariants, non-negativity, integer values, and implied constraints to maintain consistency between person and household tables.

Generate Detailed Histograms for Lower Geographic Levels. At each subsequent geographic level, a similar optimization problem is solved, minimizing the differences between that level's set of noisy histograms and the output detailed histogram for that level, while matching invariants and meeting other constraints. For these lower geographic levels, the optimization problem also includes an additional constraint in the form of the detailed histogram from the parent level. Effectively, the lower-level optimization problems assign geographic identifiers to the synthetic microdata defined by the national-level detailed histogram in such a way as to best match the noisy lower-level queries.

[17] The optimization problem is actually solved in two stages, one to enforce non-negativity and optimize over the set of queries and a second stage to produce an integer-valued detailed histogram.

Generate Final Synthetic Microdata and Cross-Tabulations for Publication. The final detailed histogram is at the block level. This histogram is then transformed into synthetic microdata records, each with a full set of characteristics and a block identifier, constituting the Microdata Detail File (MDF). The Bureau's tabulation system then reads in the MDF and constructs the PL94-171 and DHC tables for specified geographic levels.

3 Data Usability Insights from the DDP

Publication of the 2010 Demonstration Data product allowed data users to compare statistics from the DDP with those from Summary File 1 (SF1), which was produced using traditional disclosure avoidance techniques. In doing so, users discovered several problematic aspects with respect to the accuracy of the DDP [4,20,26,28–31]. Many users have concluded that if 2020 decennial census data were published based on the DAS as implemented in producing the DDP it would be unusable for their needs. Below, we explore several of the issues that seem to contribute to these data usability problems by examining the particularly surprising inaccuracies in occupancy rates in the DDP.[18]

The DDP data showed 310 of the 3,221 counties in the United States and Puerto Rico as having occupancy rates of 100% (i.e., no vacant housing units).[19] In the SF1 data, no counties had 100% occupancy rates. These discrepancies represent known error in the DDP data because counts of both total housing units and occupied housing units were invariant to the block level in SF1. These errors can be traced to particular aspects of the DAS algorithm design and policy decisions, including the design of the household queries and allocation of the PLB among them and the invariants and constraints applied during the optimization process.

3.1 DDP Calculation of Occupancy Rates

First, it is important to understand how occupancy rates were calculated for the DDP. The Census Bureau's Data Stewardship Executive Policy (DSEP) committee made two critical recommendations that impacted the occupied housing unit counts: (1) the count of housing units be invariant for all geographic levels in the census hierarchy; (2) the count of occupied housing units will be subject to disclosure avoidance [16]. Housing unit characteristics were not directly included in the DAS queries, but the count of occupied housing units is, by definition, equal to the count of households. The count of vacant housing units is the difference between total housing units and households.

[18] The Census Bureau fielded so many questions about occupancy rates that they added a question to their FAQ [7]. The answer mentions that Census would look into the issue and post answers or updates. As of 2020-05-22, no answers or updates have been posted.

[19] Readers interested in learning more about the discrepancy should watch Beth Jarosz' presentation at the December 2019 CNStat workshop on the 2010 Demonstration Data Product [20].

3.2 Household Query Design and PLB Allocation

The combination of the design of the household queries and PLB allocations is likely to have resulted in a low signal-to-noise ratio in the noisy household histograms. Counts of households are, by definition, equal to or less than counts of persons. (In 2010 the national average household size was 2.68.) Except for the detailed histogram, the household histograms generally include many more cells than the person histograms. This means that CEF counts in many of the household histogram cells are quite small. In the SF1 data, the median count of households in counties is approximately 10,000. As an example, if 10,000 households were distributed evenly over the 384 cells in the *sex of householder * household type * elderly* histogram, each cell would have a count of about 26. Of course, most households will be in the *male-no elderly* cells, so other cells will have even smaller counts. With a scale parameter of 83.33, 50% of the draws of noise from the Laplace distribution for this histogram are expected to have absolute values greater than 57, easily doubling the original count or dropping it below zero. The situation is similar or worse for the other household histograms, meaning the information content of each histogram is largely obscured by the noise injected.

The set of household queries and allocation of PLB over the queries further contributes to the low signal-to-noise ratio. Overall, the household queries received an initial PLB allocation of 2.0, compared to the person queries' allocation of 4.0. No single household query received more than a 0.25 fractional allocation of the 2.0 PLB. This means that none of the counts or relationships are particularly well preserved. Furthermore, most of the household variables appear in just one or two of the queries (see Appendix A for details). Variables that appear in multiple queries provide redundancy in the set of histograms passed into the optimization problem that helps to pull the solution back toward the true values. Without this redundancy, the original signal tends to get lost in the noise.

3.3 Optimization and Constraints

The non-negativity constraint and block-level total housing unit invariant used as constraints in the DAS optimization ultimately result in the observed occupancy rate errors. The non-negativity constraint requires that every cell in the final detailed histogram be non-negative. As described above, many of the cells in the noisy household histograms will be negative, especially for geographic units with smaller numbers of households. Returning these cells to zero effectively adds households to these small places, resulting in positive bias. Dividing counties into quintiles by SF1 household count, counties in the three lowest quintiles consistently have more households in the DDP data than in SF1 (Fig. 2).

The invariant number of housing units down to the block level implies an upper-bound constraint on the number of households. Each geographic unit must have no more households than it has housing units. With the low signal-to-noise ratio in the noisy histograms, especially at the block level, this constraint is the strongest signal present in the optimization problem. Many geographic units therefore receive a number of households equal to the number of housing units,

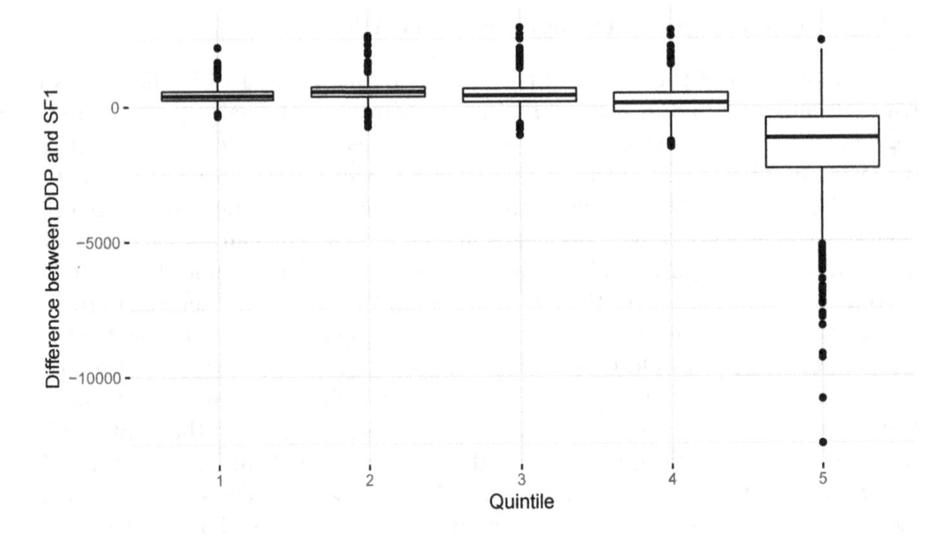

Fig. 2. Boxplots of county-level differences in household counts between DDP and SF1. Quintiles are computed from the invariant SF1 counts.

resulting in 100% occupancy rates. This is especially true for geographic units with smaller numbers of households that are affected by positive bias due to the non-negativity constraint.

While this combination of factors is especially problematic for the occupancy rate results in the DDP, the issues are not limited to this particular case. Researchers analyzing the DDP data in comparison to the SF1 data have found evidence of related issues in many aspects of the data. The issue of scale-independent noise affects all of the millions of cells with small counts in both the person and household histograms, making counts of many population subsets unreliable. The combination of the non-negativity constraint and population invariants consistently leads to bias increasing counts of small subgroups and small geographic units and decreasing counts of larger subgroups and geographic units.

4 Conclusion

Adopting a disclosure control technique based on differential privacy "marks a sea change for the way that official statistics are produced and published" [18]. It is critical that data users understand this new technique so they can judge whether the published data are fit to use. As the occupancy rate example demonstrated, the algorithm that generated the 2010 Demonstration Data Product Data produced highly problematic and biased data. Users must also be aware of the policy decisions required by the technique so that they may participate effectively in the decision-making process.

A Privacy Loss Budget Allocations

Table 2 lists the 7 person and 6 household queries that received direct allocations of the privacy loss budget. The allocations are shown in the PLB_{frac} column. The bold rows are the queries with the largest PLB allocation.

The $Scale_{nation}$ and $Scale_{county}$ columns list the scale factors used to generate the statistical distributions from which noise injection values are drawn. The $Scale_{nation}$ values are used for the nation and state histograms, and the $Scale_{county}$ values are used for the county, tract group, tract, block group, and block histograms.

The $Hist_{size}$ column lists the number of cells in the particular query. This is the number of cells on each row of the histogram (i.e., for each geographic unit). The value is generated by multiplying together the number of categories for each variable in a query. For example, the *Sex * Age (64 year bins)* query has two categories for *sex* and two categories for *age* giving a histogram size of 4. Category counts for each variable are listed in frequently asked question 11 in [7].

Table 2. Privacy loss budget allocations and scale parameters for 2010 DDP queries.

Query	$Hist_{size}$	PLB_{frac}	$Scale_{nation}$	$Scale_{county}$
Person				
Detailed person	467,712	0.10	25.0	41.67
Household/group quarters	8	0.20	12.5	20.83
Voting age * Hisp * Race * Citizen	**504**	**0.50**	**5.0**	**8.33**
Sex * Age (single year bins)	232	0.05	50.0	83.33
Sex * Age (4 year bins)	58	0.05	50.0	83.33
Sex * Age (16 year bins)	16	0.05	50.0	83.33
Sex * Age (64 year bins)	4	0.05	50.0	83.33
Household				
Detailed household	96,768	0.20	25.0	41.67
$\mathbf{HH_{Hisp} * HH_{race} * HH_{size} * HH_{type}}$	**2,688**	**0.25**	**20.0**	**33.33**
$\mathbf{HH_{sex} * HH_{Hisp} * HH_{race} * HH_{type}}$	**672**	**0.25**	**20.0**	**33.33**
$HH_{Hisp} * HH_{type} * HH_{multi}$	28	0.10	50.0	83.33
$HH_{sex} * HH_{type} * HH_{elderly}$	384	0.10	50.0	83.33
$HH_{sex} * HH_{age} * HH_{type}$	432	0.10	50.0	83.33

The variable names in the household queries are described in Table 3.

Table 3. Variable names and descriptions.

Variable name	Variable description
HH_{sex}	Sex of householder
HH_{race}	Race of householder
HH_{Hisp}	Hispanic/Latino origin of householder
HH_{size}	Household size
HH_{type}	Houehold type
HH_{multi}	Presence of three or more generations in household
$HH_{elderly}$	Presence of persons age 60+, 65+, or 75+

B Top-Down Algorithm flow diagram

Figure 3 depicts the flow of data through the noise injection and optimization steps of the Census Bureau's Top-Down Algorithm.

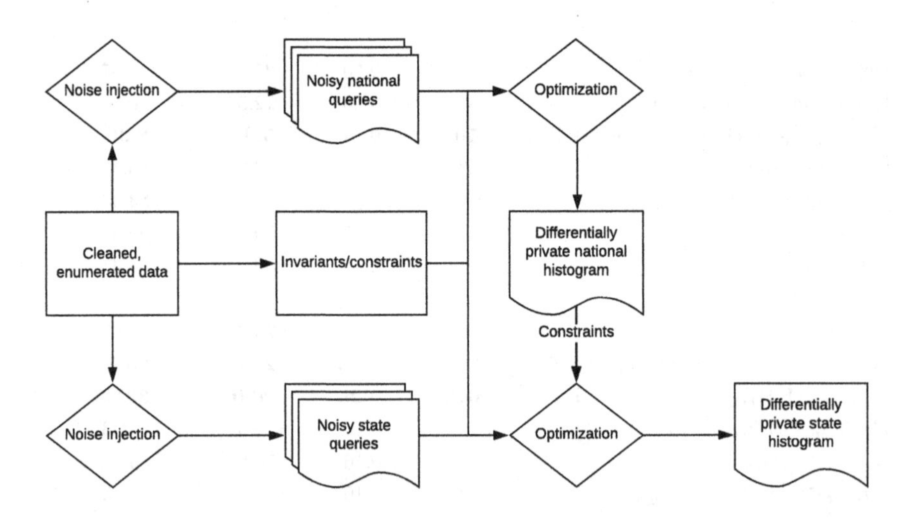

Fig. 3. Census DAS optimization flow diagram.

References

1. Abowd, J.: Disclosure avoidance for block level data and protection of confidentiality in public tabulations, December 2018. https://www2.census.gov/cac/sac/meetings/2018-12/abowd-disclosure-avoidance.pdf

2. Abowd, J.: Protecting the Confidentiality of America's Statistics: Adopting Modern Disclosure Avoidance Methods at the Census Bureau (2018). https://www.census.gov/newsroom/blogs/research-matters/2018/08/protecting_the_confi.html
3. Abowd, J., Garfinkel, S.: Disclosure Avoidance and the 2018 Census Test: Release of the Source Code (2019). https://www.census.gov/newsroom/blogs/research-matters/2019/06/disclosure_avoidance.html
4. Akee, R.: Population counts on American Indian Reservations and Alaska Native Villages, with and without the application of differential privacy. In: Workshop on 2020 Census Data Products: Data Needs and Privacy Considerations, Washington, DC, December 2019
5. Bambauer, J., Muralidhar, K., Sarathy, R.: Fool's gold: an illustrated critique of differential privacy. Vanderbilt J. Entertain. Technol. Law **16**, 55 (2014)
6. Boyd, D.: Balancing data utility and confidentiality in the 2020 US census. Technical report, Data and Society, New York, NY, December 2019
7. Census Bureau: Frequently Asked Questions for the 2010 Demonstration Data Products. https://www.census.gov/programs-surveys/decennial-census/2020-census/planning-management/2020-census-data-products/2010-demonstration-data-products/faqs.html
8. Census Bureau: Standard hierarchy of census geographic entities. Technical report, US Census Bureau, Washington DC, July 2010
9. Census Bureau: 2010 Demonstration Data Products (2019). https://www.census.gov/programs-surveys/decennial-census/2020-census/planning-management/2020-census-data-products/2010-demonstration-data-products.html
10. Census Bureau: 2010 demonstration P.L. 94–171 redistricting summary file and demographic and housing demonstration file: technical documentation. Technical report, US Department of Commerce, Washington, DC, October 2019
11. Census Bureau: 2020 Census 2010 Demonstration Data Products Disclosure Avoidance System. US Census Bureau (2019)
12. Census Bureau: 2020 census 2010 demonstration data products disclosure avoidance system: design specification, version 1.4. Technical report, US Census Bureau, Washington DC (2019)
13. Cormode, G.: Building blocks of privacy: differentially private mechanisms. Technical report, Rutgers University, Brunswick, NJ (nd)
14. Dinur, I., Nissim, K.: Revealing information while preserving privacy. In: Proceedings of the Twenty-Second ACM SIGMOD-SIGACT-SIGART Symposium on Principles of Database Systems. PODS 2003, pp. 202–210. ACM, New York (2003). https://doi.org/10.1145/773153.773173
15. Dwork, C., McSherry, F., Nissim, K., Smith, A.: Calibrating noise to sensitivity in private data analysis. In: Halevi, S., Rabin, T. (eds.) TCC 2006. LNCS, vol. 3876, pp. 265–284. Springer, Heidelberg (2006). https://doi.org/10.1007/11681878_14
16. Fontenot Jr., A.: 2010 demonstration data products - design parameters and global privacy-loss budget. Technical report 2019.25, US Census Bureau, Washington, DC, October 2019
17. Garfinkel, S., Abowd, J.M., Martindale, C.: Understanding database reconstruction attacks on public data. ACM Queue **16**(5), 28–53 (2018)
18. Garfinkel, S.L., Abowd, J.M., Powazek, S.: Issues encountered deploying differential privacy. In: Proceedings of the 2018 Workshop on Privacy in the Electronic Society. WPES 2018, pp. 133–137. ACM, New York (2018). https://doi.org/10.1145/3267323.3268949

19. Jarmin, R.: Census Bureau Adopts Cutting Edge Privacy Protections for 2020 Census (2019). https://www.census.gov/newsroom/blogs/random-samplings/2019/02/census_bureau_adopts.html
20. Jarosz, B.: Importance of decennial census for regional planning in California. In: Workshop on 2020 Census Data Products: Data Needs and Privacy Considerations, Washington, DC, December 2019
21. Leclerc, P.: The 2020 decennial census topdown disclosure limitation algorithm: a report on the current state of the privacy loss-accuracy trade-off. In: Workshop on 2020 Census Data Products: Data Needs and Privacy Considerations, Washington DC, December 2019
22. Leclerc, P.: Guide to the census 2018 end-to-end test disclosure avoidance algorithm and implementation. Technical report, US Census Bureau, Washington DC, July 2019
23. Leclerc, P.: Reconstruction of person level data from data presented in multiple tables. In: Challenges and New Approaches for Protecting Privacy in Federal Statistical Programs: A Workshop, Washington, DC, June 2019
24. McKenna, L.: Disclosure avoidance techniques used for the 1970 through 2010 decennial censuses of population and housing. Technical report 18–47, US Census Bureau, Washington, DC (2018)
25. McKenna, L.: Disclosure avoidance techniques used for the 1960 through 2010 decennial censuses of population and housing public use microdata samples. Technical report, US Census Bureau, Washington, DC (2019)
26. Nagle, N.: Implications for municipalities and school enrollment statistics. In: Workshop on 2020 Census Data Products: Data Needs and Privacy Considerations, Washington, DC, December 2019
27. Reiter, J.P.: Differential privacy and federal data releases. Ann. Rev. Stat. Appl. **6**(1), 85–101 (2019). https://doi.org/10.1146/annurev-statistics-030718-105142
28. Sandberg, E.: Privatized data for Alaska communities. In: Workshop on 2020 Census Data Products: Data Needs and Privacy Considerations, Washington, DC, December 2019
29. Santos-Lozada, A.: Differential privacy and mortality rates in the United States. In: Workshop on 2020 Census Data Products: Data Needs and Privacy Considerations, Washington, DC, December 2019
30. Santos-Lozada, A.R., Howard, J.T., Verdery, A.M.: How differential privacy will affect our understanding of health disparities in the United States. Proc. Natl. Acad. Sci. (2020). https://doi.org/10.1073/pnas.2003714117
31. Spielman, S., Van Riper, D.: Geographic review of differentially private demonstration data. In: Workshop on 2020 Census Data Products: Data Needs and Privacy Considerations, Washington, DC, December 2019
32. Wood, A., et al.: Differential privacy: a primer for a non-technical audience. Vanderbilt J. Entertain. Technol. Law **21**(1), 209–276 (2018)

Author Index

Printed in the United States
By Bookmasters